Phase Type Distributions

**Stochastic Models in Computer Science and Telecommunication Networks Set**

coordinated by
Bruno Sericola

Volume 2

# Phase Type Distributions

*Theory and Application*

András Horváth
Miklós Telek

ISTE

WILEY

First published 2024 in Great Britain and the United States by ISTE Ltd and John Wiley & Sons, Inc.



ISTE Ltd
27-37 St George's Road
London SW19 4EU
UK

www.iste.co.uk

John Wiley & Sons, Inc.
111 River Street
Hoboken, NJ 07030
USA

www.wiley.com



Library of Congress Control Number: 2024941676

British Library Cataloguing-in-Publication Data
A CIP record for this book is available from the British Library
ISBN 978-1-84821-945-8

# Contents

# Introduction

## I.1. The benefits and the limitations of the exponential distribution

During the development of the theory of stochastic processes, many specific models got introduced for describing different real life phenomena. With respect to the contents of this book, the introduction of the Poisson process to model phone calls occurring in a period of time by A. K. Erlang played a seminal role [BRO 60a]. A. K. Erlang also recognized that the inter-arrival time of consecutive call arrivals is exponentially distributed.

Starting from the first specific stochastic processes, various generalizations were introduced to extend their modeling power. One of the simplest and analytically tractable general stochastic models, which got very popular in many application fields, is the class of Markov chains. In continuous time, a characteristic property of Markov chains is that, at any point in time, the next event occurs in an exponentially distributed amount of time, while in discrete time, the next event occurs in a geometrically distributed number of steps. Both of these distributions enjoy the memoryless property which means that, at any point in time, the remaining time to the next event is independent of the time elapsed since the last one.

Markov chains are very efficient modeling tools for such real life processes where the inter-event times are exponentially (geometrically) distributed, and, as observed by A. K. Erlang in the case of phone call arrivals, many practically important real systems enjoy this property. Unfortunately, it does not apply to all real systems. In spite of this, due to the theoretical simplicity and computational tractability of Markov chains, in many cases, real life processes are modeled by Markov chains also when the inter-event times of the real process are significantly different from being exponentially distributed. One reason for this is that the stochastic behavior of systems with non-exponentially (non-geometrically) distributed inter-event times is often too complex to analyze.

## I.2. Phase-type distributions

In order to have models that provide both the analytical tractability of Markov chains and the possibility to describe non-exponentially distributed inter-event times, simple stochastic models have been introduced using the exponential distribution as a building block. It was recognized already by A. K. Erlang [BRO 60b] that the sum of two independent exponentially distributed random variables has a distribution that is different from exponential. A generalization of this observation leads to apply serial and parallel compositions of exponential random variables to approximate non-exponential inter-event time distributions. This approach is referred to as *the method of phases* in [KLE 75], where its applicability was questioned due to the lack of a general theorem which unifies the use of special (e.g. serial and parallel) structures.

The general theory of such distributions, which exploits the whole flexibility associated with a Markov chain of a finite number of states (referred to as phases), has been introduced by M. Neuts in [NEU 75], and the associated distribution got widely referred to as continuous phase type (continuous phase type (PH)) distributions.

The introduction of PH distributions resulted in a fertile research in the stochastic modeling community. Many essential questions have been investigated in order to understand the applicability of PH distributions regarding both:

– basic properties of PH distributions, such as,

- what is the density function and the moments of a PH distribution;

- whether the Markov chain based description of a PH distribution is unique;

- how many parameters identify a PH distribution of $n$ phases;

– their use for approximation of a general non-negative distribution, for example,

- are there efficient procedures to find a PH distribution that closely approximates the general one;

- what are the appropriate measures of "goodness" of the approximation.

The complete summary of the literature on PH distributions in a single book is an infeasible task. In this book, we provide a large portion of the obtained results focusing mostly on the application of PH distributions in approximating general non-negative distributions. To this end, we collect qualitative properties of PH distributions that imply limitations in distribution fitting. For example, we present bounds on the moments of PH distributions, the behavior of the tail of PH distributions, etc. These qualitative properties help to set proper expectations about the quality of distribution approximation. For example, if the moments of a distribution are out of the moment bounds of the considered class of PH distributions, then accurate distribution approximation is infeasible even with the best approximation procedure.

We discuss two distribution approximation approaches which are referred to as *matching* and *fitting*. Matching takes a set of characterizing parameters (e.g. a set of moments) and composes a PH distribution with exactly the same parameters, if possible. If the selected parameters properly captures the distribution, then the approximation has good precision but it is still not exact in general.

Fitting is instead an optimization problem. It is based on a distance measure between two distributions and applies an optimization procedure to set the parameters of the PH distribution such that the distance measure is minimized.

Apart of these main lines of the book, we point out to various extensions which improve the applicability of PH distributions in different fields. As an example, we discuss the efficient simulation of PH distributions.

Besides the favorable properties of PH distributions in distribution approximation, another essential ingredient of their success and popularity is a set of efficient computational procedures to evaluate the behavior of stochastic models which contain PH distributions. A part of these procedures is referred to as matrix analytic methods [NEU 81c]. The treatment of these methods is out of the scope of this book.

## I.3. The structure of the book

After the introduction, in Chapter 1, we collect most of the mathematical background which we use in later chapters. Readers who are familiar with the analysis of Markov chains might skip this chapter and return only when a piece of background information becomes of interest where it is used.

Chapters 2 and 3 introduce continuous time and discrete time PH distributions, respectively, and investigate their basic properties, like the distribution function, moments, etc. Additionally, many other useful properties are studied that characterize the flexibility of these classes of PH distributions.

Chapter 4 gives an outlook toward more general classes of distributions that have similar applicability as PH distributions but lack a stochastic interpretation based on the associated Markov chains.

The application of PH distributions started with special combinations of exponentially distributed random variables. Still today, certain subclasses of PH distributions play an important role in many applications. Chapter 5 summarizes the most frequently used subclasses and their properties.

Chapters 6 and 7 address the approximation of positive distributions with PH distributions. The first one discusses the matching approaches while the second one discusses the fitting methods and their respective properties.

When PH distributions are used in complex stochastic models whose complexity require discrete event simulation-based evaluation, the efficient simulation of PH distributions is of essential importance. Chapter 8 is devoted to this subject.

The analytical tractability and flexibility of PH distributions resulted in a line of research where various functions of PH-distributed random variables are applied to approximate distributions with special properties. The final chapter, Chapter 9, summarizes the related results.

Along the text, many numerical examples and related figures help the intuitive understanding of the presented materials. For numerical experimentation with PH distributions, readers can use many program packages that deals with PH distributions in general, for example, BuTools [BUT 15], SMCSolver [BIN 06] and KPC-Toolbox [CAS 08]. There are also a large number of more specific software tools to perform PH distribution fitting tasks, for example, HyperStar [REI 13] and PH-fit [HOR 02]. A short description of some tools is given in Appendix 1.

The previous literature on PH distributions also contains excellent textbooks that overview many properties of PH distributions. Some of these textbooks restrict their attention to a specific PH distribution related matters, for example, [BLA 17] which deals with matrix exponential distributions, and many other textbooks, in contrast to the present one, consider also the stochastic processes built from PH distributions (e.g. [NEU 81c], [BUC 14], [HE 14], [LAK 19]).

The basic properties of PH distributions are well summarized in each of these textbooks and we provide them also here for the sake of completeness. For what concerns the more involved properties, we present many specific features of PH distributions which have not been described in previous textbooks, for example, the eigenvalue structure in sections 2.8 and 3.7, the steepest increase property in section 2.9 and the properties of moments in Chapter 6.

# 1

# Mathematical Background

In this section, we collect some basic mathematical concepts. We do not intend to provide an exhaustive description of these concepts but to collect only the properties which we will use later on.

## 1.1. Basic properties of random variables

If $X$ is a random variable (RV) on a subset of $\mathbb{R}$, then its cumulative distribution function (CDF) is defined as $F_X(x) = \mathrm{P}(X < x)$. $F_X(x)$ is a monotone increasing function.

If $F_X(x)$ is continuous, then $X$ is said to be a continuous RV. A continuous RV is characterized by its probability density function (PDF) which is defined as:

$$f_X(x) = \frac{d}{dx} F_X(x).$$

Further properties of a continuous RV are:

$$
\begin{aligned}
m_k &= \mathbb{E}(X^k) = \int_{x=-\infty}^{\infty} t^k f_X(x)dx, && k\text{th moment} \\
D^2 &= m_2 - m_1^2, && \text{variance} \\
SCV &= \frac{m_2 - m_1^2}{m_1^2}, && \text{squared coefficient of variation}
\end{aligned}
\qquad [1.1]
$$

$$
\begin{aligned}
\lambda_X(x) &= \frac{f_X(x)}{1 - F_X(x)}, && \text{hazard function} \\
F_\tau(x) &= \mathrm{P}(X - \tau < x | X > \tau) = \frac{F_X(x+\tau) - F_X(\tau)}{1 - F_X(\tau)}, && \text{remaining lifetime}
\end{aligned}
$$

where the squared coefficient of variation (SCV) is a normalized version of the variance and the remaining lifetime distribution describes the distribution of $X - \tau$ (how much $X$ is greater than $\tau$) if $X$ is larger than $\tau$.

If $F_X(x)$ is piecewise constant with positive jumps between the constant intervals, then $X$ is said to be a discrete RV. A discrete random variable is characterized by its probability mass function (PMF), which describes the points, $x_i$, where $F_X(x)$ has a jump and the probabilities, $p_i$, which are the sizes of the jumps at $x_i$. That is $P(X = x_i) = p_i$. Based on the PMF, the moments of $X$ are obtained as:

$$m_k = \mathbb{E}(X^k) = \sum_i x_i^k p_i.$$

The variance and the SCV are obtained from $m_k$ in the same way as for continuous RVs.

If $F_X(x)$ is neither continuous nor piecewise constant, then $X$ is said to be a mixed RV and its analysis is based on its CDF.

## 1.2. Moments of random variables and related quantities

For a (non-defective) random variable, $m_0 = 1$. Besides the mean $m_1$, the most often used parameters related to the moments are the variance and the SCV [1.1].

In this work, we use several variants of the moments for notational convenience:

– the central moments

$$c_k = \mathbb{E}((X - m_1)^k) = \sum_{j=0}^{k} \binom{k}{j} m_j (-m_1)^{k-j}, \qquad [1.2]$$

– the factorial moments

$$f_k = \mathbb{E}(X(X-1)\dots(X-k+1)), \qquad [1.3]$$

– the reduced moments

$$r_k = \frac{m_k}{k!}, \qquad [1.4]$$

– the normalized moments

$$n_k = \frac{m_k}{m_{k-1} m_1}. \qquad [1.5]$$

There are some essential relations among moments [STI 94]. For any random variable:

$$\det \begin{pmatrix} m_0 & m_1 & m_2 & \cdots & m_{k-1} & m_k \\ m_1 & m_2 & \iddots & \iddots & \iddots & m_{k+1} \\ m_2 & \iddots & \iddots & \iddots & \iddots & \vdots \\ \vdots & \iddots & \iddots & \iddots & \iddots & \vdots \\ m_{k-1} & \iddots & \iddots & \iddots & \iddots & m_{2k-1} \\ m_k & m_{k+1} & \cdots & \cdots & m_{2k-1} & m_{2k} \end{pmatrix} \geq 0 \qquad [1.6]$$

for $k \geq 0$. The matrix composed of the moments according to [1.6] is referred to as the Hankel matrix. For $k = 1$, we have $\det \begin{pmatrix} m_0 & m_1 \\ m_1 & m_2 \end{pmatrix} = m_2 m_0 - m_1^2 \geq 0$, that is, the variance is non-negative. Also, for any non-negative random variable:

$$\det \begin{pmatrix} m_1 & m_2 & m_3 & \cdots & m_{k-1} & m_k \\ m_2 & m_3 & \iddots & \iddots & \iddots & m_{k+1} \\ m_3 & \iddots & \iddots & \iddots & \iddots & \vdots \\ \vdots & \iddots & \iddots & \iddots & \iddots & \vdots \\ m_{k-1} & \iddots & \iddots & \iddots & \iddots & m_{2k-2} \\ m_k & m_{k+1} & \cdots & \cdots & m_{2k-2} & m_{2k-1} \end{pmatrix} \geq 0 \qquad [1.7]$$

for $k \geq 1$.

## 1.3. Laplace transformation

The Laplace transform (LT) of function $f$ is defined as:

$$f^*(s) = \int_{x=0}^{\infty} e^{-sx} f(x) dx. \qquad [1.8]$$

COMMENT 1.1.– *If* $\int_{x=0}^{\infty} |f(x)|dx \leq \infty$, *then the integral in* [1.8] *converges for* $\mathcal{R}(s) \geq 0$ *(where* $\mathcal{R}(s)$ *denotes the real part of s) because for* $\mathcal{R}(s) \geq 0$, *we have:*

$$\int_{x=0}^{\infty} |e^{-sx} f(x)| dx \leq \int_{x=0}^{\infty} |e^{-sx}| \cdot |f(x)| dx \leq \int_{x=0}^{\infty} |f(x)| dx. \qquad [1.9]$$

The LT of a non-negative RV $X$ is defined based on its PDF and denoted as:

$$f_X^*(s) = \mathbb{E}(e^{-sX}) = \int_{x=0}^{\infty} e^{-sx} f_X(x) dx.$$

The LT can be efficiently used in the analysis of non-negative RVs, for example:

– the $n$th moment of $X$ can be obtained as $(-1)^n \frac{d^n}{ds^n} f_X^*(s)\Big|_{s=0}$ because:

$$(-1)^n \frac{d^n}{ds^n} f_X^*(s)\Big|_{s=0} = (-1)^n \frac{d^n}{ds^n} \int_{x=0}^{\infty} e^{-sx} f(x)dx\Big|_{s=0}$$
$$= (-1)^n \int_{x=0}^{\infty} (-x)^n e^{-sx} f(x)dx\Big|_{s=0} = \int_{x=0}^{\infty} x^n f(x)dx = \mathbb{E}(X^n), \quad [1.10]$$

– when $X$ and $Y$ are independent RVs the LT of $X + Y$ is:

$$f_{X+Y}^*(s) = \mathbb{E}\left(e^{-s(X+Y)}\right) = \mathbb{E}\left(e^{-sX} e^{-sY}\right)$$
$$= \mathbb{E}\left(e^{-sX}\right) \mathbb{E}\left(e^{-sY}\right) = f_X^*(s) f_Y^*(s). \quad [1.11]$$

## 1.4. $z$ transform

The $z$ transform [1] of the sequence $\{a_0, a_1, \ldots\}$ is defined as:

$$A(z) = \sum_{k=0}^{\infty} z^k a_k. \quad [1.12]$$

If $\sum_{k=0}^{\infty} |a_k|$ is finite, then the summation in [1.12] converges for $|z| \leq 1$.

If $T$ is a RV with support on the set of natural numbers with PMF $a_k = \mathrm{P}(X = k)$, then its $z$ transform is defined as:

$$T(z) = \mathbb{E}(z^T) = \sum_{k=0}^{\infty} z^k a_k.$$

Analogously to the LT, the $z$ transform can be used for the analysis of discrete RVs:

– the $n$th factorial moment of $T$ can be computed as:

$$\frac{d^n}{dz^n} T(z)\Big|_{z=1} = \frac{d^n}{dz^n} \sum_{k=0}^{\infty} z^k a_k\Big|_{z=1}$$
$$= \sum_{k=0}^{\infty} k(k-1)\ldots(k-n+1) z^{k-n} a_k\Big|_{z=1}$$
$$= \mathbb{E}(X(X-1)\ldots(X-n+1)) = f_n, \quad [1.13]$$

1 The $z$ transform, used in engineering terminology, is identical with the momentum generating function, which is more commonly used in probability theory and statistics.

– when $X$ and $Y$ are independent RVs the $z$ transform of $Z = X + Y$ is:

$$\begin{aligned} Z(z) &= \mathbb{E}\left(z^{X+Y}\right) = \mathbb{E}\left(z^X z^Y\right) \\ &= \mathbb{E}\left(z^X\right)\mathbb{E}\left(z^Y\right) = X(z)Y(z). \end{aligned} \qquad [1.14]$$

## 1.5. Matrix functions of quadratic matrices

Matrix functions of quadratic matrices are defined by the series expansion of the function. If $\boldsymbol{Q}$ is a quadratic matrix and $F$ is an analytic function with series expansion $F(x) = \sum_{i=0}^{\infty} a_i x^i$, then:

$$F(\boldsymbol{Q}) = \sum_{i=0}^{\infty} a_i \boldsymbol{Q}^i.$$

The most often used matrix function $e^{\boldsymbol{Q}t}$ is defined through its series expansion as:

$$e^{\boldsymbol{Q}t} = \sum_{i=0}^{\infty} \frac{t^i}{i!} \boldsymbol{Q}^i.$$

## 1.6. Matrix inverse

The matrix inverse of quadratic matrices can be introduced and computed in several different ways. We use the relation that element $(i, j)$ of the inverse of matrix $\boldsymbol{M}$ is:

$$(\boldsymbol{M}^{-1})_{ij} = \frac{\det_{ji}(\boldsymbol{M})}{\det(\boldsymbol{M})}, \qquad [1.15]$$

where $\det(\boldsymbol{M})$ is the determinant of matrix $\boldsymbol{M}$ and $\det_{ij}(\boldsymbol{M})$ denotes the determinant of the matrix obtained by removing row $i$ and column $j$ from matrix $\boldsymbol{M}$ multiplied by $(-1)^{i+j}$.

## 1.7. Eigenvalues and the characteristic polynomial

The eigenvalues of a real quadratic matrix $\boldsymbol{Q}$ of size $n \times n$ are the roots of the degree $n$ characteristic polynomial $C_{\boldsymbol{Q}}(x) = \det(x\boldsymbol{I} - \boldsymbol{Q})$ where $\boldsymbol{I}$ is the identity matrix.

LEMMA 1.2.– *For matrix $\boldsymbol{M}$ of size $n \times n$, row vector $\boldsymbol{\beta} = \begin{pmatrix} b_0 & b_1 & \dots & b_{n-1} \end{pmatrix}$ and column vector $\mathbf{e}_n = \begin{pmatrix} 0 & 0 & \dots & 0 & 1 \end{pmatrix}^T$ of size $n$, we have:*

$$\boldsymbol{\beta}(x\boldsymbol{I} - \boldsymbol{M})^{-1}\mathbf{e}_n = \frac{b_{n-1}x^{n-1} + b_{n-2}x^{n-2} + \ldots + b_1 x + b_0}{x^n + a_{n-1}x^{n-1} + \ldots + a_1 x + a_0}, \qquad [1.16]$$

*where matrix $\boldsymbol{M}$ is:*

$$\boldsymbol{M} = \begin{pmatrix} 0 & 1 & & \\ & 0 & \ddots & \\ & & \ddots & 1 \\ -a_0 & -a_1 & \dots & -a_{n-1} \end{pmatrix}.$$

*Proof.* The characteristic polynomial of matrix $\boldsymbol{M}$ is $C_{\boldsymbol{M}}(x) = \det(x\boldsymbol{I} - \boldsymbol{M}) = x^n + a_{n-1}x^{n-1} + \ldots + a_1 x + a_0$.

Applying [1.15] for element $(i, n)$ of $(x\boldsymbol{I} - \boldsymbol{M})^{-1}$ for $1 \le i \le n-1$, we have:

$$(x\boldsymbol{I} - \boldsymbol{M})^{-1}_{in} = \frac{1}{C_{\boldsymbol{M}}(x)} \det{}_{ni} \begin{pmatrix} x & -1 & & \\ & x & \ddots & \\ & & \ddots & -1 \\ a_0 & a_1 & \dots & x + a_{n-1} \end{pmatrix}$$

$$= \frac{(-1)^{i+n}}{C_{\boldsymbol{M}}(x)} \det \begin{pmatrix} \boldsymbol{A} & \boldsymbol{0} \\ \boldsymbol{0} & \boldsymbol{B} \end{pmatrix} = \frac{(-1)^{i+n}}{C_{\boldsymbol{M}}(x)} \det(\boldsymbol{A}) \det(\boldsymbol{B})$$

$$= \frac{(-1)^{i+n}}{C_{\boldsymbol{M}}(x)} x^{i-1}(-1)^{n-i} = \frac{x^{i-1}}{C_{\boldsymbol{M}}(x)},$$

where $\boldsymbol{A}$ of size $(i-1) \times (i-1)$ and $\boldsymbol{B}$ of size $(n-i) \times (n-i)$ are:

$$\boldsymbol{A} = \begin{pmatrix} x & -1 & & \\ & x & \ddots & \\ & & \ddots & -1 \\ & & & x \end{pmatrix} \quad \text{and} \quad \boldsymbol{B} = \begin{pmatrix} -1 & & & \\ x & -1 & & \\ & \ddots & \ddots & \\ & & x & -1 \end{pmatrix}.$$

Finally, multiplying $(x\boldsymbol{I} - \boldsymbol{M})^{-1}$ by the row vector $\boldsymbol{\beta}$ from the left and by the column vector $\mathbf{e}_n$ from the right, we obtain the statement of the lemma. □

## 1.8. Spectral decomposition

For a real quadratic matrix $\boldsymbol{Q}$ of size $n \times n$, the spectral decomposition (also referred to as the Jordan decomposition) is a factorization:

$$\boldsymbol{Q} = \boldsymbol{B}^{-1} \boldsymbol{\Lambda} \boldsymbol{B},$$

where $\boldsymbol{\Lambda}$ is either a diagonal matrix with potentially complex diagonal elements containing the eigenvalues (in this case $\boldsymbol{Q}$ is referred to as diagonalizable), or $\boldsymbol{\Lambda}$ is a block diagonal matrix, referred to as the Jordan matrix, in the form:

$$\boldsymbol{\Lambda} = \begin{pmatrix} \boldsymbol{J}_{n_1}(\lambda_1) & & & \\ & \boldsymbol{J}_{n_2}(\lambda_2) & & \\ & & \ddots & \\ & & & \boldsymbol{J}_{n_k}(\lambda_k) \end{pmatrix}, \qquad [1.17]$$

whose diagonal blocks are called Jordan blocks. The size $u \times u$ Jordan block associated with eigenvalue $\lambda$ is:

$$\boldsymbol{J}_u(\lambda) = \begin{pmatrix} \lambda & 1 & & \\ & \lambda & \ddots & \\ & & \ddots & 1 \\ & & & \lambda \end{pmatrix}. \qquad [1.18]$$

If all eigenvalues of a real valued square matrix are non-zero, then the matrix is referred to as non-singular or full rank and its inverse exists.

If $\boldsymbol{Q}$ is diagonalizable, then:

$$\boldsymbol{Q} = \boldsymbol{B}^{-1}\boldsymbol{\Lambda}\boldsymbol{B} = \sum_{\ell=1}^{n} \lambda_\ell \boldsymbol{u}_\ell \boldsymbol{v}_\ell,$$

where $\lambda_\ell$ is the $\ell$th diagonal element of $\boldsymbol{\Lambda}$ referred to as its $\ell$th eigenvalue, $\boldsymbol{u}_\ell$ is the $\ell$th column of $\boldsymbol{B}^{-1}$ referred to as its $\ell$th right eigenvector, and $\boldsymbol{v}_\ell$ is the $\ell$th row of $\boldsymbol{B}$ referred to as its $\ell$th left eigenvector. The spectral decomposition is not unique, for example, a permutation of $\lambda_\ell, \boldsymbol{u}_\ell, \boldsymbol{v}_\ell$ results in the same matrix $\boldsymbol{Q}$. Also the multiplication of $\boldsymbol{u}_\ell$ and the division of $\boldsymbol{v}_\ell$ by the same non-zero constant leaves the results invariant. When $\boldsymbol{Q}$ is real, the complex eigenvalues appear in complex conjugate pairs and the associated left and right eigenvectors are complex conjugates as well. The eigenvectors associated with real eigenvalues are real (or can be transformed to be real by the multiplication of $\boldsymbol{u}_\ell$ and the division of $\boldsymbol{v}_\ell$ by the same non-zero constant).

From $\boldsymbol{B}\boldsymbol{B}^{-1} = \boldsymbol{I}$, it also follows for the rows of $\boldsymbol{B}$ and the columns of $\boldsymbol{B}^{-1}$ that $\boldsymbol{v}_k \boldsymbol{u}_\ell = \delta_{k\ell}$ where $\delta_{k\ell}$ is the Kronecker delta ($\delta_{k\ell} = 1$ if $k = \ell$ and zero otherwise). This implies that the eigenvectors associated with different eigenvalues are orthogonal.

The $\sum_{\ell=1}^{n} \lambda_\ell \boldsymbol{u}_\ell \boldsymbol{v}_\ell$ representation of a diagonalizable matrix is also referred to as dyad decomposition because $\boldsymbol{u}_\ell \boldsymbol{v}_\ell$ is a dyadic matrix. The matrix functions of diagonalizable matrices are easy to obtain via dyad decomposition as:

$$F(\boldsymbol{Q}) = \sum_{\ell=1}^{n} F(\lambda_\ell) \boldsymbol{u}_\ell \boldsymbol{v}_\ell.$$

It is because, assuming $F(x) = \sum_{i=0}^{\infty} a_i x^i$ is the series expansion of $F(x)$, we can write:

$$F(\boldsymbol{Q}) = \sum_{i=0}^{\infty} a_i \boldsymbol{Q}^i = \sum_{i=0}^{\infty} a_i \left( \sum_{\ell=1}^{n} \lambda_\ell \boldsymbol{u}_\ell \boldsymbol{v}_\ell \right)^i = \sum_{i=0}^{\infty} a_i \sum_{\ell=1}^{n} \lambda_\ell^i \boldsymbol{u}_\ell \boldsymbol{v}_\ell$$

$$= \sum_{\ell=1}^{n} \sum_{i=0}^{\infty} a_i \lambda_\ell^i \boldsymbol{u}_\ell \boldsymbol{v}_\ell = \sum_{\ell=1}^{n} F(\lambda_\ell) \boldsymbol{u}_\ell \boldsymbol{v}_\ell.$$

If $\boldsymbol{Q}$ is non-diagonalizable, then the dyad decomposition cannot be used directly. In those cases, the powers of the Jordan blocks play a role in the analysis of matrix functions. For example, the $k$th power of a Jordan block with eigenvalue $\lambda$ is:

$$\begin{pmatrix} \lambda & 1 & & \\ & \lambda & 1 & \\ & & \ddots & \ddots \\ & & & \lambda \end{pmatrix}^k = \begin{pmatrix} \lambda^k & \lambda^{(k-1)^+} & \lambda^{(k-2)^+} & \cdots \\ & \lambda^k & \lambda^{(k-1)^+} & \cdots \\ & & \ddots & \vdots \\ & & & \lambda^k \end{pmatrix}, \qquad [1.19]$$

where $(k)^+$ denotes $\max(k, 0)$.

The following theorems provide relations between matrices with identical Jordan matrices or matrix blocks.

THEOREM 1.3.– *If the Jordan matrix of matrices $\boldsymbol{R}$ and $\boldsymbol{S}$ (both of size $n \times n$) are identical, that is, with appropriately ordered eigenvalues the Jordan decomposition of $\boldsymbol{R}$ and $\boldsymbol{S}$ are $\boldsymbol{R} = \boldsymbol{G}^{-1}\boldsymbol{\Lambda}\boldsymbol{G}$ and $\boldsymbol{S} = \boldsymbol{H}^{-1}\boldsymbol{\Lambda}\boldsymbol{H}$, then there exists a non-singular matrix $\boldsymbol{W}$ such that $\boldsymbol{R} = \boldsymbol{W}^{-1}\boldsymbol{S}\boldsymbol{W}$.*

*Proof.* With $\boldsymbol{W} = \boldsymbol{H}^{-1}\boldsymbol{G}$, we have:

$$\boldsymbol{W}^{-1}\boldsymbol{S}\boldsymbol{W} = \boldsymbol{G}^{-1}\boldsymbol{H}\boldsymbol{H}^{-1}\boldsymbol{\Lambda}\boldsymbol{H}\boldsymbol{H}^{-1}\boldsymbol{G} = \boldsymbol{G}^{-1}\boldsymbol{\Lambda}\boldsymbol{G} = \boldsymbol{R}. \qquad \square$$

THEOREM 1.4.– *If the Jordan matrix of matrix $\boldsymbol{R}$ of size $n \times n$ is $\boldsymbol{\Lambda}_R$ and with appropriately ordered eigenvalues, the Jordan matrix of matrix $\boldsymbol{S}$ of size $m \times m$ with $m > n$ is:*

$$\boldsymbol{\Lambda}_S = \begin{pmatrix} \boldsymbol{\Lambda}_R & \boldsymbol{0} \\ \boldsymbol{0} & \boldsymbol{\Lambda}_{\bar{R}} \end{pmatrix},$$

*then there exists a matrix $\boldsymbol{W}$ of size $n \times m$ such that $\boldsymbol{R}\boldsymbol{W} = \boldsymbol{W}\boldsymbol{S}$.*

*Proof.* The block structure of the Jordan decomposition of $\boldsymbol{S}$ is:

$$\boldsymbol{S} = \boldsymbol{H}^{-1}\boldsymbol{\Lambda}_S\boldsymbol{H} = (\boldsymbol{U}_1\ \boldsymbol{U}_2) \begin{pmatrix} \boldsymbol{\Lambda}_R & \boldsymbol{0} \\ \boldsymbol{0} & \boldsymbol{\Lambda}_{\bar{R}} \end{pmatrix} \begin{pmatrix} \boldsymbol{V}_1 \\ \boldsymbol{V}_2 \end{pmatrix}$$

where $U_1$ is of size $m \times n$ and the rest of the block dimensions follow. From $HH^{-1} = I$, we have $V_1U_1 = I$ and $V_1U_2 = 0$. Using this block structure, we have:

$$\begin{aligned} G^{-1}V_1S &= G^{-1}V_1 \left(U_1 \ U_2\right) \begin{pmatrix} \Lambda_R & 0 \\ 0 & \Lambda_{\bar{R}} \end{pmatrix} \begin{pmatrix} V_1 \\ V_2 \end{pmatrix} \\ &= G^{-1} \left(I \ 0\right) \begin{pmatrix} \Lambda_R & 0 \\ 0 & \Lambda_{\bar{R}} \end{pmatrix} \begin{pmatrix} V_1 \\ V_2 \end{pmatrix} = G^{-1} \left(\Lambda_R \ 0\right) \begin{pmatrix} V_1 \\ V_2 \end{pmatrix} \\ &= G^{-1}\Lambda_R V_1, \end{aligned}$$

and

$$RG^{-1}V_1 = G^{-1}\Lambda_R G G^{-1}V_1 = G^{-1}\Lambda_R V_1.$$

That is, $W = G^{-1}V_1$. □

In the case of diagonalizable matrices, theorem 1.3 implies that the eigenvalues of matrix $R$ and matrix $S$ are identical, while theorem 1.4 means that the eigenvalues of matrix $R$ is a subset of the eigenvalues of matrix $S$.

DEFINITION 1.5.– *The Jordan decomposition $Q = B^{-1}\Lambda B$ is referred to as* normalized *if $B\mathbb{1} = \mathbb{1}$, where $\mathbb{1}$ denotes the column vector of ones.*

Not all real matrices have a normalized Jordan decomposition but the generator matrices associated with PH distributions have normalized Jordan decomposition as shown in [TEL 07, theorem 5]. We note that multiplying $B\mathbb{1} = \mathbb{1}$ with $B^{-1}$ from the left implies $B^{-1}\mathbb{1} = \mathbb{1}$.

COROLLARY 1.6.– *If the Jordan decomposition of $R$ and $S$ are normalized in theorem 1.3, then the transformation matrix provided in the proof of the theorem satisfies $W\mathbb{1} = \mathbb{1}$.*

*Proof.* If the Jordan decomposition of $R$ and $S$ are normalized, then $G\mathbb{1} = \mathbb{1}$, $H^{-1}\mathbb{1} = \mathbb{1}$ and consequently:

$$W\mathbb{1} = H^{-1}G\mathbb{1} = H^{-1}\mathbb{1} = \mathbb{1}.$$ □

COROLLARY 1.7.– *If the Jordan decomposition of $R$ and $S$ are normalized in theorem 1.4, then the transformation matrix satisfies $W\mathbb{1} = \mathbb{1}$.*

*Proof.* If the Jordan decomposition of $R$ and $S$ are normalized, then $G^{-1}\mathbb{1} = \mathbb{1}$, $H\mathbb{1} = \mathbb{1}$, and consequently $V_1\mathbb{1} = \mathbb{1}$. Using it, we have:

$$W\mathbb{1} = G^{-1}V_1\mathbb{1} = G^{-1}\mathbb{1} = \mathbb{1}.$$ □

If matrices $\boldsymbol{R}$ and $\boldsymbol{S}$ satisfy the condition of theorem 1.4, then the proof of the theorem provides a constructive way to compute $\boldsymbol{W}$ using the Jordan decomposition of $\boldsymbol{R}$ and $\boldsymbol{S}$, which is a computationally sensitive task. If the Jordan decomposition of $\boldsymbol{R}$ and $\boldsymbol{S}$ can be normalized, then $\boldsymbol{W}$ can also be obtained by solving the linear system of equations $\boldsymbol{RW} = \boldsymbol{WS}$ and $\boldsymbol{W}\mathbb{1} = \mathbb{1}$.

## 1.9. Ordinary differential equation of vector functions

Let $\boldsymbol{p}(t) = \big(p_1(t) \; \dots \; p_n(t)\big)$ be a row vector function of size $n$ and $\boldsymbol{Q}$ be a matrix of size $n \times n$. We often make use of the ordinary differential equation (ODE):

$$\frac{d}{dt}\boldsymbol{p}(t) = \boldsymbol{p}(t)\boldsymbol{Q}, \qquad [1.20]$$

with initial condition $\boldsymbol{p}(0) = \boldsymbol{\pi}$. The scalar version of this ODE is:

$$\frac{d}{dt}p_i(t) = \sum_{j=1}^{n} p_j(t)\boldsymbol{Q}_{ji},$$

which indicates that the derivative of the vector function is taken element-wise.

LEMMA 1.8.– *The solution of* [1.20] *is:*

$$\boldsymbol{p}(t) = \boldsymbol{\pi} e^{\boldsymbol{Q}t}. \qquad [1.21]$$

*Proof.* To show that [1.21] satisfies [1.20], we write:

$$\frac{d}{dt}e^{\boldsymbol{Q}t} = \sum_{i=0}^{\infty} \frac{d}{dt}\frac{t^i}{i!}\boldsymbol{Q}^i = \sum_{i=1}^{\infty} \frac{t^{i-1}}{(i-1)!}\boldsymbol{Q}^i = \sum_{i=1}^{\infty} \frac{t^{i-1}}{(i-1)!}\boldsymbol{Q}^{i-1}\boldsymbol{Q} = e^{\boldsymbol{Q}t}\boldsymbol{Q}$$

$$= \boldsymbol{Q}\sum_{i=1}^{\infty} \frac{t^{i-1}}{(i-1)!}\boldsymbol{Q}^{i-1} = \boldsymbol{Q}e^{\boldsymbol{Q}t}.$$

□

The derivation in the proof also indicates that $e^{\boldsymbol{Q}t}$ and $\boldsymbol{Q}$ commute.

If $\boldsymbol{Q}$ is diagonalizable with dyad decomposition $\boldsymbol{Q} = \sum_{\ell=1}^{n} \lambda_\ell \boldsymbol{u}_\ell \boldsymbol{v}_\ell$, then:

$$e^{\boldsymbol{Q}t} = \sum_{i=0}^{\infty} \frac{t^i}{i!}\boldsymbol{Q}^i = \sum_{i=0}^{\infty} \frac{t^i}{i!}\sum_{\ell=1}^{n} \lambda_\ell^i \boldsymbol{u}_\ell \boldsymbol{v}_\ell = \sum_{\ell=1}^{n} e^{\lambda_\ell t}\boldsymbol{u}_\ell \boldsymbol{v}_\ell.$$

If an eigenvalue has positive real part, then the related $e^{\lambda_\ell t}$ function diverges as $t$ tends to infinity. If all eigenvalues have negative real parts, then all $e^{\lambda_\ell t}$ functions converge to zero as $t$ tends to infinity. If all eigenvalues have negative real parts except one which is zero, then all $e^{\lambda_\ell t}$ functions converge to zero except the one with zero eigenvalue which is constant.

As a result, if all eigenvalues of $\boldsymbol{Q}$ have negative real parts, then $\int_{t=0}^{\infty} e^{\boldsymbol{Q}t} dt$ converges and:

$$\int_{t=0}^{\infty} e^{\boldsymbol{Q}t} dt = -\boldsymbol{Q}^{-1} \tag{1.22}$$

which is easily obtained from the dyad decomposition if $\boldsymbol{Q}$ is diagonalizable. In general, it follows from:

$$\begin{aligned}\int_{t=0}^{\infty} e^{\boldsymbol{Q}t} dt &= \lim_{T\to\infty} \sum_{i=0}^{\infty} \boldsymbol{Q}^i \int_{t=0}^{T} \frac{t^i}{i!} dt \\ &= \lim_{T\to\infty} \sum_{i=0}^{\infty} \boldsymbol{Q}^i \frac{T^{i+1}}{(i+1)!} = \boldsymbol{Q}^{-1}\Big(\underbrace{\lim_{T\to\infty} e^{\boldsymbol{Q}T}}_{\to \boldsymbol{0}} - \boldsymbol{I}\Big).\end{aligned}$$

If at least one eigenvalue of $\boldsymbol{Q}$ has non-negative real parts, then $\int_{t=0}^{\infty} e^{\boldsymbol{Q}t} dt$ does not converge.

## 1.10. Exponential distribution

The exponential distribution plays an important role in the theory of Markov chains. If random variable $X$ is exponentially distributed with parameter $\lambda$, then:

$$F_X(t) = \mathrm{P}(X < t) = 1 - \lambda e^{-\lambda t}, \qquad \text{cumulative distribution function}$$

$$f_X(t) = \frac{d}{dt} F_X(t) = \lambda e^{-\lambda t}, \qquad \text{probability density function}$$

$$\lambda_X(t) = \frac{f_X(t)}{1 - F_X(t)} = \lambda, \qquad \text{hazard function}$$

$$f_X^*(s) = \mathbb{E}\big(e^{-sX}\big) = \int_{t=0}^{\infty} e^{-st} f_X(t) dt = \frac{\lambda}{s+\lambda}, \qquad \text{Laplace transform}$$

$$m_i = \mathbb{E}\big(X^i\big) = \int_{t=0}^{\infty} t^i f_X(t) dt = \frac{i!}{\lambda^i}, \qquad n\text{th moment}$$

$$D^2 = m_2 - m_1^2 = \frac{1}{\lambda^2}, \qquad \text{variance}$$

$$SCV = \frac{m_2 - m_1^2}{m_1^2} = 1, \qquad \text{squared coefficient of variation}$$

$$r_n = \frac{m_n}{n!} = \frac{1}{\lambda^n}, \qquad n\text{th reduced moment}$$

$$F_\tau(t) = \text{P}(X - \tau < t | X > \tau) =$$
$$= \frac{F_X(t+\tau) - F_X(\tau)}{1 - F_X(\tau)} = 1 - \lambda e^{-\lambda t}. \text{ CDF of the remaining lifetime}$$

The fact that the remaining lifetime of an exponentially distributed RV is independent of $\tau$ and is identical with its original distribution is referred to as *memoryless property*. The constant hazard rate of exponential distribution is due to this memoryless property.

## 1.11. Erlang distribution

The Erlang distribution is closely related to the exponential distribution. The sum of $n$ independent, identically distributed exponential random variables, $Y = \sum_{i=1}^{n} X_i$, where $X_i$ is exponentially distributed with parameter $\lambda$, is Erlang distributed with shape parameter $n$ and rate parameter $\lambda$. We denote this distribution as Erlang($n, \lambda$). If the rate is not defined, we denote the Erlang distribution with shape parameter $n$ as Erlang($n$).

The PDF, the LT and the moments of the Erlang($n, \lambda$) distribution are, respectively:

$$f(t) = \frac{\lambda^n t^{n-1}}{(n-1)!} e^{-\lambda t}, \qquad [1.23]$$

$$f^*(s) = \left( \frac{\lambda}{s+\lambda} \right)^n, \qquad [1.24]$$

$$\mathbb{E}(Y^k) = \frac{(n+k-1)!}{(n-1)!\lambda^k}, \qquad [1.25]$$

where the LT can be obtained using [1.11].

## 1.12. Discrete time Markov chain

The are several excellent textbooks introducing Markov chains (e.g. [TRI 01, BOL 06, LAK 19]). In this and the next sections, we summarize only those properties which are used later in this book for completeness.

Let $X_k$ denote the state of a finite state discrete time Markov chain (DTMC) after $k$ steps of the process where the state space is composed by $n$ states, that is $X_k \in \{1, 2, \ldots, n\}$. For $i \in \{1, 2, \ldots, n\}$, we denote the state probabilities as $p_i(k) = \Pr(X_k = i)$, which is the probability that the DTMC is in state $i$ after $k$ steps. The row vector of the state probabilities is $\boldsymbol{p}(k) = \begin{pmatrix} p_1(k) & \ldots & p_n(k) \end{pmatrix}$. The probability that the Markov chain moves from state $i$ to state $j$ in the $k$th step is denoted by $\Pi_{ij}(k)$ where $\Pi_{ij}(k) = \Pr(X_k = j | X_{k-1} = i)$. We are interested in

Markov chains where this probability does not depend on the time, that is, the process is time inhomogeneous. In this case, $\Pi_{ij}(k) = \Pi_{ij}$ for $\forall k$. The matrix composed of the state transition probabilities, $\Pi_{ij}$, is denoted by $\mathbf{\Pi}$ and referred to as the one-step state transition probability matrix. The transient behavior of the DTMC with transition probability matrix $\mathbf{\Pi}$ satisfies:

$$p_i(k) = \sum_{j=1}^{n} p_j(k-1)\mathbf{\Pi}_{ji},$$

which is the law of total probability for the state of the process at time $k-1$. For the row vector of the state probabilities, we can write:

$$\boldsymbol{p}(k) = \boldsymbol{p}(k-1)\mathbf{\Pi},$$

from which

$$\boldsymbol{p}(k) = \boldsymbol{p}(0)\mathbf{\Pi}^k$$

where $\boldsymbol{p}(0) = \boldsymbol{\pi}$ denotes the initial distribution of the process.

The transition probability matrix $\mathbf{\Pi}$ and initial distribution $\boldsymbol{\pi}$ have the following properties. The elements of $\boldsymbol{\pi}$ and $\mathbf{\Pi}$ are probabilities, that is, non-negative reals not greater than one. Each row of $\mathbf{\Pi}$ and also $\boldsymbol{\pi}$ describes a probability distribution, that is, $\sum_i \pi_i = 1$ and $\sum_j \Pi_{ij} = 1\ \forall i$. With the introduction of the column vector of ones, $\mathbb{1}$, these summations can be written in the following matrix and vector forms, $\mathbf{\Pi}\mathbb{1} = \mathbb{1}$ and $\boldsymbol{\pi}\mathbb{1} = 1$. From $\mathbf{\Pi}\mathbb{1} = \mathbb{1}$, we have that 1 is always an eigenvalue of $\mathbf{\Pi}$ (and $\mathbb{1}$ is an associated right eigenvector). A matrix with these properties (non-negative with row sum equal to one) is also referred to as stochastic matrix. The eigenvalues of a stochastic matrix are on the closed unit disk of the complex plain, that is, the absolute value of any eigenvector is not greater than one.

A DTMC is said to be *irreducible* if any state $j$ is reachable from any state $i$ by a finite non-zero $\Pi_{ik_1}, \Pi_{k_1k_2}, \ldots, \Pi_{k_{m-2}k_{m-1}}, \Pi_{k_{m-1}j}$ transition probability sequence through states $k_1, \ldots, k_{m-1}$. For an irreducible finite state DTMC, the multiplicity of the eigenvalue 1 is one.

The stationary distribution of an irreducible finite DTMC with generator $\mathbf{\Pi}$, $\boldsymbol{p} \triangleq \lim_{k\to\infty} \boldsymbol{p}(k)$, if exists, can be computed as the unique solution of the linear system of equations:

$$\boldsymbol{p}\mathbf{\Pi} = \boldsymbol{p}, \quad \boldsymbol{p}\mathbb{1} = 1, \qquad [1.26]$$

where $\boldsymbol{p}$ is indeed the normalized left eigenvector associated with eigenvalue one. If $\lim_{k\to\infty} \boldsymbol{p}(k)$ does not exist (due to some periodic behavior which results in eigenvalues on the unit cycle, e.g., $-1$), still the limit $\lim_{K\to\infty} \sum_{k=0}^{K-1} \boldsymbol{p}(k)/K$ exists.

If a DTMC is not irreducible, then it is composed of a number of transient communicating blocks and an absorbing block [KEL 78]. A *communicating block* is a maximal subset of DTMC states in which any state is reachable from any other state of the subset. A *transient* communicating block is such that:

– if the process leaves the block, it never returns to the block,

– if the process is in the block, then the absorbing block is reached after a finite number of steps.

The absorbing block is such that once the process enters the block, it never leaves the block.

The return times of state $i$ of the DTMC are the number of steps, $k$, for which $\Pr(X_k = i | X_0 = i) > 0$. Each communicating block has a cycle period. If the cycle period of a block is $m$, then the return times of the states in the block are integer multiples of $m$. If $m = 1$, then the block is said to be aperiodic; if $m > 1$, then it is said to be periodic.

## 1.13. Continuous time Markov chain

Let $X(t)$ denote the state of a finite state continuous time Markov chain (CTMC) at time $t$ with state space composed of $n$ states. We denote the state probabilities as $p_i(t) = \Pr(X(t) = i)$, which is the probability that the CTMC is in state $i$ at time $t$, and the row vector of the state probabilities as $\boldsymbol{p}(t) = \begin{pmatrix} p_1(t) & \dots & p_n(t) \end{pmatrix}$. The transient behavior of a CTMC with generator $\boldsymbol{Q}$ and initial distribution $\boldsymbol{\pi}$ satisfies the ordinary differential equation:

$$\frac{d}{dt}\boldsymbol{p}(t) = \boldsymbol{p}(t)\boldsymbol{Q} \quad \text{with initial condition} \quad \boldsymbol{p}(0) = \boldsymbol{\pi},$$

whose solution is the matrix exponential function:

$$\boldsymbol{p}(t) = \boldsymbol{\pi} e^{\boldsymbol{Q}t} = \boldsymbol{\pi} \sum_{i=0}^{\infty} \frac{t^i}{i!} \boldsymbol{Q}^i. \qquad [1.27]$$

The properties of $\boldsymbol{Q}$ and $\boldsymbol{\pi}$ are as follows. The elements of $\boldsymbol{\pi}$ are probabilities, that is, non-negative numbers not greater than one, and their sum is one, that is, $\sum_i \pi_i = \boldsymbol{\pi}\mathbb{1} = 1$. The off-diagonal elements of $\boldsymbol{Q}$ are transition intensities, that is, non-negative numbers. The diagonal elements of $\boldsymbol{Q}$ are such that each row sum is zero, that is, $\sum_j Q_{ij} = 0 \; \forall i$ (or $\boldsymbol{Q}\mathbb{1} = \boldsymbol{0}$ where $\boldsymbol{0}$ is a column vector of zeros). It follows that the diagonal elements are non-positive. Hereinafter, the sizes of vectors $\mathbb{1}$ and $\boldsymbol{0}$ are defined by the context such that the dimensions in the expressions are compatible. From $\boldsymbol{Q}\mathbb{1} = \boldsymbol{0}$, we have that 0 is always an eigenvalue of $\boldsymbol{Q}$ (and $\mathbb{1}$ is an associated right eigenvector).

A CTMC is said to be *irreducible* if any state $j$ is reachable from any state $i$ by a finite non-zero $Q_{ik_1}, Q_{k_1k_2}, \ldots, Q_{k_{m-2}k_{m-1}}, Q_{k_{m-1}j}$ sequence. For an irreducible finite state Markov chain, the multiplicity of the zero eigenvalue of $\boldsymbol{Q}$ is one.

The stationary distribution of an irreducible, finite state CTMC with generator $\boldsymbol{Q}$, $\boldsymbol{p} \triangleq \lim_{t\to\infty} \boldsymbol{p}(t)$, can be computed as the unique solution of the linear system of equations:

$$\boldsymbol{p}\boldsymbol{Q} = \boldsymbol{0}, \quad \boldsymbol{p}\mathbb{1} = 1, \qquad [1.28]$$

where $\boldsymbol{p}$ is indeed the normalized left eigenvector associated with eigenvalue zero.

## 1.14. Kronecker algebra

For a compact description of complicated matrix operations, Kronecker product (denoted as $\otimes$) and Kronecker summation (denoted as $\oplus$) are defined as follows:

$$\boldsymbol{A} \otimes \boldsymbol{B} = \begin{pmatrix} A_{11}\boldsymbol{B} & \ldots & A_{1m}\boldsymbol{B} \\ \vdots & & \vdots \\ A_{n1}\boldsymbol{B} & \ldots & A_{nm}\boldsymbol{B} \end{pmatrix},$$

$$\boldsymbol{A} \oplus \boldsymbol{B} = \boldsymbol{A} \otimes \boldsymbol{I} + \boldsymbol{I} \otimes \boldsymbol{B}.$$

The Kronecker summation is applicable only for quadratic matrices but with potentially different sizes, while the Kronecker product operation is applicable also for non-quadratic matrices of equal or different sizes.

If $\boldsymbol{A}$ and $\boldsymbol{B}$ are quadratic matrices of size $n$ and $m$ and $\alpha_i$ and $\beta_j$ are their respective eigenvalues, then the eigenvalues of $\boldsymbol{A} \otimes \boldsymbol{B}$ are $\alpha_i\beta_j$ and the eigenvalues of $\boldsymbol{A} \oplus \boldsymbol{B}$ are $\alpha_i + \beta_j$.

The Kronecker product introduces a kind of commutativity. For example, if $\boldsymbol{A}$ and $\boldsymbol{B}$ do not commute, then $\boldsymbol{AB} \neq \boldsymbol{BA}$ but we still have:

$$(\boldsymbol{A} \otimes \boldsymbol{I})(\boldsymbol{I} \otimes \boldsymbol{B}) = \boldsymbol{A} \otimes \boldsymbol{B} = (\boldsymbol{I} \otimes \boldsymbol{B})(\boldsymbol{A} \otimes \boldsymbol{I}).$$

One of the main applications of Kronecker operators is the solution of linear equations like the Sylvester equation:

$$\boldsymbol{AX} + \boldsymbol{XB} = \boldsymbol{C} \qquad [1.29]$$

where the $\boldsymbol{A}$, $\boldsymbol{B}$ and $\boldsymbol{C}$ are known matrices and $\boldsymbol{X}$ is an unknown matrix. The vec operator transforms a matrix to a column vector by listing its columns one after the other. Applying the vec operator on the left side of the Sylvester equation and using $\text{vec}(\boldsymbol{KLM}) = (\boldsymbol{M}^T \otimes \boldsymbol{K})\text{vec}(\boldsymbol{L})$ gives:

$$\begin{aligned}\text{vec}(\boldsymbol{AX}+\boldsymbol{XB}) &= \text{vec}(\boldsymbol{AXI}+\boldsymbol{IXB}) = \text{vec}(\boldsymbol{AXI}) + \text{vec}(\boldsymbol{IXB}) \\ &= \left(\boldsymbol{I}^T \otimes \boldsymbol{A} + \boldsymbol{B}^T \otimes \boldsymbol{I}\right)\text{vec}(\boldsymbol{X}) = \left(\boldsymbol{B}^T \oplus \boldsymbol{A}\right)\text{vec}(\boldsymbol{X}).\end{aligned}$$

Based on this manipulation, the Sylvester equation in [1.29] takes the form of the regular linear equation:

$$\left(\boldsymbol{B}^T \oplus \boldsymbol{A}\right)\text{vec}(\boldsymbol{X}) = \text{vec}(\boldsymbol{C}).$$

Another main application of Kronecker operators is the description of independent Markov processes. Let $X_k$ and $Y_k$ be two independent irreducible DTMCs with one step state transition probability matrix $\boldsymbol{\Pi_X}$ of size $n_X$ and $\boldsymbol{\Pi_Y}$ of size $n_Y$, respectively. The joint process $(X_k, Y_k)$ is also a DTMC over a state space of size $n_X n_Y$. If the state space of the joint process is numbered according to the lexicographical order, then the state transition probability matrix of the joint process is $\boldsymbol{\Pi} = \boldsymbol{\Pi_X} \otimes \boldsymbol{\Pi_Y}$.

Furthermore, the transient behavior of the joint DTMC is:

$$\begin{aligned}\boldsymbol{p}(k) &= (\boldsymbol{p}_X(0) \otimes \boldsymbol{p}_Y(0))\,(\boldsymbol{\Pi_X} \otimes \boldsymbol{\Pi_Y})^k = (\boldsymbol{p}_X(0) \otimes \boldsymbol{p}_Y(0))\left({\boldsymbol{\Pi_X}}^k \otimes {\boldsymbol{\Pi_Y}}^k\right) \\ &= \left(\boldsymbol{p}_X(0){\boldsymbol{\Pi_X}}^k \otimes \boldsymbol{p}_Y(0){\boldsymbol{\Pi_Y}}^k\right) = \boldsymbol{p}_X(k) \otimes \boldsymbol{p}_Y(k).\end{aligned}$$

When $\boldsymbol{\Pi_X}$ and $\boldsymbol{\Pi_Y}$ are such that 1 is a unique eigenvalue of both of them and all other eigenvalues are inside the unit disk ($X_k$ and $Y_k$ are irreducible DTMCs without periodic behavior), then 1 is a unique eigenvalue of $\boldsymbol{\Pi_X} \otimes \boldsymbol{\Pi_Y}$ and all other eigenvalues are inside the unit disk. The stationary distribution of the joint DTMC is the solution of $\boldsymbol{p} = \boldsymbol{p\Pi}$, which is $\boldsymbol{p} = \boldsymbol{p}_X \otimes \boldsymbol{p}_Y$ where $\boldsymbol{p}_X$ and $\boldsymbol{p}_Y$ denote the stationary distribution of $X_k$ and $Y_k$ satisfying $\boldsymbol{p}_X\boldsymbol{\Pi_X} = \boldsymbol{p}_X$ and $\boldsymbol{p}_Y\boldsymbol{\Pi_Y} = \boldsymbol{p}_Y$. It is because:

$$\boldsymbol{p\Pi} = (\boldsymbol{p}_X \otimes \boldsymbol{p}_Y)\,(\boldsymbol{\Pi_X} \otimes \boldsymbol{\Pi_Y}) = (\boldsymbol{p}_X\boldsymbol{\Pi_X} \otimes \boldsymbol{p}_Y\boldsymbol{\Pi_Y}) = (\boldsymbol{p}_X \otimes \boldsymbol{p}_Y) = \boldsymbol{p}.$$

Now, let $X(t)$ and $Y(t)$ be two independent irreducible CTMCs with generator matrix $\boldsymbol{Q_X}$ of size $n_X$ and $\boldsymbol{Q_Y}$ of size $n_Y$, respectively. The joint process $(X(t), Y(t))$ is also a CTMC over a state space of size $n_X n_Y$. Applying a lexicographical state ordering the generator matrix of the joint process is $\boldsymbol{Q_X} \oplus \boldsymbol{Q_Y}$. That is, $\boldsymbol{Q} = \boldsymbol{Q_X} \oplus \boldsymbol{Q_Y}$ is such that its diagonal elements are non-positive, its off-diagonal elements are non-negative and $(\boldsymbol{Q_X} \otimes \boldsymbol{Q_Y})\mathbb{1} = \boldsymbol{0}$.

The transient behavior of the joint CTMC is:

$$\begin{aligned}\boldsymbol{p}(t) &= \boldsymbol{p}(0)e^{\boldsymbol{Q}t} = (\boldsymbol{p}_X(0) \otimes \boldsymbol{p}_Y(0))\,e^{(\boldsymbol{Q_X} \oplus \boldsymbol{Q_Y})t} \\ &= (\boldsymbol{p}_X(0) \otimes \boldsymbol{p}_Y(0))\left(e^{\boldsymbol{Q_X}t} \otimes e^{\boldsymbol{Q_Y}t}\right) \\ &= \boldsymbol{p}_X(0)e^{\boldsymbol{Q_X}t} \otimes \boldsymbol{p}_Y(0)e^{\boldsymbol{Q_Y}t} = \boldsymbol{p}_X(t) \otimes \boldsymbol{p}_Y(t),\end{aligned}$$

where we used:

$$\begin{aligned} e^{(\boldsymbol{Q_X} \oplus \boldsymbol{Q_Y})t} &= \sum_{i=0}^{\infty} \frac{t^i}{i!} (\boldsymbol{Q_X} \oplus \boldsymbol{Q_Y})^i = \sum_{i=0}^{\infty} \frac{t^i}{i!} \sum_{j=0}^{i} \binom{i}{j} (\boldsymbol{Q_X}^j \otimes \boldsymbol{Q_Y}^{i-j}) \\ &= \sum_{j=0}^{\infty} \frac{t^j}{j!} \sum_{i=j}^{\infty} \frac{t^{i-j}}{(i-j)!} (\boldsymbol{Q_X}^j \otimes \boldsymbol{Q_Y}^{i-j}) = \sum_{j=0}^{\infty} \frac{t^j}{j!} \boldsymbol{Q_X}^j \otimes \sum_{i=0}^{\infty} \frac{t^i}{i!} \boldsymbol{Q_Y}^i \\ &= e^{\boldsymbol{Q_X} t} \otimes e^{\boldsymbol{Q_Y} t}. \end{aligned}$$

The binomial expansion in the second step is based on the mentioned commutativity of $(\boldsymbol{Q_X} \otimes \boldsymbol{I})$ and $(\boldsymbol{I} \otimes \boldsymbol{Q_Y})$.

When $\boldsymbol{Q_X}$ and $\boldsymbol{Q_Y}$ are such that zero is a unique eigenvalue of both of them and all other eigenvalues have negative real parts ($X(t)$ and $Y(t)$ are irreducible CTMCs), then zero is a unique eigenvalue of $\boldsymbol{Q_X} \oplus \boldsymbol{Q_Y}$ and all other eigenvalues have negative real parts. Based on a similar reasoning as for the DTMC case, the stationary distribution of the joint CTMC is the solution of $\boldsymbol{0} = \boldsymbol{pQ}$, which is $\boldsymbol{p} = \boldsymbol{p}_X \otimes \boldsymbol{p}_Y$ where $\boldsymbol{p}_X$ and $\boldsymbol{p}_Y$ denote the stationary distribution of $X(t)$ and $Y(t)$.

# 2

# Continuous Phase Type Distributions

We introduce first the set of continuous phase type distributions, which are defined based on CTMCs, before discussing their discrete counterparts, the discrete phase type distributions, which are defined based on DTMCs. It is in reverse order compared to the introduction of DTMCs and CTMCs in section 1.12 and 1.13, respectively, but it better follows the history of the development of these distributions. This order is also motivated by the fact that several properties of continuous phase type distributions are simpler than the corresponding ones of discrete phase type distributions.

Due to the fact that the majority of the literature related to phase type distributions is devoted to continuous phase type distributions, which are commonly abbreviated as continuous phase type (PH). Hereinafter PH refers to continuous phase type distributions and DPH to discrete phase type distributions.

## 2.1. Definition and basic properties

PH distributions are defined by the behavior of a CTMC, which is often referred to as the background Markov chain.

Let $X(t)$ be a CTMC with $n$ transient states and one absorbing state, meaning that the absorbing state is reachable from all transient states (by a sequence of state transitions), and when the CTMC moves to the absorbing state, it remains there forever. Let $\boldsymbol{\pi}$ be the initial distribution of the CTMC, that is $\pi_i = \mathrm{P}(X(0) = i)$. Without loss of generality, we number the states of the CTMC such that states $1, \ldots, n$ are transient and state $(n + 1)$ is the absorbing state. The generator matrix of such a CTMC has the structure:

$$\boldsymbol{Q} = \begin{pmatrix} \boldsymbol{A} & \boldsymbol{a} \\ \boldsymbol{0} & 0 \end{pmatrix},$$

where $\boldsymbol{A}$ is a square matrix of size $n$ and $\boldsymbol{a}$ is a column vector of size $n$. Since the rows of the generator matrix sum to zero ($\boldsymbol{Q}\mathbb{1} = \mathbf{0}$), the elements of $\boldsymbol{a}$ can be computed from $\boldsymbol{A}$, that is $\boldsymbol{a} = -\boldsymbol{A}\mathbb{1}$.

We denote the first $n$ elements of the initial vector $\pi$ by $\boldsymbol{\alpha}$, that is:

$$\pi = \begin{pmatrix} \boldsymbol{\alpha} & \pi_{n+1} \end{pmatrix}.$$

Vector $\boldsymbol{a}$ completely defines the initial vector, since element $(n+1)$ of $\pi$ satisfies $\pi_{n+1} = 1 - \boldsymbol{\alpha}\mathbb{1}$, due to the fact that $\pi\mathbb{1} = 1$. $\boldsymbol{\alpha}$ defines the initial probabilities of the transient states and $\pi_{n+1}$ is the probability that the CTMC starts from the absorbing state. With the help of this CTMC, we are ready to define PH distributions.

DEFINITION 2.1.– *The time to reach the absorbing state of a CTMC with $n$ transient states and one absorbing state:*

$$T = \min\{t : X(t) = n+1, t \geq 0\},$$

*is* phase type distributed.

From definition 2.1, $T$ is a non-negative real. The name of PH distribution comes the fact that the states of the absorbing CTMC has been referred to as *phases* (e.g. in [KLE 75]), and the set of distributions that can be constructed from such phases are referred to as phase type distributions.

Throughout this document, we assume that the CTMC does not start from the absorbing state, that is $\pi_{n+1} = 0$ and $\boldsymbol{\alpha}\mathbb{1} = 1$. Consequently, the distribution of $T$ is such that it has no probability mass at zero and consequently $T$ has a continuous distribution on $\mathbb{R}^+$.

DEFINITION 2.2.– *If the initial probability vector and the generator matrix of the absorbing CTMC associated with the transient states are $\boldsymbol{\alpha}$ and $\boldsymbol{A}$, then PH($\boldsymbol{\alpha}, \boldsymbol{A}$) denotes the resulting PH distribution.*

DEFINITION 2.3.– *We refer to the size of vector $\boldsymbol{\alpha}$ (and square matrix $\boldsymbol{A}$) as the* size *of PH($\boldsymbol{\alpha}, \boldsymbol{A}$).*

DEFINITION 2.4.– *We denote the set of size $n$ PH distributions with PH($n$) and similar notations are used for the later discussed sets of distributions associated with a matrix representation of size $n$.*

In the majority of the cases, the size of PH($\boldsymbol{\alpha}, \boldsymbol{A}$) characterizes its inherent rank (which is defined later in definition 2.6), but unfortunately, there are cases when it is not true. For a precise discussion of those cases, we need two different definitions for the size and the rank of PH($\boldsymbol{\alpha}, \boldsymbol{A}$).

To evaluate the distribution of random variable $T$, we relate it with the transient behavior of the CTMC with initial distribution $\pi$ and generator matrix $\boldsymbol{Q}$:

$$F_T(t) = \ P(T < t) = \ P(X(t) = n+1) = \boldsymbol{\pi} e^{\boldsymbol{Q}t} \mathbf{e}_{n+1}^T,$$

where $\mathbf{e}_{n+1}$ is the $(n+1)$st unit row vector (the vector with all zero elements except in position $(n+1)$ which is 1). The second equality states that the time to absorption has occurred before time $t$ if and only if the Markov chain is in state $n+1$ at time $t$.

This straight forward description of the distribution of $T$ is not widely used due to the redundancy of matrix $\boldsymbol{Q}$ and vector $\boldsymbol{\pi}$. Matrix $\boldsymbol{A}$ and the initial vector associated with the transient states ($\boldsymbol{\alpha}$) define all information about the distribution of $T$ and the analytical description based on $\boldsymbol{\alpha}$ and $\boldsymbol{A}$ is more commonly used in practice because it is much simpler to use in complex stochastic models. To obtain the distribution based on $\boldsymbol{\alpha}$ and $\boldsymbol{A}$, we detail the block structure of matrix $\boldsymbol{Q}$ in the computation:

$$\begin{aligned} F_T(t) &= \ P(T < t) = \ P(X(t) = n+1) = 1 - \sum_{i=1}^{n} P(X(t) = i) = \\ &= 1 - (\boldsymbol{\alpha} 0) e^{\boldsymbol{Q}t} \begin{pmatrix} \mathbb{1} \\ 0 \end{pmatrix} = 1 - (\boldsymbol{\alpha} 0) \sum_{i=0}^{\infty} \frac{t^i}{i!} \begin{pmatrix} \boldsymbol{A} & \boldsymbol{a} \\ \mathbf{0} & 0 \end{pmatrix}^i \begin{pmatrix} \mathbb{1} \\ 0 \end{pmatrix} \\ &= 1 - (\boldsymbol{\alpha} 0) \sum_{i=0}^{\infty} \frac{t^i}{i!} \begin{pmatrix} \boldsymbol{A}^i & \bullet \\ \mathbf{0} & 0 \end{pmatrix} \begin{pmatrix} \mathbb{1} \\ 0 \end{pmatrix} = 1 - \boldsymbol{\alpha} \sum_{i=0}^{\infty} \frac{t^i}{i!} \boldsymbol{A}^i \mathbb{1} = 1 - \boldsymbol{\alpha} e^{\boldsymbol{A}t} \mathbb{1}, \end{aligned} \quad [2.1]$$

where $\bullet$ indicates an irrelevant matrix block whose elements are multiplied by zero. The PDF of $T$ can be obtained from the derivative of its CDF as follows:

$$\begin{aligned} f_T(t) &= \frac{d}{dt} F_T(t) = \frac{d}{dt} \left( 1 - \boldsymbol{\alpha} \sum_{i=0}^{\infty} \frac{t^i}{i!} \boldsymbol{A}^i \mathbb{1} \right) = -\boldsymbol{\alpha} \sum_{i=0}^{\infty} \frac{d}{dt} \frac{t^i}{i!} \boldsymbol{A}^i \mathbb{1} \qquad [2.2] \\ &= -\boldsymbol{\alpha} \sum_{i=1}^{\infty} \frac{t^{i-1}}{(i-1)!} \boldsymbol{A}^{i-1} \boldsymbol{A} \mathbb{1} = -\boldsymbol{\alpha} e^{\boldsymbol{A}t} \boldsymbol{A} \mathbb{1} = \boldsymbol{\alpha} e^{\boldsymbol{A}t} \boldsymbol{a}, \end{aligned}$$

where we used $\boldsymbol{a} = -\boldsymbol{A}\mathbb{1}$ in the last step. The stochastic interpretation of this PDF is:

$$\begin{aligned} f_T(t) &= \lim_{\Delta \to 0} \frac{1}{\Delta} P(t \leq T < t + \Delta) \\ &= \lim_{\Delta \to 0} \frac{1}{\Delta} P(X(t) \neq n{+}1, X(t+\Delta) = n{+}1) \end{aligned}$$

$$= \lim_{\Delta \to 0} \frac{1}{\Delta} \sum_{i=1}^{n} \mathrm{P}(X(t) = i, X(t+\Delta) = n+1)$$

$$= \lim_{\Delta \to 0} \frac{1}{\Delta} \sum_{i=1}^{n} \underbrace{\mathrm{P}(X(t) = i)}_{[\boldsymbol{\alpha} e^{\boldsymbol{A}t}]_i} \underbrace{\mathrm{P}(X(t+\Delta) = n+1 | X(t) = i)}_{\sim a_i \Delta} = \boldsymbol{\alpha} e^{\boldsymbol{A}t} \boldsymbol{a}, \quad [2.3]$$

where the first under braced term is the probability that the CTMC stays in one of the transient states at time $t$ and the second under braced term is the probability that the CTMC undergoes a state transition to the absorbing state in $(t, t+\Delta)$.

The hazard rate function $\lambda_T(t)$ of $T$ is:

$$\lambda_T(t) = \frac{f_T(t)}{1 - F_T(t)} = \frac{\boldsymbol{\alpha} e^{\boldsymbol{A}t} \boldsymbol{a}}{\boldsymbol{\alpha} e^{\boldsymbol{A}t} \mathbb{1}} \quad [2.4]$$

and its stochastic interpretation is:

$$\lambda_T(t) = \lim_{\Delta \to 0} \frac{1}{\Delta} \mathrm{P}(t \le T < t + \Delta | T > t)$$

$$= \lim_{\Delta \to 0} \frac{1}{\Delta} \mathrm{P}(X(t+\Delta) = n+1 | X(t) \ne n+1)$$

$$= \frac{\lim_{\Delta \to 0} \frac{1}{\Delta} \mathrm{P}(X(t) \ne n+1, X(t+\Delta) = n+1)}{\mathrm{P}(X(t) \ne n+1)} = \frac{f_T(t)}{1 - F_T(t)}.$$

Before computing the remaining properties of PH distributions, we need to classify the eigenvalues of $\boldsymbol{A}$. The $i, j$ element of matrix $e^{\boldsymbol{A}t}$ contains the transition probability $\mathrm{P}(X(t) = j | X(0) = i)$. As $t$ tends to infinity $e^{\boldsymbol{A}t}$ tends to zero, because states $1, \ldots, n$ are transient states, which are eventually left by the CTMC for the absorbing state. It means that the eigenvalues of $\boldsymbol{A}$ have negative real parts and, as a consequence, $\boldsymbol{A}$ is non-singular.

The Laplace transform of $T$ is:

$$f_T^*(s) = \mathbf{E}(e^{-sT}) = \int_{t=0}^{\infty} e^{-st} f_T(t) dt = \int_{t=0}^{\infty} e^{-st} \boldsymbol{\alpha} e^{\boldsymbol{A}t} \boldsymbol{a} dt$$

$$= \boldsymbol{\alpha} \int_{t=0}^{\infty} e^{(-s\boldsymbol{I} + \boldsymbol{A})t} dt \boldsymbol{a} = \boldsymbol{\alpha} (s\boldsymbol{I} - \boldsymbol{A})^{-1} \boldsymbol{a}, \quad [2.5]$$

where we used [1.22]. We note that the integral is guaranteed to converge for $\mathcal{R}(s) \ge 0$ according to remark 1.1, because $\boldsymbol{\alpha} \int_{t=0}^{\infty} e^{\boldsymbol{A}t} dt \boldsymbol{a} = \int_{t=0}^{\infty} f_T(t) dt = 1 < \infty$.

To compute the $k$th moment of $T$, we use the integral relation:

$$\left[t^k e^{\boldsymbol{A}t}\right]_0^{\infty} = \int_{t=0}^{\infty} k t^{k-1} e^{\boldsymbol{A}t} dt + \int_{t=0}^{\infty} t^k e^{\boldsymbol{A}t} \boldsymbol{A} dt, \quad [2.6]$$

whose left-hand side (LHS) is zero because the eigenvalues of $\boldsymbol{A}$ have negative real parts. Multiplying [2.6] by $(-\boldsymbol{A})^{-1}$, we get:

$$\int_{t=0}^{\infty} t^k e^{\boldsymbol{A}t} dt = k \int_{t=0}^{\infty} t^{k-1} e^{\boldsymbol{A}t} dt (-\boldsymbol{A})^{-1}.$$

Using this relation, the $k$th moment of $T$ is:

$$\begin{aligned}\mathbb{E}(T^k) &= \int_{t=0}^{\infty} t^k f_T(t) dt = \boldsymbol{\alpha} \int_{t=0}^{\infty} t^k e^{\boldsymbol{A}t} dt (-\boldsymbol{A}) \mathbb{1} = k\boldsymbol{\alpha} \int_{t=0}^{\infty} t^{k-1} e^{\boldsymbol{A}t} dt \mathbb{1} \\ &= k(k-1)\boldsymbol{\alpha} \int_{t=0}^{\infty} t^{k-2} e^{\boldsymbol{A}t} dt (-\boldsymbol{A})^{-1} \mathbb{1} = \ldots = k! \boldsymbol{\alpha} (-\boldsymbol{A})^{-k} \mathbb{1}.\end{aligned} \qquad [2.7]$$

The same moment expressions can be obtained from the LT of $T$ and [1.10], using:

$$\frac{d}{ds}(s\boldsymbol{I} - \boldsymbol{A})^{-k} = (-k)(s\boldsymbol{I} - \boldsymbol{A})^{-k-1}.$$

The Taylor series expansion of $f_T^*(s)$:

$$\begin{aligned}f_T^*(s) &= \sum_{i=0}^{\infty} s^i \frac{d^i}{ds^i} f_T^*(s)\Big|_{s=0} = \boldsymbol{\alpha} \sum_{i=0}^{\infty} s^i \frac{d^i}{ds^i} (s\boldsymbol{I} - \boldsymbol{A})^{-1}\Big|_{s=0} \boldsymbol{a} \\ &= \boldsymbol{\alpha} \sum_{i=0}^{\infty} s^i (-1)^i i! (s\boldsymbol{I} - \boldsymbol{A})^{-(i+1)}\Big|_{s=0} (-\boldsymbol{A}) \mathbb{1} = \boldsymbol{\alpha} \sum_{i=0}^{\infty} s^i (-1)^i i! (-\boldsymbol{A})^{-i} \mathbb{1}\end{aligned} \qquad [2.8]$$

also indicates the relation between the sequence of the moments [2.7] and $f_T^*(s)$.

The four basic properties of PH distributions (CDF, PDF, Laplace transform and moments) have several interesting consequences, some of which we summarize below.

## 2.2. Stochastic meaning of $(-A)^{-1}$

The matrix $(-\boldsymbol{A})^{-1}$ has an important stochastic meaning. Let $T_{ij}$ be the time spent in transient state $j$ before moving to the absorbing state starting from state $i$. For $\mathbb{E}(T_{ij})$, we have:

$$\mathbb{E}(T_{ij}) = \frac{\delta_{ij}}{-A_{ii}} + \sum_{k,k\neq i} \frac{A_{ik}}{-A_{ii}} \mathbb{E}(T_{kj}),$$

where $\delta_{ij}$ is the Kronecker delta. The first term of the LHS is the time spent in state $j$ while the Markov chain is in the initial state, and the second term is the time spent in state $j$ during later visits to $j$. Multiplying both sides by $-A_{ii}$ and adding $\mathbb{E}(T_{ij})\,A_{ii}$ gives:

$$0 = \delta_{ij} + \sum_k A_{ik}\mathbb{E}(T_{kj}),$$

whose matrix form is:

$$\mathbf{0} = \boldsymbol{I} + \boldsymbol{A}\overline{\mathbf{T}} \quad\longrightarrow\quad \overline{\mathbf{T}} = (-\boldsymbol{A})^{-1},$$

where $\overline{\mathbf{T}}$ is the matrix composed of the elements $\mathbb{E}(T_{ij})$. Consequently, element $(i,j)$ of matrix $(-\boldsymbol{A})^{-1}$ is $\mathbb{E}(T_{ij})$, which is a non-negative real number.

## 2.3. Rational Laplace transform

The Laplace transform $f_T^*(s)$ is a rational function (i.e. both the numerator and the denominator are polynomials in $s$) whose numerator is of degree at most $(n-1)$ and denominator is of degree at most $n$. This is because:

$$f_T^*(s) = \boldsymbol{\alpha}(s\boldsymbol{I} - \boldsymbol{A})^{-1}\boldsymbol{a} = \sum_i\sum_j \alpha_i((s\boldsymbol{I} - \boldsymbol{A})^{-1})_{ij}a_j$$

$$= \sum_i\sum_j \alpha_i \frac{\det_{ji}(s\boldsymbol{I} - \boldsymbol{A})}{\det(s\boldsymbol{I} - \boldsymbol{A})} a_j = \frac{\sum_i\sum_j \alpha_i a_j \det_{ji}(s\boldsymbol{I} - \boldsymbol{A})}{\det(s\boldsymbol{I} - \boldsymbol{A})},$$

where $\det(s\boldsymbol{I} - \boldsymbol{A})$ is a degree $n$ polynomial and the numerator is the sum of at most degree $(n-1)$ polynomials (i.e. the numerator is an at most degree $n-1$ polynomial). Normalization of the fraction such that the coefficient of $s^n$ in the denominator is 1 provides the normalized form:

$$f_T^*(s) = \frac{\varphi_{n-1}s^{n-1} + \varphi_{n-2}s^{n-2} + \ldots + \varphi_1 s + \varphi_0}{s^n + \varrho_{n-1}s^{n-1} + \varrho_{n-2}s^{n-2} + \ldots + \varrho_1 s + \varrho_0}, \qquad [2.9]$$

where noting that $\int_{x=0}^{\infty} f_T(x)dx = f_T^*(0) = 1$, we obtain $\varphi_0 = \varrho_0$.

EXAMPLE 2.5.– *If* $\boldsymbol{\alpha} = (\alpha_1\ \alpha_2)$, $\boldsymbol{A} = \begin{pmatrix} A_{11} & A_{12} \\ A_{21} & A_{22} \end{pmatrix}$ *and* $\boldsymbol{a} = \begin{pmatrix} a_1 \\ a_2 \end{pmatrix}$, *then:*

$$(s\boldsymbol{I} - \boldsymbol{A})^{-1} = \frac{1}{\det(s\boldsymbol{I} - \boldsymbol{A})} \begin{pmatrix} \det_{11}(s\boldsymbol{I} - \boldsymbol{A}) & \det_{21}(s\boldsymbol{I} - \boldsymbol{A}) \\ \det_{12}(s\boldsymbol{I} - \boldsymbol{A}) & \det_{22}(s\boldsymbol{I} - \boldsymbol{A}) \end{pmatrix}$$

$$= \frac{1}{\det(s\boldsymbol{I} - \boldsymbol{A})} \begin{pmatrix} s - A_{22} & A_{12} \\ A_{21} & s - A_{11} \end{pmatrix},$$

*with* $\det(s\boldsymbol{I}-\boldsymbol{A}) = s^2-(A_{11}+A_{22})+A_{11}A_{22}-A_{12}A_{21}$. *The term* $\det(s\boldsymbol{I}-\boldsymbol{A})$ *is a degree 2 polynomial of s and* $\det_{ij}(s\boldsymbol{I}-\boldsymbol{A})$ *for* $i,j \in \{1,2\}$ *is a degree 1 polynomial of s at most. For* $f_T^*(s)$, *we have:*

$$f_T^*(s) = \boldsymbol{\alpha}(s\boldsymbol{I}-\boldsymbol{A})^{-1}\boldsymbol{a} = \frac{\sum_{i=1}^{2}\sum_{j=1}^{2}\alpha_i \det_{ji}(s\boldsymbol{I}-\boldsymbol{A})a_j}{\det(s\boldsymbol{I}-\boldsymbol{A})}$$
$$= \frac{\alpha_1(s-A_{22})a_1+\alpha_1A_{12}a_2+\alpha_2A_{21}a_1+\alpha_2(s-A_{11})a_2}{s^2-(A_{11}+A_{22})s+A_{11}A_{22}-A_{12}A_{21}}$$
$$= \frac{(\alpha_1a_1+\alpha_2a_2)s+\alpha_1A_{21}a_2+\alpha_2A_{12}a_1-\alpha_1A_{22}a_1-\alpha_2A_{11}a_2}{s^2-(A_{11}+A_{22})s+A_{11}A_{22}-A_{12}A_{21}}.$$

We define the rank of PH($\boldsymbol{\alpha}$, $\boldsymbol{A}$) based on the rational Laplace transform as follows.

DEFINITION 2.6.– *The order of the denominator of the rational function in* [2.9] *after all possible simplifications is the* rank *of PH(*$\boldsymbol{\alpha}$, $\boldsymbol{A}$*).*

In example 2.5, an automatic expansion of the Laplace transform of PH($\boldsymbol{\alpha}$, $\boldsymbol{A}$) of size 2 according to [2.9] results in a rational function of order 2 (without simplification). In general, if the order $n$ rational function cannot be simplified (which is commonly the case), then both the size and the rank of PH($\boldsymbol{\alpha}$, $\boldsymbol{A}$) are $n$.

In some special cases, given a size $n$ PH($\boldsymbol{\alpha}$, $\boldsymbol{A}$), the numerator and the denominator in [2.9] have common factors, which can be eliminated. When the common factors are eliminated, the order of the resulting rational function is lower than $n$. In these cases, the rank of PH($\boldsymbol{\alpha}$, $\boldsymbol{A}$) is less than its size.

EXAMPLE 2.7.– *We compute the rank of PH(*$\boldsymbol{\gamma}$, $\boldsymbol{G}$*) with*

$$\boldsymbol{\gamma} = \begin{pmatrix} 1-\frac{\lambda}{\mu} & \frac{\lambda}{\mu} \end{pmatrix} \text{ and } \boldsymbol{G} = \begin{pmatrix} -\lambda & \lambda \\ 0 & -\mu \end{pmatrix}$$

*according to definition 2.6 as follows:*

$$f_T^*(s) = \boldsymbol{\gamma}(s\boldsymbol{I}-\boldsymbol{G})^{-1}(-\boldsymbol{G})\mathbb{1} = \boldsymbol{\gamma}\begin{pmatrix} s+\lambda & -\lambda \\ 0 & s+\mu \end{pmatrix}^{-1}(-\boldsymbol{G})\mathbb{1}$$
$$= \begin{pmatrix} 1-\frac{\lambda}{\mu} & \frac{\lambda}{\mu} \end{pmatrix}\begin{pmatrix} \frac{1}{s+\lambda} & \frac{\lambda}{s+\lambda}\frac{1}{s+\mu} \\ 0 & \frac{1}{s+\mu} \end{pmatrix}\begin{pmatrix} 0 \\ \mu \end{pmatrix}$$
$$= \begin{pmatrix} \frac{\mu-\lambda}{\mu} & \frac{\lambda}{\mu} \end{pmatrix}\begin{pmatrix} \frac{\lambda}{s+\lambda}\frac{\mu}{s+\mu} \\ \frac{\mu}{s+\mu} \end{pmatrix}$$
$$= \frac{\mu-\lambda}{\mu}\cdot\frac{\lambda}{s+\lambda}\frac{\mu}{s+\mu}+\frac{\lambda}{\mu}\cdot\frac{\mu}{s+\mu} = \frac{\mu-\lambda}{s+\mu}\frac{\lambda}{s+\lambda}+\frac{\lambda}{s+\mu}$$
$$= \frac{(\mu-\lambda)\lambda}{(s+\mu)(s+\lambda)}+\frac{\lambda(s+\lambda)}{(s+\mu)(s+\lambda)} = \frac{\lambda(s+\mu)}{(s+\mu)(s+\lambda)} = \frac{\lambda}{s+\lambda}.$$

*That is the rank of PH($\gamma, \boldsymbol{G}$) is 1, even though the size of PH($\gamma, \boldsymbol{G}$) is 2.*

### 2.4. Decomposition of matrix exponential functions

The PDF of a PH distribution can be decomposed into a sum of functions. To this end, let $\boldsymbol{A} = \boldsymbol{B}^{-1}\boldsymbol{\Lambda}\boldsymbol{B}$ be the spectral decomposition [1] of $\boldsymbol{A}$ and let $\boldsymbol{u} = \boldsymbol{\alpha}\boldsymbol{B}^{-1}$ and $\boldsymbol{v} = \boldsymbol{B}\mathbb{1}$. Then,

$$f_T(t) = \boldsymbol{\alpha} e^{\boldsymbol{A}t}\boldsymbol{a} = -\boldsymbol{\alpha} e^{\boldsymbol{A}t}\boldsymbol{A}\mathbb{1} = -\boldsymbol{\alpha}\boldsymbol{B}^{-1}e^{\boldsymbol{\Lambda}t}\boldsymbol{\Lambda}\boldsymbol{B}\mathbb{1} = -\boldsymbol{u}e^{\boldsymbol{\Lambda}t}\boldsymbol{\Lambda}\boldsymbol{v}\,. \qquad [2.10]$$

Two cases can be distinguished.

– The eigenvalues of $\boldsymbol{A}$ are different (i.e. they all have algebraic multiplicity equal to one) and, consequently, $\boldsymbol{\Lambda}$ is a diagonal matrix. In this case, $f_T(t)$ is a sum of exponential functions because:

$$f_T(t) = -\boldsymbol{u}e^{\boldsymbol{\Lambda}t}\boldsymbol{\Lambda}\boldsymbol{v} = -\sum_i u_i v_i \lambda_i e^{\lambda_i t} = -\sum_i c_i \lambda_i e^{\lambda_i t}, \qquad [2.11]$$

where $c_i = u_i v_i$ is a constant coefficient of the exponential function.

Here, the eigenvalues ($\lambda_i$) as well as the associated coefficients ($c_i$) can be either real or complex conjugate pairs, but all eigenvalues have negative real parts. In the case of a real eigenvalue $\lambda_i$, the associated $c_i$ constant can be any real number, positive, negative or zero. If $c_i$ is zero, then the $\lambda_i$ eigenvalue does not contribute to the distribution. If $c_i$ is negative, then the $\lambda_i$ eigenvalue contributes with a negative exponential function and the rest of the exponential functions have to be such that they sum up the overall density function to be non-negative. Figure 2.1 presents a simple example of [2.11] in the case of a density function with real eigenvalues.

For a complex conjugate pair of eigenvalues, we have:

$$c_i\lambda_i e^{\lambda_i t} + \bar{c}_i\bar{\lambda}_i e^{\bar{\lambda}_i t} = 2|c_i\lambda_i| e^{\mathcal{R}(\lambda_i)t}\cos(\mathcal{I}(\lambda_i)t - \varphi_i),$$

where $c_i\lambda_i = |c_i\lambda_i|e^{\Im\varphi_i}$, $\mathcal{R}(\lambda_i)$ and $\mathcal{I}(\lambda_i)$ are the real and the imaginary part of $\lambda_i$ and $\Im$ is the imaginary unit. Figure 2.2 demonstrates the shape of $e^{\mathcal{R}(\lambda_i)t}\cos(\mathcal{I}(\lambda_i)t - \varphi_i)$, and Figure 2.3 presents a simple example with complex eigenvalues.

– There are eigenvalues of $\boldsymbol{A}$ with algebraic multiplicity larger than one, and hence, $\boldsymbol{\Lambda}$ contains real Jordan blocks. Based on [1.19], the matrix exponential of a Jordan block is:

$$\exp\left(\begin{pmatrix} \lambda & 1 & & \\ & \lambda & 1 & \\ & & \ddots & \ddots \\ & & & \lambda \end{pmatrix} t\right) = \begin{pmatrix} e^{\lambda t} & te^{\lambda t} & \frac{1}{2!}t^2e^{\lambda t} & \frac{1}{3!}t^3e^{\lambda t} \\ & e^{\lambda t} & te^{\lambda t} & \frac{1}{2!}t^2e^{\lambda t} \\ & & \ddots & \ddots \\ & & & e^{\lambda t} \end{pmatrix}.$$

1 The case of different Jordan blocks with identical eigenvalue is not considered here, because it collapses to the same exponential function.

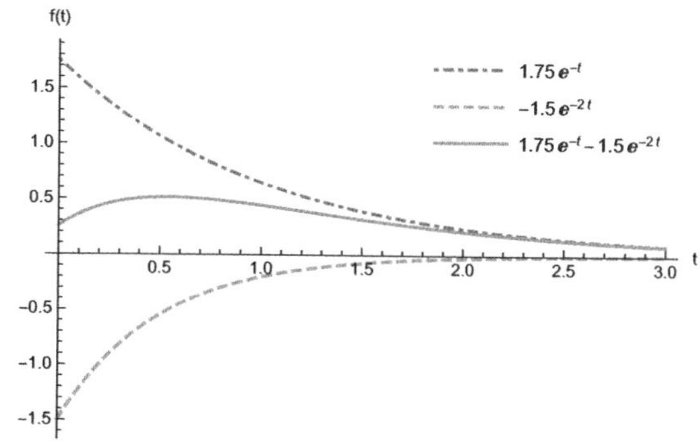


**Figure 2.1.** *PDF of PH($\alpha$, $A$) with real eigenvalues where* $\alpha = (0.75\ 0.25)$ *and* $A = \begin{pmatrix} -2 & 2 \\ 0 & -1 \end{pmatrix}$, *from which* $f_T(t) = 1.75e^{-t} - 1.5e^{-2t}$. *For a color version of this figure, see www.iste.co.uk/horvath/phase.zip*

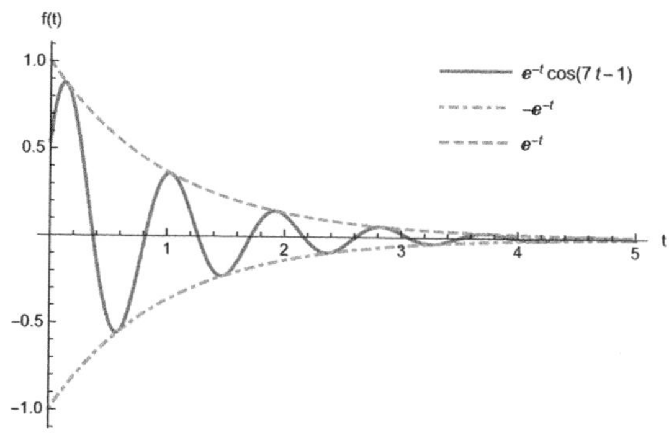


**Figure 2.2.** *The shape of the* $e^{\mathcal{R}(\lambda_i)t}\cos(\mathcal{I}(\lambda_i)t - \varphi_i)$ *function. For a color version of this figure, see www.iste.co.uk/horvath/phase.zip*

Consequently, in the case of real eigenvalues the density function takes the form:

$$f_T(t) = \sum_{i=1}^{\#\lambda} \sum_{j=1}^{\#\lambda_i} c_{ij}(-\lambda_i)^j t^{j-1} e^{\lambda_i t}, \qquad [2.12]$$

where $\#\lambda$ is the number of different eigenvalues and $\#\lambda_i$ is the multiplicity of $\lambda_i$. The $te^{-t}$ curve in Figure 2.4 demonstrates the shape of this function for $j = 2$.

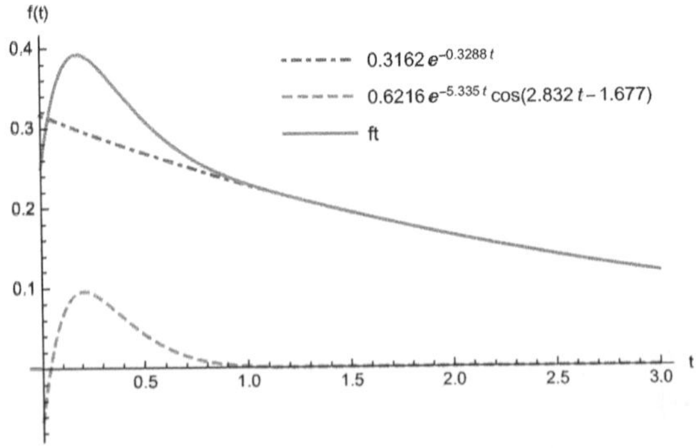


**Figure 2.3.** *PDF of PH($\alpha$, $A$) with complex eigenvalues, where* $\alpha = (0.75\ 0.25\ 0)$ *and* $A = \begin{pmatrix} -4 & 4 & 0 \\ 0 & -4 & 3 \\ 3 & 0 & -3 \end{pmatrix}$, *from which* $f_T(t) = 0.3162e^{-0.3288t} + 0.6216e^{-5.335t}\cos(2.832t - 1.677)$. *For a color version of this figure, see www.iste.co.uk/horvath/phase.zip*

Similar to the previous case, the eigenvalues ($\lambda_i$) as well as the associated coefficients ($c_{i,j}$) can be real or complex conjugate pairs. For a complex conjugate pair of eigenvalues, we have:

$$c_{i,j}(-\lambda_i)^j t^{j-1} e^{\lambda_i t} + \bar{c}_{i,j}(-\bar{\lambda}_i)^j t^{j-1} e^{\bar{\lambda}_i t}$$
$$= 2|c_{i,j}(-\lambda_i)^j| t^{j-1} e^{\mathcal{R}(\lambda_i)t} \cos(\mathcal{I}(\lambda_i)t - \varphi_{i,j}),$$

where $c_{i,j}(-\lambda_i)^j = |c_{i,j}(-\lambda_i)^j| e^{\Im \varphi_{i,j}}$. Figure 2.4 demonstrates an example of this function with $j = 2$.

As a result, the density function of a PH distribution has the form:

$$f_T(t) = \sum_{i=1}^{\#\lambda_R} \sum_{j=1}^{\#\lambda_i^R} c_{ij}^R (-\lambda_i^R)^j t^{j-1} e^{\lambda_i^R t} \tag{2.13}$$
$$+ \sum_{i=1}^{\#\lambda_C} \sum_{j=1}^{\#\lambda_i^C} 2|c_{i,j}^C (-\lambda_i^C)^j| t^{j-1} e^{\mathcal{R}(\lambda_i^C)t} \cos(\mathcal{I}(\lambda_i^C)t - \varphi_{i,j}),$$

where $\#\lambda_R$ and $\#\lambda_C$ are the number of different real and complex conjugate pair of eigenvalues, and $\#\lambda_i^R$ and $\#\lambda_i^C$ are the multiplicity of the $i$th real and complex eigenvalues $\lambda_i^R$ and $\lambda_i^C$.

Another way to identify the rank of PH($\alpha$, $A$) (see definition 2.6) is through the decomposition of the matrix exponential function, which results in the following alternative definition.

DEFINITION 2.8.– *If* $\boldsymbol{A}$ *is diagonalizable, then the* rank *of PH(*$\boldsymbol{\alpha}, \boldsymbol{A}$*) is the number of the exponential terms in* [2.11] *with non-zero* $c_i$ *coefficients.*

*If* $\boldsymbol{A}$ *is non-diagonalizable, then the* rank *of PH(*$\boldsymbol{\alpha}, \boldsymbol{A}$*) is characterized by the highest order non-zero* $c_{ij}$ *coefficients associated with the eigenvalues in* [2.12]*. That is, the* rank *of PH(*$\boldsymbol{\alpha}, \boldsymbol{A}$*) is* $\sum_{i=1}^{\#\lambda} \#\hat{\lambda}_i$*, where* $\#\hat{\lambda}_i = \max_j(c_{ij} \neq 0)$.

EXAMPLE 2.9.– *The rank of PH(*$\boldsymbol{\gamma}, \boldsymbol{G}$*) from example 2.7 can also be computed based on definition 2.8. The spectral decomposition of matrix* $\boldsymbol{G}$ *is:*

$$\boldsymbol{G} = \begin{pmatrix} -\lambda & \lambda \\ 0 & -\mu \end{pmatrix} = \begin{pmatrix} \frac{\mu}{\mu-\lambda} & \frac{-\lambda}{\mu-\lambda} \\ 0 & 1 \end{pmatrix} \begin{pmatrix} -\lambda & 0 \\ 0 & -\mu \end{pmatrix} \begin{pmatrix} \frac{\mu-\lambda}{\mu} & \frac{\lambda}{\mu} \\ 0 & 1 \end{pmatrix}$$

*from which*

$$e^{t\boldsymbol{G}} = \begin{pmatrix} \frac{\mu}{\mu-\lambda} & \frac{-\lambda}{\mu-\lambda} \\ 0 & 1 \end{pmatrix} \begin{pmatrix} e^{-\lambda t} & 0 \\ 0 & e^{-\mu t} \end{pmatrix} \begin{pmatrix} \frac{\mu-\lambda}{\mu} & \frac{\lambda}{\mu} \\ 0 & 1 \end{pmatrix}$$

*and for* $f_T^*(t)$ *we have*

$$\begin{aligned} f_T^*(t) &= \boldsymbol{\gamma} e^{t\boldsymbol{G}}(-\boldsymbol{G})\mathbb{1} \\ &= \begin{pmatrix} 1 - \frac{\lambda}{\mu} & \frac{\lambda}{\mu} \end{pmatrix} \begin{pmatrix} \frac{\mu}{\mu-\lambda} & \frac{-\lambda}{\mu-\lambda} \\ 0 & 1 \end{pmatrix} \begin{pmatrix} e^{-\lambda t} & 0 \\ 0 & e^{-\mu t} \end{pmatrix} \begin{pmatrix} \frac{\mu-\lambda}{\mu} & \frac{\lambda}{\mu} \\ 0 & 1 \end{pmatrix} \begin{pmatrix} 0 \\ \mu \end{pmatrix} \\ &= (1\ 0) \begin{pmatrix} e^{-\lambda t} & 0 \\ 0 & e^{-\mu t} \end{pmatrix} \begin{pmatrix} \lambda \\ \mu \end{pmatrix} = 1\lambda e^{-\lambda t} + 0\mu e^{-\mu t} = \lambda e^{-\lambda t}. \end{aligned}$$

*That is, the rank of PH(*$\boldsymbol{\gamma}, \boldsymbol{G}$*) is* 1.

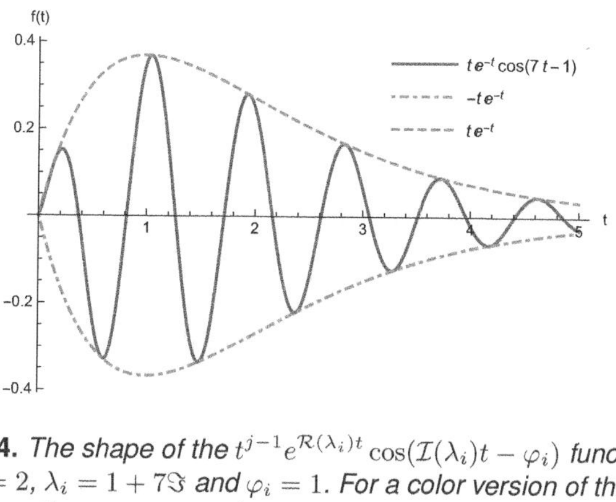


**Figure 2.4.** *The shape of the* $t^{j-1}e^{\mathcal{R}(\lambda_i)t}\cos(\mathcal{I}(\lambda_i)t - \varphi_i)$ *function with* $j = 2$, $\lambda_i = 1 + 7\Im$ *and* $\varphi_i = 1$. *For a color version of this figure, see www.iste.co.uk/horvath/phase.zip*

## 2.5. Similarity transformation

In general, infinitely many Markov chains can represent the same PH distribution. Utilizing the results of theorem 1.3 and 1.4, the following sections introduce representations with identical and different sizes, respectively.

### 2.5.1. *Similarity transformation with identical sizes*

The following *similarity transformation* generates representations with identical sizes.

Let $\boldsymbol{T}$ be a non-singular square matrix with unit row sums ($\boldsymbol{T}\mathbb{1} = \mathbb{1}$), whose size is the same as the size of $\boldsymbol{A}$. The vector–matrix pairs $(\boldsymbol{\alpha}, \boldsymbol{A})$ and $(\boldsymbol{\alpha T}, \boldsymbol{T}^{-1}\boldsymbol{AT})$ are two different vector–matrix representations of the same PH distribution, since:

$$F_{\mathcal{T}}(t) = 1 - \boldsymbol{\alpha T} e^{\boldsymbol{T}^{-1}\boldsymbol{AT}t}\mathbb{1} = 1 - \boldsymbol{\alpha T T}^{-1} e^{\boldsymbol{A}t}\boldsymbol{T}\mathbb{1} = 1 - \boldsymbol{\alpha} e^{\boldsymbol{A}t}\mathbb{1}. \qquad [2.14]$$

DEFINITION 2.10.– *We refer to the:*

$$(\boldsymbol{\alpha}, \boldsymbol{A}) \quad \longrightarrow \quad (\boldsymbol{\alpha T}, \boldsymbol{T}^{-1}\boldsymbol{AT}),$$

*transformation as* similarity transformation *with similarity matrix* $\boldsymbol{T}$.

DEFINITION 2.11.– *Since infinitely many Markov chains can describe a given PH distribution, the particular Markov chain which is used for describing a PH distribution is referred to as* representation *and the PH($\boldsymbol{\alpha}$, $\boldsymbol{A}$) notation makes the representation explicit.*

EXAMPLE 2.12.– *For PH($\boldsymbol{\alpha}$, $\boldsymbol{A}$) defined by initial vector $\boldsymbol{\alpha}$ and generator matrix $\boldsymbol{A}$ with:*

$$\boldsymbol{\alpha} = \begin{pmatrix} 0.1 & 0.2 & 0.7 \end{pmatrix}, \; \boldsymbol{A} = \begin{pmatrix} -5 & 2 & 1 \\ 0 & -4 & 3 \\ 0 & 0 & -2 \end{pmatrix},$$

*we can apply a similarity transformation with matrix:*

$$\boldsymbol{T} = \begin{pmatrix} \frac{2}{5} & \frac{1}{5} & \frac{2}{5} \\ \frac{9}{20} & \frac{7}{20} & \frac{1}{5} \\ \frac{3}{10} & \frac{3}{10} & \frac{2}{5} \end{pmatrix},$$

*which gives*

$$\boldsymbol{\gamma} = \boldsymbol{\alpha T} = \begin{pmatrix} 0.34 & 0.3 & 0.36 \end{pmatrix}, \; \boldsymbol{G} = \boldsymbol{T}^{-1}\boldsymbol{AT} = \begin{pmatrix} -2 & 2 & 0 \\ 0 & -4 & 4 \\ 0 & 0 & -5 \end{pmatrix}.$$

*That is, PH($\boldsymbol{\alpha}, \boldsymbol{A}$) and PH($\boldsymbol{\gamma}, \boldsymbol{G}$) are two different representations of the same PH distribution, since they represent two different Markov chains with absorbing state, which result in the same distribution for the time to absorption.*

The similarity transformation maintains several matrix properties. For example, $\boldsymbol{A}$ and $\boldsymbol{T}^{-1}\boldsymbol{A}\boldsymbol{T}$ have the same characteristic polynomial, determinant, trace, eigenvalues (with the same algebraic and geometric multiplicities) and Jordan normal form (up to a permutation of the Jordan blocks). The Jordan decomposition in [2.10] is also a similarity transformation and it indicates that the same matrix exponential function is obtained from different, but similar matrix representations.

### 2.5.2. *Similarity transformation with different sizes*

Vector matrix pairs with different sizes can also represent the same distribution. Let matrix $\boldsymbol{V}$ of size $m \times n$ be such that $\boldsymbol{V}\mathbb{1} = \mathbb{1}$. The vector–matrix pairs $(\boldsymbol{\alpha}, \boldsymbol{A})$ of size $n$ and $(\boldsymbol{\gamma}, \boldsymbol{G})$ of size $m$ are two different vector–matrix pairs representing the same PH distribution if $\boldsymbol{A}\boldsymbol{V} = \boldsymbol{V}\boldsymbol{G}$ and $\boldsymbol{\alpha}\boldsymbol{V} = \boldsymbol{\gamma}$ because:

$$F_T(t) = 1 - \boldsymbol{\gamma} e^{\boldsymbol{G}t}\mathbb{1} = 1 - \boldsymbol{\alpha}\boldsymbol{V} e^{\boldsymbol{G}t}\mathbb{1} = 1 - \boldsymbol{\alpha} e^{\boldsymbol{A}t}\boldsymbol{V}\mathbb{1} = 1 - \boldsymbol{\alpha} e^{\boldsymbol{A}t}\mathbb{1} \quad [2.15]$$

in this case.

EXAMPLE 2.13.– *The simplest example for the similarity of a size 2 PH representation with a size 1 PH representation (which is the exponential distribution) is according to examples 2.7 and 2.9, where we transform PH($\boldsymbol{\alpha}, \boldsymbol{A}$) with:*

$$\boldsymbol{\alpha} = (1),\ \boldsymbol{A} = (-\lambda)\,, \quad [2.16]$$

*with the transformation matrix* $\boldsymbol{V} = \begin{pmatrix} 1 - \frac{\lambda}{\mu} & \frac{\lambda}{\mu} \end{pmatrix}$. *It gives:*

$$\boldsymbol{\gamma} = \boldsymbol{\alpha}\boldsymbol{V} = \begin{pmatrix} 1 - \frac{\lambda}{\mu} & \frac{\lambda}{\mu} \end{pmatrix},$$

*and from* $\boldsymbol{A}\boldsymbol{V} = \boldsymbol{V}\boldsymbol{G}$ *we have:*

$$\boldsymbol{G} = \begin{pmatrix} x_1 & x_2 \\ \frac{(x_1+\lambda)(\lambda-\mu)}{\lambda} & x_2 - \lambda - \frac{x_2\mu}{\lambda} \end{pmatrix},$$

*which is an equivalent, but redundant representation of the exponential distribution with two parameters $x_1$ and $x_2$. For $x_1 = -\lambda$ and $x_2 = \lambda$, the matrix $\boldsymbol{G}$ simplifies to:*

$$\boldsymbol{G} = \begin{pmatrix} -\lambda & \lambda \\ 0 & -\mu \end{pmatrix},$$

*which is the most often used size 2 representation of the exponential distribution because it is closely related to the departure process of M/M/1 queues [BUR 56].*

The $\boldsymbol{AV} = \boldsymbol{VG}$ relation is a generalized similarity transformation for matrices of different size. If $\boldsymbol{A}$ is of size $m \times m$, $\boldsymbol{G}$ is of size $n \times n$ and $m < n$, then the eigenvalues of $\boldsymbol{A}$ are eigenvalues of $\boldsymbol{G}$ with at least the same algebraic and geometric multiplicities as in theorem 1.4.

When $\boldsymbol{\alpha V} = \boldsymbol{\gamma}$, $\boldsymbol{AV} = \boldsymbol{VG}$, $\boldsymbol{A}$ is of size $m \times m$, $\boldsymbol{G}$ is of size $n \times n$ and $m < n$, then $\boldsymbol{\gamma}$ is orthogonal to those eigenvectors of $\boldsymbol{G}$ which are associated with the extra eigenvalues of $\boldsymbol{G}$ compared to the ones of $\boldsymbol{A}$. As a result, $1 - \boldsymbol{\gamma} e^{\boldsymbol{G}t}\mathbb{1}$ and $1 - \boldsymbol{\alpha} e^{\boldsymbol{A}t}\mathbb{1}$ are composed of the same exponential functions in [2.10].

### 2.5.3. *Full rank representation*

The previous section shows that vector–matrix pairs of possibly different sizes can represent the same distribution. In practice, the following case is of essential interest.

DEFINITION 2.14.– *A size $n$ PH($\boldsymbol{\alpha}, \boldsymbol{A}$) is* full rank *if its rank is equal to $n$.*

According to this definition, in example 2.13, the PH($\boldsymbol{\alpha}, \boldsymbol{A}$) representation of the exponential distribution in [2.16] is full rank, while its PH($\boldsymbol{\gamma}, \boldsymbol{G}$) representation is not full rank. The main difference between the full rank and the non-full rank representation is that the effect of some of the eigenvalues of the non-full rank representation (eigenvalue $-\mu$ in the case of $\boldsymbol{G}$) is eliminated by associated zero coefficients in [2.11].

## 2.6. Closure properties

The class of PH distributions is closed for several frequently applied operations on random variables. Let $X$ and $Y$ be two random variables with PH distribution represented by ($\boldsymbol{\alpha}, \boldsymbol{A}$) and ($\boldsymbol{\beta}, \boldsymbol{B}$), respectively. The following operations with $X$ and $Y$ results in another PH distribution.

*Summation*: $Z = X + Y$ is PH distributed with representation PH($\boldsymbol{\gamma}, \boldsymbol{G}$) where:

$$\boldsymbol{\gamma} = \begin{pmatrix} \boldsymbol{\alpha} & \boldsymbol{0} \end{pmatrix} \text{ and } \boldsymbol{G} = \begin{pmatrix} \boldsymbol{A} & \boldsymbol{a\beta} \\ \boldsymbol{0} & \boldsymbol{B} \end{pmatrix}.$$

Intuitively, this representation of $Z$ means that $Z$ is PH distributed such that it starts according to the initial distribution of $X$, it goes through the transient phases of $X$ and instead of moving to the absorbing state of $X$ the process moves to one of the transient state of $Y$ with initial distribution $\boldsymbol{\beta}$. From this point, it goes through the transient phases of $Y$ and moves to an absorbing state. In case the process goes through the transient states of $Y$ first, we have an alternative representation for the same distribution:

$$\begin{pmatrix} \boldsymbol{\beta} & \boldsymbol{0} \end{pmatrix} \text{ and } \begin{pmatrix} \boldsymbol{B} & \boldsymbol{b\alpha} \\ \boldsymbol{0} & \boldsymbol{A} \end{pmatrix}.$$

*Mixture*: If $Z$ takes the value of $X$ with probability $p$ and the value of $Y$ with probability $1-p$, that is,

$$Z = \begin{cases} X & \text{with probability } p, \\ Y & \text{with probability } (1-p), \end{cases}$$

then $Z$ is PH($\boldsymbol{\gamma}, \boldsymbol{G}$) with

$$\boldsymbol{\gamma} = \begin{pmatrix} p\boldsymbol{\alpha} & (1-p)\boldsymbol{\beta} \end{pmatrix} \text{ and } \boldsymbol{G} = \begin{pmatrix} \boldsymbol{A} & \mathbf{0} \\ \mathbf{0} & \boldsymbol{B} \end{pmatrix}.$$

In this PH structure, the initial vector determines if the process goes through the transient states of $X$ or the ones of $Y$. Again the role of $\boldsymbol{A}$ and $\boldsymbol{B}$ can be interchanged.

In the above two cases, if the size of $\boldsymbol{A}$ and $\boldsymbol{B}$ are $n$ and $m$, respectively, then the size of $\boldsymbol{G}$ is $n+m$.

*Minimum*: The minimum of $X$ and $Y$ is also PH distributed with representation PH($\boldsymbol{\gamma}, \boldsymbol{G}$) where:

$$\boldsymbol{\gamma} = \boldsymbol{\alpha} \otimes \boldsymbol{\beta} \text{ and } \boldsymbol{G} = \boldsymbol{A} \oplus \boldsymbol{B}.$$

Intuitively, this representation means that two independent transient Markov chains run in parallel and the minimum occurs when the joint process leaves the product space of the transient states, that is, one of them reaches the absorbing state. If the size of $\boldsymbol{A}$ and $\boldsymbol{B}$ are $n$ and $m$, then $\boldsymbol{G}$ is of size $nm$.

*Maximum*: The maximum of $X$ and $Y$ is also PH distributed with representation PH($\boldsymbol{\gamma}, \boldsymbol{G}$) where:

$$\boldsymbol{\gamma} = \begin{pmatrix} \boldsymbol{\alpha} \otimes \boldsymbol{\beta} & \mathbf{0} & \mathbf{0} \end{pmatrix} \text{ and } \boldsymbol{G} = \begin{pmatrix} \boldsymbol{A} \oplus \boldsymbol{B} & \boldsymbol{a} \otimes \mathbf{I} & \mathbf{I} \otimes \boldsymbol{b} \\ \mathbf{0} & \boldsymbol{B} & \mathbf{0} \\ \mathbf{0} & \mathbf{0} & \boldsymbol{A} \end{pmatrix}.$$

The stochastic process representing the Markov chain until the maximum of $X$ and $Y$ has three parts: i) both $X$ and $Y$ are in a transient state (represented by the first diagonal matrix block of $\boldsymbol{G}$); ii) $Y$ is in a transient state and $X$ is in the absorbing state (in this case, $Y$ still evolves according to the transient generator $\boldsymbol{B}$, but $X$ does not change state any more, as it is represented by the second diagonal matrix block of $\boldsymbol{G}$); iii) $X$ is in a transient state and $Y$ is in the absorbing state (in this case, $X$ still evolves according to the transient generator $\boldsymbol{A}$, but $Y$ does not change state any more, as it is represented by the third diagonal matrix block of $\boldsymbol{G}$). If the size of $\boldsymbol{A}$ and $\boldsymbol{B}$ are $n$ and $m$, then $\boldsymbol{G}$ is of size $nm+n+m$.

Also when it comes to calculate the so-called *equilibrium distribution*, the class of PH distributions is closed. In a renewal process with independent and identically distributed inter-event times $\tau_i$, the time from a random time instance (e.g. arrival of an independent observer) to the next event is referred to as residual time or forward recurrence time and its distribution as equilibrium distribution. If the CDF of the inter-event time of the renewal process is $F(x)$ and its mean is $\mathbb{E}(\tau)$, then the PDF of the equilibrium distribution is:

$$f_{EQ}(t) = \frac{1 - F(t)}{\mathbb{E}(\tau)},$$

whose complementary CDF is given in [KUL 11, Eq. 8.40].

If the inter-event time of the renewal process is PH($\boldsymbol{\alpha}, \boldsymbol{A}$) distributed then the PDF of its residual time is:

$$f_{EQ}(t) = \frac{1 - F(t)}{\mathbb{E}(\tau)} = \frac{\boldsymbol{\alpha} e^{\boldsymbol{A}t}\mathbb{1}}{\boldsymbol{\alpha}(-\boldsymbol{A})^{-1}\mathbb{1}} = \frac{\boldsymbol{\alpha}(-\boldsymbol{A})^{-1}e^{\boldsymbol{A}t}(-\boldsymbol{A})\mathbb{1}}{\boldsymbol{\alpha}(-\boldsymbol{A})^{-1}\mathbb{1}}$$

$$= \frac{\boldsymbol{\alpha}(-\boldsymbol{A})^{-1}}{\boldsymbol{\alpha}(-\boldsymbol{A})^{-1}\mathbb{1}} e^{\boldsymbol{A}t}\boldsymbol{a},$$

which means that the residual time is PH($\boldsymbol{\gamma}, \boldsymbol{G}$) distributed where:

$$\boldsymbol{\gamma} = \frac{\boldsymbol{\alpha}(-\boldsymbol{A})^{-1}}{\boldsymbol{\alpha}(-\boldsymbol{A})^{-1}\mathbb{1}} \text{ and } \boldsymbol{G} = \boldsymbol{A}.$$

## 2.7. Positive density on $(0, \infty)$

THEOREM 2.15.– *If $T$ is PH($\boldsymbol{\alpha}, \boldsymbol{A}$) distributed, then its density is strictly positive on* $(0, \infty)$.

*Proof.* According to [2.3], the PDF can be written as $f_T(t) = \sum_{i=1}^{n} \mathrm{P}(X(t) = i)a_i$. Without loss of generality, we assume that the $a_i$ rates are positive for $i \leq k$ and $a_i = 0$ for $i > k$, where $k$ should be at least 1. In this case, $f_T(t) = 0$ if and only if $\mathrm{P}(X(t) \in \{1, \ldots, k\}) = 0$. If there exists $i \leq k$ such that $\alpha_i > 0$, then $\mathrm{P}(X(t) = i) > 0$ for all $t > 0$, because the CTMC can stay in the initial state until time $t$ with positive probability.

That is, $f_T(t) = 0$ only when the CTMC starts from $\{k+1, \ldots, n\}$ and it cannot reach the set $\{1, \ldots, k\}$ in time $t$. If the CTMC cannot reach $\{1, \ldots, k\}$ in time $t$, then $\{1, \ldots, k\}$ is not reachable from $\{k+1, \ldots, n\}$ and consequently the absorbing state is not reachable from the initial state, which is in conflict with the assumption that the states in $\{1, \ldots, n\}$ are transient states. □

## 2.8. Eigenvalue structure

If $T$ is size $n$ PH($\boldsymbol{\alpha}, \boldsymbol{A}$) distributed, then the eigenvalues of $\boldsymbol{A}$, $\lambda_i$, satisfy some simple conditions. It was already mentioned that $\lambda_i$ must have negative real parts. Further constraints on the location of the eigenvalues are summarized by the following theorems.

The eigenvalue with the maximal real part is referred to as *dominant eigenvalue* and is denoted by $\hat{\lambda}$. Let $\hat{a} = \max_{i,j} |A_{ij}|$, be the maximal element of matrix $|\boldsymbol{A}|$. We refer to $\hat{a}$ as the maximal rate in $\boldsymbol{A}$ and note that $\hat{\lambda}$ and $\hat{a}$ are real and $\hat{\lambda} < 0$ and $\hat{a} > 0$.

THEOREM 2.16.– *On the complex plain, the eigenvalues of* $\boldsymbol{A}$ *lie in the closed circle around* $-\hat{a}$ *with radius* $\hat{\lambda} + \hat{a}$ *(see Figure 2.5).*

*Proof.* We transform matrix $\boldsymbol{A}$ to a non-negative matrix and apply the Perron–Frobenius theorem.

The eigenvalues of $\boldsymbol{B} = \boldsymbol{A}/\hat{a} + \boldsymbol{I}$ are $\gamma_i = \lambda_i/\hat{a} + 1$, where $\lambda_i$ denote the $i$th eigenvalue of $\boldsymbol{A}$. By its definition, $\boldsymbol{B}$ is a non-negative ($\boldsymbol{B}_{ij} \geq 0$), sub-stochastic ($\boldsymbol{B}\mathbb{1} \leq \mathbb{1}$) matrix, whose Perron–Frobenius eigenvalue $\hat{\gamma} = \hat{\lambda}/\hat{a} + 1$ is real, less than 1 and the eigenvalues of $\boldsymbol{B}$ lie on the closed circle around the origin with radius $\hat{\gamma}$, that is, $|\gamma_i| \leq \hat{\gamma}$. □

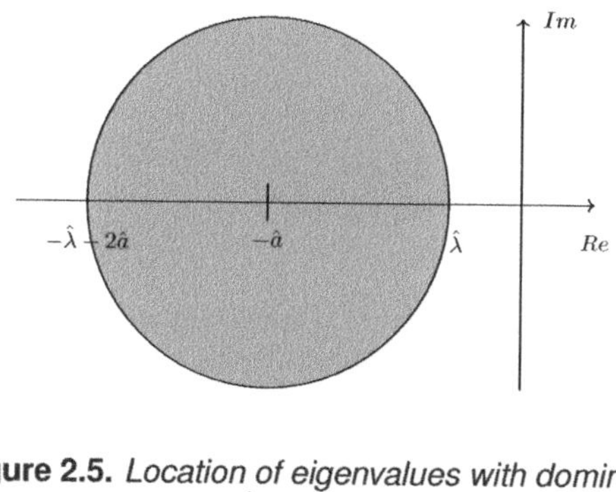


**Figure 2.5.** *Location of eigenvalues with dominant eigenvalue $\hat{\lambda}$ and maximal rate $\hat{a}$*

THEOREM 2.17.– *The eigenvalue with the maximal real part, denoted by* $\hat{\lambda}$, *is real and any* $\lambda_i$ *eigenvalue of* $\boldsymbol{A}$ *of size* $n$, *with real part* $\mathcal{R}(\lambda_i)$ *and imaginary part* $\mathcal{I}(\lambda_i)$, *satisfies (see Figure 2.6):*

$$|\mathcal{I}(\lambda_i)| \leq (\hat{\lambda} - \mathcal{R}(\lambda_i)) \tan\left(\frac{\pi}{2} - \frac{\pi}{n}\right) \qquad [2.17]$$

*and the PH distribution whose complex eigenvalue achieves equality has a cyclic structure of size $n$.*

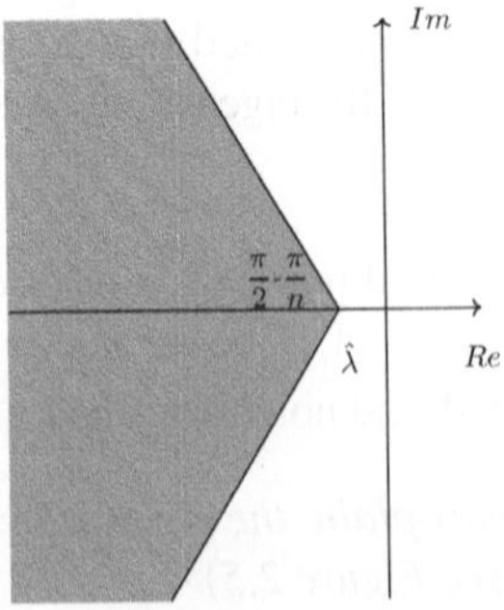


**Figure 2.6.** *Location of eigenvalues with dominant eigenvalue $\hat{\lambda}$ and size $n$*

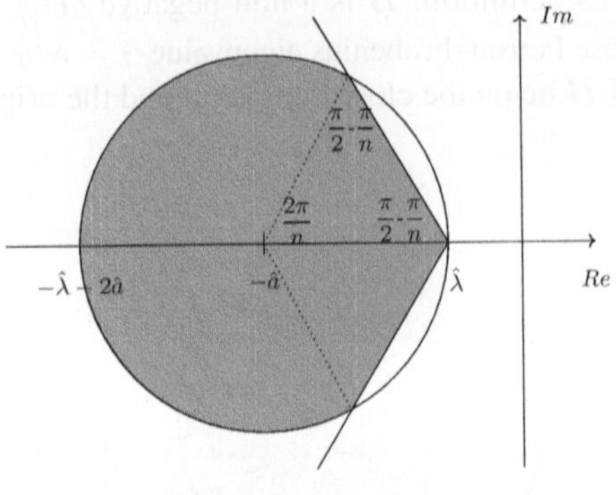


**Figure 2.7.** *Location of eigenvalues with dominant eigenvalue $\hat{\lambda}$, maximal rate $\hat{a}$ and size $n$*

*Proof.* A direct consequence of theorem 2.16 is that the eigenvalue with the maximal real part is unique, that is, there is no complex eigenvalue with the same real part as $\hat{\lambda}$.

Similar to the previous proof, we transform matrix $\boldsymbol{A}$ to the non-negative $\boldsymbol{B} = \boldsymbol{A}/\hat{a} + \boldsymbol{I}$, whose eigenvalues can be obtained from $\lambda_i$ as $\gamma_i = \lambda_i/\hat{a} + 1$.

According to [KEL 78, theorem 1], if the spectral radius of a non-negative matrix $\boldsymbol{B}$ is $\hat{\gamma}$ and $\gamma_i = \mu + \nu\Im$ is one of its eigenvalues, then:

$$\mu + |\nu| \tan\left(\frac{\pi}{m}\right) \leq \hat{\gamma}, \qquad [2.18]$$

where $m$ is the length of the longest simple circuit of $\boldsymbol{B}$. The longest simple circuit of $\boldsymbol{B}$ is the largest cycle period among the communication classes of the DTMC (see section 1.12). [2.18] can be written as:

$$|\nu| \leq (\hat{\gamma} - \mu) \tan\left(\frac{\pi}{2} - \frac{\pi}{m}\right).$$

Substituting $m = n$, $\mu = \mathcal{R}(\gamma_i)$, $\nu = \mathcal{I}(\gamma_i)$, $\hat{\gamma} = \hat{\lambda}/\hat{a} + 1$ and $\gamma_i = \lambda_i/\hat{a} + 1$, we obtain [2.17]. Finally, we note that the length of the longest simple circuit of $\boldsymbol{B}$ is the same as for $\boldsymbol{A}$ and $m \leq n$. In [2.17], the equality holds only when $m = n$, that is, $\boldsymbol{A}$ has a cyclic structure of size $n$. □

EXAMPLE 2.18.– *For example, for the size* 5 *cyclic PH distribution*[2] *with generator:*

$$\boldsymbol{A} = \begin{pmatrix} -\varphi & \varphi & & & \\ & -\varphi & \varphi & & \\ & & -\varphi & \varphi & \\ & & & -\varphi & \varphi \\ z\varphi & & & & -\varphi \end{pmatrix},$$

*where* $0 \leq z < 1$*, we have* $\hat{a} = \varphi$ *and:*

$$\boldsymbol{B} = \boldsymbol{A}/\hat{a} + \boldsymbol{I} = \begin{pmatrix} 0 & 1 & & & \\ & 0 & 1 & & \\ & & 0 & 1 & \\ & & & 0 & 1 \\ z & & & & 0 \end{pmatrix},$$

*whose characteristic polynomial is* $\det(x\boldsymbol{I} - \boldsymbol{B}) = x^5 - z$ *and the eigenvalues are* $\gamma_i = e^{(\log(z)+\Im i 2\pi)/5}$ *for* $i = 1, ..., 5$*, which are the fifth roots of unity multiplied by* $z$*. Consequently, the eigenvalues of* $\boldsymbol{A}$ *are* $\lambda_i = \varphi(e^{(\log(z)+\Im i2\pi)/5} - 1)$ *with* $i = 1, ..., 5$*. For* $z = 0.5$*,* $\gamma_i$ *are* $0.870551$*,* $0.269015 + 0.827943\Im$*,* $0.269015 - 0.827943\Im$*,* $-0.70429 + 0.511697\Im$ *and* $-0.70429 - 0.511697\Im$ *as depicted in Figure 2.8.*

Theorems 2.16 and 2.17 present necessary conditions for the location of the eigenvalues, whose intersection is depicted in Figure 2.6. A necessary and sufficient condition for the location of the eigenvalues is provided in [BEN 04] based on the results of Karpelevich [KAR 51] and its refinement in [ITO 97], which we summarize below.

Let $\Theta_n$ be the set of points in the complex plane that are eigenvalues of non-negative $n \times n$ matrices with spectral radius 1.

2 Later this structure is discussed as the feedback Erlang distribution.

THEOREM 2.19.– *[ITO 97] $\Theta_n$ is a proper subset of the closed unit disk, it is symmetric to the real axis, and intersects the unit circle at the ith root of unity for $i = 1, \ldots, n$. The boundary of $\Theta_n$ consists of curvilinear arcs connecting the consecutive intersecting points. Let the endpoints of a curvilinear arc be $e^{2\pi\Im a_1/b_1}$ and $e^{2\pi\Im a_2/b_2}$ (where $a_1, b_1$ and $a_2, b_2$ are relative primes, $0 \leq a_1 \leq b_1$, $0 \leq a_2 \leq b_2$, $1 \leq b_1 \leq b_2 \leq n$). Each of these arc is given by the following parametric equation:*

$$\lambda^{b_2}(\lambda^{b_1} - x)^{\lfloor n/b_1 \rfloor} = (1 - x)^{\lfloor n/b_1 \rfloor} \lambda^{b1 \lfloor n/b_1 \rfloor},$$

*where $x$ runs over the $(0, 1)$ interval.*

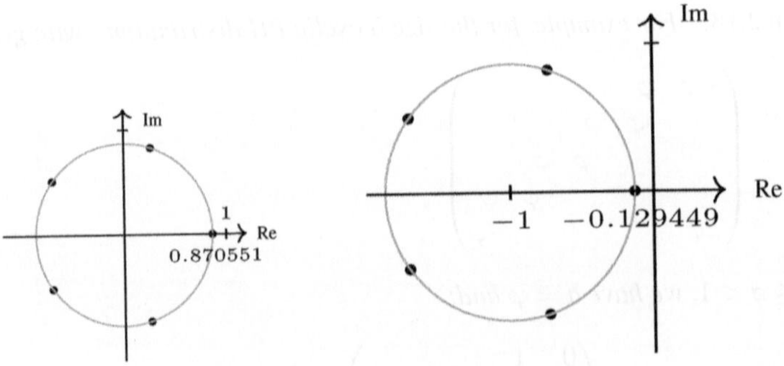


**Figure 2.8.** *Location of eigenvalues $\gamma_i$ and $\lambda_i$ for the size 5 cyclic PH distribution with $z = 0.5$ and $\varphi = 1$*

Figure 2.9 depicts $\Theta_n$ for $n = 2, 3, 4, 5$.

Let $\boldsymbol{A}$ be a PH generator of size $n$ with dominant eigenvalue $\hat{\lambda}$ and maximal rate $\hat{a}$, and let $\Omega_n$ be the set of points in the complex plane that are eigenvalues of $\boldsymbol{A}$.

THEOREM 2.20.– *The points of $\Omega_n$ can be obtained from the points of $\Theta_n$ by a linear transformation:*

$$\lambda_i \in \Omega_n \text{ if and only if } \frac{\lambda_i + \hat{a}}{\hat{\lambda} + \hat{a}} \in \Theta_n.$$

*Proof.* $\boldsymbol{A}/\hat{a} + \boldsymbol{I}$ is a non-negative matrix, whose Perron–Frobenius eigenvalue is $\hat{\lambda}/\hat{a} + 1$. Dividing $\boldsymbol{A}/\hat{a} + \boldsymbol{I}$ by this Perron–Frobenius eigenvalue, we obtain that $\boldsymbol{U} = \frac{1}{\hat{\lambda}+\hat{a}}(\boldsymbol{A} + \hat{a}\boldsymbol{I})$ is a non-negative matrix with spectral radius 1, and its eigenvalues are in $\Theta_n$ according to theorem 2.19. The eigenvalues of $\boldsymbol{U}$ are obtained from $\lambda_i$ by $\frac{\lambda_i+\hat{a}}{\hat{\lambda}+\hat{a}}$. □

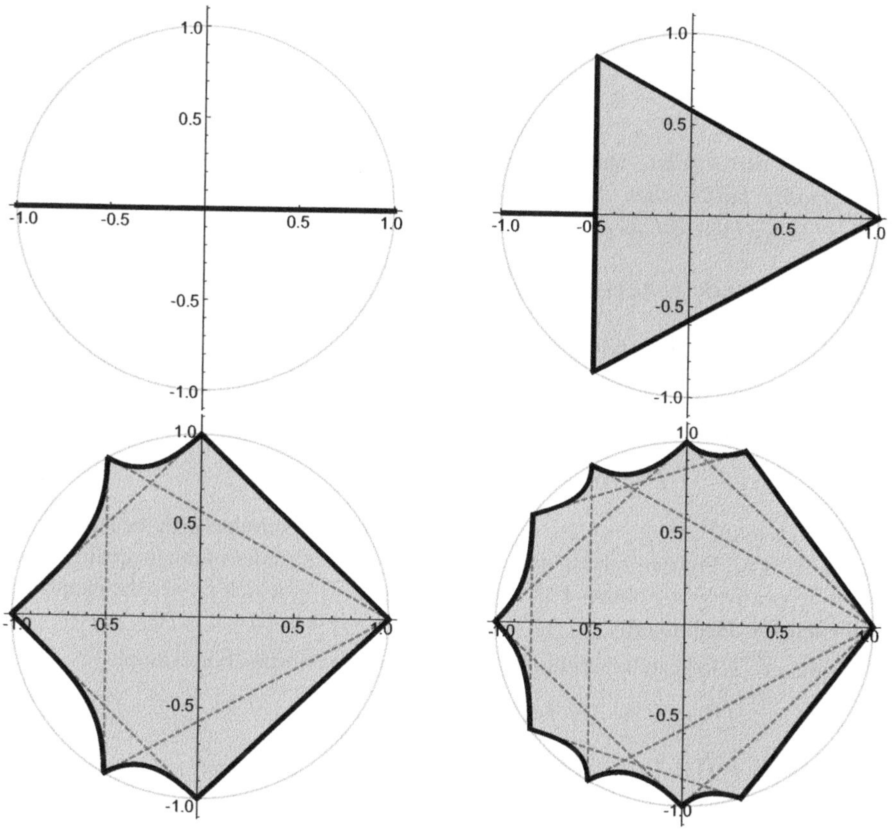

**Figure 2.9.** *Location of the eigenvalues of PH distributions with size* 2, 3, 4 *and* 5

## 2.9. Steepest increase property

In [O'CI 99], O'Cinneide published the "steepest increase" conjecture which was proved partially by O'Cinneide in [O'CI 99] and completely by Yao in [YAO 02]. The steepest increase lemma states that the PDF of a size $n$ PH$(\boldsymbol{\alpha}, \boldsymbol{A})$ distribution satisfies:

$$\frac{f'(t)}{f(t)} \leq \frac{n-1}{t} \quad \text{for } t > 0,$$

or equivalently, $f(t)/t^{n-1}$ is decreasing in $t$, that is:

$$\frac{d}{dt}\frac{f(t)}{t^{n-1}} \leq 0 \quad \text{for } t > 0.$$

Next, we present a "sharp" form of the lemma.

LEMMA 2.21.– *The PDF of a size n PH($\alpha$, $\mathbf{A}$) satisfies:*

$$\frac{f'(t)}{f(t)} \leq \frac{n-1}{t} - \lambda, \qquad [2.19]$$

*where $\lambda$ is the absolute value of the dominant eigenvalue of matrix $\mathbf{A}$. In [2.19], the equality holds when $f(t)$ is PDF of the Erlang$(n,\lambda)$ distribution, that is $f(t) = f_{(n,\lambda)}(t) = \frac{\lambda^n t^{n-1}}{(n-1)!} e^{-\lambda t}$.*

*Proof.* In [O'CI 99, p. 744], O'Cinneide presents the inequality:

$$((n-1-\lambda)\boldsymbol{I} - \boldsymbol{A})e^{\boldsymbol{A}} \geq 0$$

for PH generator $\boldsymbol{A}$. A detailed proof of this inequality is provided by Yao in [YAO 02]. Substituting $\boldsymbol{A}$ by $\boldsymbol{A}t$ in the inequality, we obtain, for $t > 0$:

$$((n-1-\lambda t)\boldsymbol{I} - \boldsymbol{A}t)e^{\boldsymbol{A}t} \geq 0, \qquad [2.20]$$

which leads to $e^{\boldsymbol{A}t}\boldsymbol{A}t \leq (n-1-\lambda t)e^{\boldsymbol{A}t}$. Note that the eigenvalue with the largest real part of $\boldsymbol{A}t$ is $\lambda t$. Pre-multiplying and post-multiplying both sides of the inequality by $\boldsymbol{\alpha}$ and $\boldsymbol{a}$, respectively, we obtain $f'(t)t \leq (n-1-\lambda t)f(t)$, which proves the inequality of the lemma. Additionally, $f_{(n,\lambda)}(t)$ satisfies $f'_{(n,\lambda)}(t)t = (n-1-\lambda t)f_{(n,\lambda)}(t)$ □

Equation [2.19] can be written in several alternative forms. For example,

$$tf'(t) \leq (n-1)f(t) - \lambda t f(t), \qquad [2.21]$$

$$tf'(t) + f(t) \leq (n-\lambda t)f(t), \qquad [2.22]$$

$$\frac{d}{dt}(tf(t)) \leq (n-\lambda t)f(t) \qquad [2.23]$$

and utilizing $\lambda \geq 0$, we also have:

$$\frac{d}{dt}(tf(t)) \leq nf(t). \qquad [2.24]$$

LEMMA 2.22.– *For $0 < t_1 < t_2$, the PDF of a PH distributed RV with generator $\boldsymbol{A}$ of size $n$ and dominant eigenvalue $-\lambda$ satisfies:*

$$\frac{f(t_2)}{f(t_1)} \leq \frac{f_{(n,\lambda)}(t_2)}{f_{(n,\lambda)}(t_1)}. \qquad [2.25]$$

*Proof.* Equation [2.21] can be written as:

$$(n-1)(\log(t))' - (\lambda t)' \geq (\log(f(t)))'. \qquad [2.26]$$

For $0 < t_1 < t_2$, integrating [2.26] from $t_1$ to $t_2$ gives:

$$\log\left(\left(\frac{t_2}{t_1}\right)^{n-1}\frac{f(t_1)}{f(t_2)}\right) \geq \lambda(t_2 - t_1).$$

We note that according to theorem 2.15, $f(t) > 0$ for $t > 0$. Then, we obtain:

$$f(t_2) \leq \frac{f(t_1)}{t_1^{n-1}} e^{\lambda t_1} t_2^{n-1} e^{-\lambda t_2} = f(t_1) \frac{f_{(n,\lambda)}(t_2)}{f_{(n,\lambda)}(t_1)}, \quad \text{for } t_2 > t_1. \qquad [2.27]$$

□

The lemma implies that if $f(t)$ is different from $f_{(n,\lambda)}(t)$, then $f(t)/f_{(n,\lambda)}(t)$ is decreasing in $t$, which is a generalization of the steepest increase conjecture [O'CI 99], which is about the monotonicity of function $f(t)/t^{n-1}$. A further consequence of lemma 2.22 and the monotonicity of $f(t)/f_{(n,\lambda)}(t)$ is the following lemma.

LEMMA 2.23.– *The CDF of a PH distribution and the associated Erlang$(n, \lambda)$ distribution satisfy the inequality:*

$$F(t) \geq F_{(n,\lambda)}(t), \qquad [2.28]$$

*for all $t > 0$.*

*Proof.* If $f(t) = f_{(n,\lambda)}(t)$, then the lemma holds true. If $f(t)$ is different from $f_{(n,\lambda)}(t)$, then $f(t)/f_{(n,\lambda)}(t)$ is decreasing in $t$, which means that there exists a $t^* > 0$ where $f(t^*) = f_{(n,\lambda)}(t^*)$ and $f(t) > f_{(n,\lambda)}(t)$ for $t < t^*$ and $f(t) < f_{(n,\lambda)}(t)$ for $t > t^*$. For $t \leq t^*$, we have:

$$F(t) = \int_{\tau=0}^{t} f(\tau) d\tau \geq \int_{\tau=0}^{t} f_{(n,\lambda)}(\tau) d\tau = F_{(n,\lambda)}(t),$$

and for $t > t^*$, we have:

$$1 - F(t) = \int_{\tau=t}^{\infty} f(\tau) d\tau \leq \int_{\tau=t}^{\infty} f_{(n,\lambda)}(\tau) d\tau = 1 - F_{(n,\lambda)}(t).$$

□

The relation in lemma 2.23 is referred to stochastic ordering [STO 83]. That is, the PH distributed RV is stochastically smaller than or equal to the associated Erlang$(n, \lambda)$ distributed RV.

# 3

# Discrete Phase Type Distributions

This chapter is devoted to the discrete counterpart of continuous phase type distributions, the discrete phase type (DPH) distributions. Some of the definitions and properties are easy to relate with the associated PH result, which we briefly discuss for completeness. Some properties are essentially different from the related PH properties, which are discussed in more details.

## 3.1. Definition and basic properties

The basic properties of DPH distributions, which we discuss first, are straightforward discrete counterparts of the related PH properties.

If the process $X_k$ with $k = 0, 1, 2, ...$ is an $n + 1$ state DTMC in which the first $n$ states are transient and the last one is absorbing, then the corresponding transition probability matrix can be partitioned as:

$$\mathbf{\Pi} = \begin{pmatrix} \boldsymbol{B} & \boldsymbol{b} \\ \mathbf{0} & 1 \end{pmatrix},$$

where $\boldsymbol{B} = (b_{ij})$ with $1 \leq i, j \leq n$ is the $n \times n$ matrix containing the transition probabilities among the transient states, $\boldsymbol{b} = (b_{i,n+1})$ with $1 \leq i \leq n$ is the column vector containing the transition probabilities from the transient states to the absorbing one, and $\mathbf{0}$ is a vector $n$ zeros. Since $\mathbf{\Pi}$ is the transition probability matrix of a DTMC whose row sum is 1, the following relation holds: $b_{i,n+1} = 1 - \sum_{j=1}^{n} b_{ij}$ for $i = 1, ..., n$. In vector form, we have $\boldsymbol{b} = \mathbb{1} - \boldsymbol{B}\mathbb{1}$. Let $\boldsymbol{\pi}$ be the initial distribution of the Markov chain, that is, $\boldsymbol{\pi}$ is the row vector of size $n + 1$ where $\pi_i = \mathrm{P}(X_0 = i)$. $\boldsymbol{\pi}$ can be decomposed into $\boldsymbol{\pi} = \begin{pmatrix} \boldsymbol{\beta} & \pi_{n+1} \end{pmatrix}$ where vector $\boldsymbol{\beta}$ of size $n$ contains the initial probabilities of the transient states and $\boldsymbol{\beta}\mathbb{1} = 1 - \pi_{n+1}$ holds.

DEFINITION 3.1.– *The time to reach the absorbing state of a discrete time Markov chain with a finite number of transient states and one absorbing state:*

$$T = \min\{k : X_k = n+1, k \geq 0\},$$

*is discrete phase type distributed [NEU 75, NEU 81a].*

COMMENT 3.2.– *As for PH distributions, we assume that the Markov chain starts from one of the transient states, that is,* $\pi_{n+1} = 0$ *and* $\beta\mathbb{1} = 1$. *Consequently,* $P(T = 0) = 0$ *and the support of* $T$ *is* $\mathbb{N}^+$.

In the case of DPH distributions, the primary interest is to compute the probability mass function, but to emphasize the similarities with PH distributions we first compute the cumulative mass function (which is also a transient measure of the Markov chain) as:

$$F_T(k) = \text{P}(T \leq k) = \text{P}(X_k = n+1) = \boldsymbol{\pi}\boldsymbol{\Pi}^k \mathbf{e}_{n+1}.$$

The calculation based on this redundant (size $n+1$) representation is not convenient, and thus the size $n$ description given by $\boldsymbol{B}$ and $\boldsymbol{\beta}$ is commonly used.

$$\begin{aligned} F_T(k) &= \text{P}(T \leq k) = \text{P}(X_k = n+1) = 1 - \sum_{i=1}^{n} \text{P}(X_k = i) \\ &= 1 - (\boldsymbol{\beta}\ 0)\, \boldsymbol{\Pi}^k \begin{pmatrix} \mathbb{1} \\ 0 \end{pmatrix} 1 - (\boldsymbol{\beta}\ 0) \begin{pmatrix} \boldsymbol{B} & \boldsymbol{b} \\ \mathbf{0} & 1 \end{pmatrix}^k = \begin{pmatrix} \mathbb{1} \\ 0 \end{pmatrix} \\ &= 1 - (\boldsymbol{\beta}\ 0) \begin{pmatrix} \boldsymbol{B}^k & \bullet \\ \mathbf{0} & 1 \end{pmatrix} \begin{pmatrix} \mathbb{1} \\ 0 \end{pmatrix} = 1 - \boldsymbol{\beta}\boldsymbol{B}^k\mathbb{1}. \end{aligned} \quad [3.1]$$

The PMF of $T$ can be obtained as follows. For $k \geq 1$:

$$\begin{aligned} f_T(k) &= \text{P}(T{=}k) = \text{P}(X_k{=}n{+}1, X_{k-1} \neq n+1) \\ &= \sum_{i=1}^{n} \text{P}(X_k{=}n{+}1, X_{k-1} = i) \\ &= \sum_{i=1}^{n} \underbrace{\text{P}(X_{k-1} = i)}_{(\boldsymbol{\beta}\boldsymbol{B}^{k-1})_i} \underbrace{\text{P}(X_k = n+1 | X_{k-1} = i)}_{b_i} = \boldsymbol{\beta}\boldsymbol{B}^{k-1}\boldsymbol{b}. \end{aligned} \quad [3.2]$$

Element $i, j$ of matrix $\boldsymbol{B}^k$ contains the probability that starting from transient state $i$, the Markov chain is in transient state $j$ after $k$ steps. Since states $1, \ldots, n$ are transient, $\boldsymbol{B}^k$ tends to zero as $k$ tends to infinity, which means that the absolute value of the eigenvalues of $\boldsymbol{B}$ are less than one. As a result, the $\sum_{k=0}^{\infty} \boldsymbol{B}^k$ sum converges. Let $\boldsymbol{M} = \sum_{k=0}^{\infty} \boldsymbol{B}^k$, then matrix $\boldsymbol{M}$ is the solution of:

$$\boldsymbol{M} = \boldsymbol{B}^0 + \sum_{k=1}^{\infty} \boldsymbol{B}^k = \boldsymbol{I} + \boldsymbol{B}\left(\sum_{k=0}^{\infty} \boldsymbol{B}^k\right) = \boldsymbol{I} + \boldsymbol{B}\boldsymbol{M},$$

from which $M = (I - B)^{-1}$. The fact that $I - B$ is non-singular follows also from the properties of the eigenvalues of $B$. Namely, denoting by $\lambda_i$, $i = 1, \ldots, n$, the eigenvalues of $B$, the eigenvalues of $I - B$ are $1 - \lambda_i$ with negative real parts.

The $z$ transform of $T$, $\mathbb{E}(z^T)$, for $|z| \leq 1$, can be computed as:

$$T(z) = \mathbb{E}(z^T) = \sum_{k=0}^{\infty} z^k f_T(k) = \sum_{k=1}^{\infty} z^k \beta B^{k-1} b$$
$$= z \sum_{k=1}^{\infty} \beta (zB)^{k-1} b = z\beta(I - zB)^{-1} b. \qquad [3.3]$$

The ordinary moments of $T$ can be computed as:

$$\mathbb{E}(T^k) = \sum_{\ell=1}^{\infty} \ell^k f_T(\ell) = \sum_{\ell=1}^{\infty} \ell^k \beta B^{\ell-1} b,$$

but to obtain a concise expression the factorial moments are computed commonly for DPH distributions:

$$f_k = \mathbb{E}(T(T-1)\ldots(T-k+1)) = \left.\frac{d^k}{dz^k} T(z)\right|_{z=1}$$
$$= \left.\frac{d^k}{dz^k} z\beta(I - zB)^{-1} b\right|_{z=1}$$
$$= k!\beta B^{k-1} \left.(I - zB)^{-k-1}\right|_{z=1} b = k!\beta B^{k-1}(I - B)^{-k-1}(I - B)\mathbb{1}$$
$$= k!\beta B^{k-1}(I - B)^{-k}\mathbb{1}. \qquad [3.4]$$

In [3.4], the $k$th derivative is obtained using $B$ and $(I - zB)^{-1}$ commute.

## 3.2. Stochastic meaning of $(I - B)^{-1}$

The matrix $(I - B)^{-1}$ has an important stochastic meaning. Let $T_{ij}$ be the time spent in transient state $j$ before reaching the absorbing one given that the Markov chain starts in state $i$. For $\mathbb{E}(T_{ij})$, we have:

$$\mathbb{E}(T_{ij}) = \sum_{k=0}^{\infty} \mathbb{E}(\mathcal{I}_{\{X_k=j\}} | X_0 = i) = \sum_{k=0}^{\infty} \mathrm{P}(X_k = j | X_0 = i)$$
$$= \left(\sum_{k=0}^{\infty} B^k\right)_{ij} = \left((I - B)^{-1}\right)_{ij}$$

Consequently, the element in position $i, j$ of $(I - B)^{-1}$ is $\mathbb{E}(T_{ij})$, which is a non-negative number.

### 3.3. Rational $z$ transform function

$T(z)$ is a rational function of $z$. Both its numerator and denominator is an at most order $n$ polynomial. This follows from:

$$T(z) = z\boldsymbol{\beta}(\boldsymbol{I} - z\boldsymbol{B})^{-1}\boldsymbol{b} = z\sum_i\sum_j \beta_i\left((\boldsymbol{I} - z\boldsymbol{B})^{-1}\right)_{ij} b_j$$

$$= z\sum_i\sum_j \beta_i \frac{\det_{ji}(\boldsymbol{I} - z\boldsymbol{B})}{\det(\boldsymbol{I} - z\boldsymbol{B})} b_j = \frac{\sum_i\sum_j z\beta_i b_j \det_{ji}(\boldsymbol{I} - z\boldsymbol{B})}{\det(\boldsymbol{I} - z\boldsymbol{B})}$$

$$= \frac{\omega_n z^n + \omega_{n-1}z^{n-1} + \ldots + \omega_1 z}{z^n + \tau_{n-1}z^{n-1} + \ldots + \tau_1 z + \tau_0} \quad \text{with} \sum_{i=1}^{n} \omega_i = 1 + \sum_{i=0}^{n-1} \tau_i, \qquad [3.5]$$

where $\det_{ji}(\boldsymbol{M})$ denotes the determinant of the matrix obtained by removing row $j$ and column $i$ of matrix $\boldsymbol{M}$ multiplied by $(-1)^{i+j}$. The determinant of $\boldsymbol{I} - z\boldsymbol{B}$ is an order $n$ polynomial of $z$, while the numerator is the sum of order $n$ polynomials of $z$ without the constant ($z^0$) term, which is at most an order $n$ polynomial of $z$ without constant term. From $\sum_{k=1}^{\infty} f_T(k) = T(1) = 1$, we have that $\sum_{i=1}^{n} \omega_i = 1 + \sum_{i=0}^{n-1} \tau_i$.

### 3.4. Decomposition of the matrix geometric function

The PMF of a DPH distribution is a matrix geometric function. Let $\boldsymbol{B} = \boldsymbol{H}^{-1}\boldsymbol{\Lambda}\boldsymbol{H}$ be the spectral decomposition of $\boldsymbol{B}$ and let $\boldsymbol{u} = \boldsymbol{\beta}\boldsymbol{H}^{-1}$ and $\boldsymbol{v} = \boldsymbol{H}\boldsymbol{b}$. Then,

$$f_T(k) = \boldsymbol{\beta}\boldsymbol{B}^{k-1}\boldsymbol{b} = \boldsymbol{\beta}\boldsymbol{H}^{-1}\boldsymbol{\Lambda}^{k-1}\boldsymbol{H}\boldsymbol{b} = \boldsymbol{u}\boldsymbol{\Lambda}^{k-1}\boldsymbol{v}\,.$$

We distinguish the following two cases:

– The eigenvalues, $\lambda_i$, of $\boldsymbol{B}$ are different (i.e. the algebraic multiplicity of each eigenvalue is one) and, consequently, $\boldsymbol{\Lambda}$ is a diagonal matrix. In this case, $f_T(k)$ is of the form:

$$f_T(k) = \boldsymbol{u}\boldsymbol{\Lambda}^{k-1}\boldsymbol{v} = \sum_i u_i v_i \lambda_i^{k-1} = \sum_i c_i \lambda_i^{k-1}, \qquad [3.6]$$

where $c_i = u_i v_i$ is a constant coefficient.

Here, the eigenvalues, $\lambda_i$, as well as the associated coefficients, $c_i$, can be real or complex conjugate pairs, but all eigenvalues are within the unit disc. In the case of a real eigenvalue, both $\lambda_i$ and the associated $c_i$ can be positive, negative or zero. If $\lambda_i$ is negative, the corresponding geometric progression has alternating sign. If $c_i$ is zero, then the eigenvalue $\lambda_i$ does not contribute to the distribution. If $c_i$ is negative and $\lambda_i$ is positive, then this eigenvalue contributes with a negative geometric progression.

For a complex conjugate pair of eigenvalues $\lambda_i = |\lambda_i| e^{\Im \alpha_i}$ and $\bar{\lambda}_i = |\lambda_i| e^{-\Im \alpha_i}$ with coefficients $c_i = |c_i| e^{\Im \varphi_i}$ and $\bar{c}_i = |c_i| e^{-\Im \varphi_i}$, we have:

$$c_i \lambda_i^k + \bar{c}_i \bar{\lambda}_i^k = |c_i||\lambda_i|^k \left( e^{\Im(\alpha_i k + \varphi_i)} + e^{-\Im(\alpha_i k + \varphi_i)} \right) = 2|c_i||\lambda_i|^k \cos(\alpha_i k + \varphi_i).$$

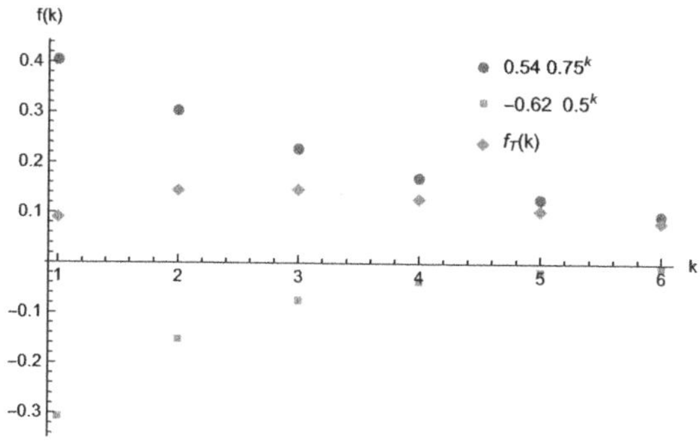


**Figure 3.1.** *PMF of the DPH($\beta$, $B$) distribution with* $\beta = \begin{pmatrix} 0.9 & 0.1 \end{pmatrix}$ *and* $B = \begin{pmatrix} 0.75 & 0.2 \\ 0 & 0.5 \end{pmatrix}$, *from which* $f_T(k) = 0.54 \cdot 0.75^k - 0.62 \cdot 0.5^k$. *For a color version of this figure, see www.iste.co.uk/horvath/phase.zip*

Figure 3.1 presents a simple example for the decomposition of the density function given in [3.6] with real eigenvalues. The geometric progressions in Figure 3.1 can be interpreted as the values of the exponential function, similar to the one in Figure 2.1, at integer points. This analogy extends to the case with complex eigenvalues, and due to that we omit further examples here. We just note that a DPH distribution with a negative real eigenvalue, which contributes with a geometric progression with alternating sign, is the discrete counterpart of the exponential function with an associated complex eigenvalue pair. For example, $\beta = \begin{pmatrix} 0.8 & 0.2 \end{pmatrix}$ and $B = \begin{pmatrix} 0.1 & 0.78 \\ 0.5 & 0.15 \end{pmatrix}$ gives $f_T(k) = 0.2016(-0.5)^k + 0.355733 \cdot 0.75^k$, which at integer points is identical to the exponential function $0.355733 e^{\log(0.75)t} + 0.2016 e^{\log(0.5)t} \cos(\pi t)$. That is, in this respect, a negative real eigenvalue of a DPH distribution is the discrete counterpart of a complex eigenvalue pair of a PH distribution.

– If there are eigenvalues of $B$ with algebraic multiplicity larger than one and, consequently, $\Lambda$ contains real Jordan blocks, then the geometric progression based decomposition of $f_T(k)$ can be obtained using [1.19]. The PMF takes the form:

$$f_T(k) = \sum_{i=1}^{\#\lambda} \sum_{j=1}^{\#\lambda_i} c_{ij} k^{j-1} \lambda_i^k,$$

where $\#\lambda$ is the number of different eigenvalues and $\#\lambda_i$ is the multiplicity of $\lambda_i$.

Similar to the previous case, the eigenvalues as well as the associated coefficients can be real or complex conjugate pairs.

As a result of all of these cases, the PMF of a DPH distribution has the following geometric progression form:

$$f_T(k) = \sum_{i=1}^{\#\lambda_R} \sum_{j=1}^{\#\lambda_i^R} c_{ij}^R k^{j-1} \lambda_i^{Rk} + \sum_{i=1}^{\#\lambda_C} \sum_{j=1}^{\#\lambda_i^C} 2|c_{i,j}^C| k^{j-1} |\lambda_i^C|^k \cos(\alpha_i^C k - \varphi_{i,j}^C) \qquad [3.7]$$

where $\#\lambda_R$ is the number of different real eigenvalues and $\#\lambda_C$ is the number of different complex conjugate eigenvalue pairs, $\lambda_i^C = |\lambda_i^C| e^{-\Im \alpha_i^C}$ and $c_{i,j} = |c_{i,j}| e^{\Im \varphi_{i,j}}$.

## 3.5. Similarity transformations

As in the case of PH distributions, infinitely many Markov chains can represent the same DPH distribution. Two Markov chains representing the same DPH can have the same size: a given representation $(\boldsymbol{\beta}, \boldsymbol{B})$ and its similarity transform $(\boldsymbol{\beta T}, \boldsymbol{T}^{-1}\boldsymbol{BT})$. Or they can have different sizes: $(\boldsymbol{\beta}, \boldsymbol{B})$ of size $n$ and $(\boldsymbol{\gamma}, \boldsymbol{G})$ of size $m$ represent the same DPH distribution if $\boldsymbol{BV} = \boldsymbol{VG}$ and $\boldsymbol{\beta V} = \boldsymbol{\gamma}$. For more details, see the corresponding results on PH distributions (section 2.5).

## 3.6. Closure properties

As the class of PH distributions, the class of DPH distributions is closed under commonly used operations on random variables. A difference is that the use of equilibrium distribution is not widespread for discrete distributions. We summarize the formulas for DPH distributions, which differ only in the use of the Kronecker operators. If $X$ and $Y$ are DPH($\boldsymbol{\alpha}, \boldsymbol{A}$) and DPH($\boldsymbol{\beta}, \boldsymbol{B}$) distributed (with $\boldsymbol{a} = (\boldsymbol{I} - \boldsymbol{A})\mathbb{1}$ and $\boldsymbol{b} = (\boldsymbol{I} - \boldsymbol{B})\mathbb{1}$), respectively, then the following operations result in another DPH distribution.

*Summation*: $Z = X + Y$ is DPH distributed with representation DPH($\boldsymbol{\gamma}, \boldsymbol{G}$) where:

$$\boldsymbol{\gamma} = (\boldsymbol{\alpha}\ \boldsymbol{0}) \text{ and } \boldsymbol{G} = \begin{pmatrix} \boldsymbol{A} & \boldsymbol{a\beta} \\ \boldsymbol{0} & \boldsymbol{B} \end{pmatrix}.$$

*Mixture*: If:

$$Z = \begin{cases} X & \text{with probability } p, \\ Y & \text{with probability } (1-p), \end{cases}$$

then $Z$ is DPH($\gamma, \boldsymbol{G}$) with:

$$\gamma = \begin{pmatrix} p\alpha & (1-p)\beta \end{pmatrix} \text{ and } \boldsymbol{G} = \begin{pmatrix} \boldsymbol{A} & \boldsymbol{0} \\ \boldsymbol{0} & \boldsymbol{B} \end{pmatrix}.$$

*Minimum*: $Z = \min(X, Y)$ is DPH($\gamma, \boldsymbol{G}$) distributed with:

$$\gamma = \alpha \otimes \beta \text{ and } \boldsymbol{G} = \boldsymbol{A} \otimes \boldsymbol{B}.$$

*Maximum*: $Z = \max(X, Y)$ is DPH($\gamma, \boldsymbol{G}$) distributed with:

$$\gamma = \begin{pmatrix} \alpha \otimes \beta & \boldsymbol{0} & \boldsymbol{0} \end{pmatrix} \text{ and } \boldsymbol{G} = \begin{pmatrix} \boldsymbol{A} \otimes \boldsymbol{B} & \boldsymbol{a} \otimes \boldsymbol{B} & \boldsymbol{A} \otimes \boldsymbol{b} \\ \boldsymbol{0} & \boldsymbol{B} & \boldsymbol{0} \\ \boldsymbol{0} & \boldsymbol{0} & \boldsymbol{A} \end{pmatrix}.$$

## 3.7. Eigenvalue structure

In the case of PH distributions, a transformation is applied to the generator in order to obtain a non-negative matrix and the properties of the eigenvalues are derived based on the transformed generator (section 2.8). In the case of DPH distributions, we can apply the related theoretical results (e.g. the Perron–Frobenius theorem) directly because the transition probability matrix is non-negative. To reduce redundancy, here we only present the final theorem with necessary and sufficient conditions.

Let $\Omega_n$ be the set of points in the complex plane that are eigenvalues of $\boldsymbol{B}$ where $\boldsymbol{B}$ is a DPH generator of size $n$ with Perron–Frobenius eigenvalue $\hat{\lambda}$ (which is the dominant eigenvalue in the DPH case). According to the Perron–Frobenius theorem, $\hat{\lambda}$ is non-negative but can be zero, for example, when:

$$\boldsymbol{B} = \begin{pmatrix} 0 & 1 & 0 \\ 0 & 0 & 1 \\ 0 & 0 & 0 \end{pmatrix},$$

which results in a DPH distribution with finite support.

THEOREM 3.3.– *Eigenvalue $\hat{\lambda} = 0$ if and only if $\Omega_n = \{0\}$. If $\hat{\lambda} > 0$, then the points of $\Omega_n$ can be obtained from the points of $\Theta_n$ (defined in theorem 2.19) by a scaling transformation:*

$$\lambda_i \in \Omega_n \textit{ if and only if } \frac{\lambda_i}{\hat{\lambda}} \in \Theta_n.$$

*Proof.* $\hat{\lambda} = 0$ if and only if 0 is an eigenvalue of $\boldsymbol{B}$ with multiplicity $n$. If $\hat{\lambda} > 0$, then $\boldsymbol{B}/\hat{\lambda}$ is a non-negative matrix with spectral radius 1 and if $\lambda_i$ is an eigenvalue of $\boldsymbol{B}$, then $\lambda_i/\hat{\lambda}$ is an eigenvalue of $\boldsymbol{B}/\hat{\lambda}$. According to theorem 2.19, the eigenvalues of $\boldsymbol{B}/\hat{\lambda}$ are in $\Theta_n$. □

4

# Matrix Exponential and Matrix Geometric Distributions

We introduced PH and DPH distributions based on the behavior of (continuous time and discrete time, respectively) Markov chains with $n$ transient states and an absorbing one. As a result, in the case of PH distributions, the vector–matrix pair representation, denoted by $(\boldsymbol{\alpha}, \boldsymbol{A})$, is such that vector $\boldsymbol{\alpha}$ is a probability vector with non-negative elements and matrix $\boldsymbol{A}$ is a generator matrix with negative diagonal and non-negative off-diagonal elements. For DPH distributions, the $(\boldsymbol{\beta}, \boldsymbol{B})$ representation is such that vector $\boldsymbol{\beta}$ is a probability vector with non-negative elements and matrix $\boldsymbol{B}$ is a transition probability matrix with non-negative elements.

Another consequence of the Markov chain–based interpretation is that the PDF of a PH distribution, $f_T(t) = \boldsymbol{\alpha} e^{\boldsymbol{A}t}\boldsymbol{a}, \forall t \in [0, \infty)$, and the PMF of a DPH distribution, $f_T(k) = \boldsymbol{\beta}\boldsymbol{B}^{k-1}\boldsymbol{b}, \forall k \in \{1, 2, ...\}$, are non-negative.

Relaxing the sign constraints for the vector and matrix elements and maintaining the sign constraint on the probability density and mass functions defines true supersets of PH and DPH distributions, namely, the matrix exponential (ME) and the matrix geometric (MG) distributions.

## 4.1. Matrix exponential distributions

DEFINITION 4.1.– *A continuous random variable $T$ with cumulative distribution function:*

$$F_T(t) = P(T < t) = 1 - \boldsymbol{\alpha} e^{\boldsymbol{A}t}\mathbb{1}, \ \forall t \in [0, \infty), \tag{4.1}$$

*where $\boldsymbol{\alpha}$ is a vector of reals of size $n$ and $\boldsymbol{A}$ is a matrix of reals of size $n \times n$, is ME distributed.*

The size of $\alpha$ and $A$ plays the same role as the number of transient states in the case of PH distributions. It directly follows from definiton 4.1 that the set of size $n$ PH distributions is a subset of the set of size $n$ ME distributions.

ME distributions share the following basic properties with PH distributions:

– matrix exponential distribution function [4.1] (see [2.1]),

– matrix exponential density function [2.2],

– matrix exponential hazard rate function [2.4],

– moment expressions [2.7],

– rational Laplace transform [2.9],

– the density function is composed of the same set of functions as in [2.13],

– non-unique representation (section 2.5),

– closure properties (section 2.6).

The main difference between the ME and the PH classes of distributions is that PH distributions have a stochastic interpretation through Markov chains. This interpretation can be used in a straightforward manner to derive properties both for PH distributions and for complex stochastic models containing PH distributions.

Another important difference is due to the sign constraints on the elements of the generator matrix of a PH distributions, which lead to significant restrictions on the eigenvalues of the generator presented in theorems 2.16, 2.17 and 2.20. Relaxing these constraints, theorems 2.16, 2.17 and 2.20 do not apply to ME distributions. Consequently, the eigenvalues of an ME distribution have more flexibility, as summarized in the following theorem.

THEOREM 4.2.– *Given an ME distribution represented by the vector–matrix pair* $(\alpha, A)$, *the eigenvalues of* $A$ *have negative real parts and among the eigenvalues with maximal real part one is real.*

*Proof.* Since $A$ is real, its eigenvalues are real or complex conjugate pairs. According to [2.13], the eigenvalues of $A$ have negative real parts because otherwise the density function does not converge to zero as $t$ tends to infinity. If there was a pair of complex eigenvalues with maximal real part such that there is no real eigenvalue with the same real part, then its contribution to the PDF (a function with alternating sign) would dominate and for large values of $t$, the PDF would have negative values. □

Figure 4.1 shows the area where the eigenvalues of ME and PH distributions can be located.

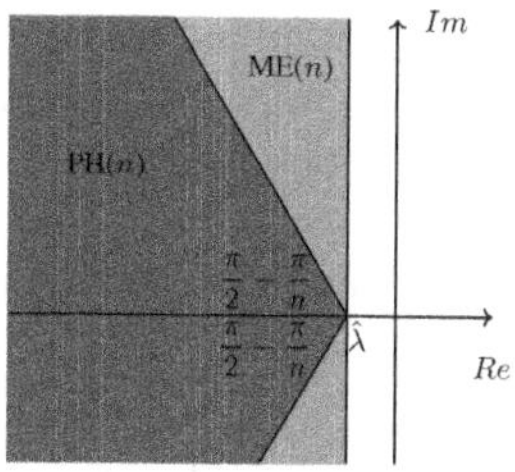


**Figure 4.1.** *Area of eigenvalues of size $n$ PH (dark gray area) and ME (gray and dark gray area) distributions with dominant eigenvalue $\hat{\lambda}$ and $n \geq 3$. The angle of the border between PH and ME distributions with the real axis is $\pi/2 - \pi/n$*

EXAMPLE 4.3.– *According to theorem 4.2, given an ME distribution with $(\boldsymbol{\alpha}, \boldsymbol{A})$, matrix $\boldsymbol{A}$ can have a pair of complex conjugate eigenvalues with the same real part as its dominant eigenvalue. For example, if $\lambda_1 = -1$, $\lambda_2 = -1 + \Im$, $\lambda_3 = -1 - \Im$, then, according to* [2.13]*:*

$$f_T(t) = ce^{-t} + 2|c^C|e^{-t}\cos(t - \varphi)$$

*where the coefficients of the real and the complex eigenvalues, $c$ and $c^C$, determine if $f_T(t)$ is non-negative for $t \geq 0$. Specifically, if $0 < 2|c^C| \leq c$, then $f_T(t)$ is non-negative and corresponds to an ME distribution with the above three eigenvalues.*

As ME distributions are more flexible than PH distributions for what concerns the eigenvalues, they are more flexible also from the point of view of their "steepest increase" and lemma 2.21 does not apply to ME distributions.

The main difficulty working with ME distributions is that not any $(\boldsymbol{\alpha}, \boldsymbol{A})$ vector–matrix pair defines a non-negative PDF (while any vector–matrix pair satisfying the sign constraints of PH distributions does). An efficient numerical method has been proposed in [REI 14] to check the non-negativity of a matrix exponential function defined by a given vector–matrix pair. General symbolic conditions are still missing. The borders of the ME($n$) class and the PH($n$) are further investigated in Chapter 5.

### 4.1.1. *Non-negative matrix exponential subclasses*

In order to avoid the difficulties related to verify the non-negativity of the PDF associated with a vector–matrix pair, $(\boldsymbol{\alpha}, \boldsymbol{A})$, one can consider subclasses of the ME family for which non-negativity is guaranteed by construction. Such subclasses are easier to define based on the PDF given in [2.13] than based on the vector–matrix pair representation. We provide two such subclasses.

A set of non-negative matrix exponential functions is given by:

$$f_r(t) = \sum_{i=1}^{n} c_i \prod_{j=1}^{k_i} (t - \tau_{ij})^2 t^{a_i} e^{\lambda_i t},$$

where $c_i \in \mathbb{R}^+$, $\lambda_i \in \mathbb{R}^-$, $k_i \in \{1, 2, ...\}$, $\tau_{ij} \in \mathbb{R}$ and $a_i$ is either 0 or 1. An ME distribution with the above PDF has real eigenvalues equal to $\lambda_1, \ldots, \lambda_n$.

A another set of non-negative matrix exponential functions can be obtained by:

$$f_c(t) = \sum_{i=1}^{n} c_i \prod_{j=1}^{k_i} \sin^2(\omega_i t - \varphi_{ij}) e^{\lambda_i t},$$

where $c_i, \omega_i \in \mathbb{R}^+$, $\lambda_i \in \mathbb{R}^-$, $k_i \in \{1, 2, ...\}$ and $\varphi_{ij} \in \mathbb{R}$. Such a PDF is associated with an ME distribution with complex conjugate eigenvalues whose real part is in $\{\lambda_1, \ldots, \lambda_n\}$ and real eigenvalues in $\{\lambda_1, \ldots, \lambda_n\}$.

In both cases, the PDF can become zero (e.g. with $n = 1$ and $t = \tau_{1j}$, in the first case, and with $n = 1$ and $t = \varphi_{1j}/\omega_1$, in the second case). Note that this cannot happen in the case of PH distributions for which the PDF is strictly positive for any $t > 0$ (see theorem 2.15).

## 4.2. Matrix geometric distributions

DEFINITION 4.4.– *A discrete random variable $T$ with probability mass function:*

$$f_T(k) = P(T = k) = \boldsymbol{\beta} \boldsymbol{B}^{k-1} (\boldsymbol{I} - \boldsymbol{B}) \mathbb{1},$$

*where $\boldsymbol{\beta}$ is a vector of reals of size $n$ and $\boldsymbol{B}$ is a matrix of reals of size $n \times n$, is MG distributed.*

Similar to the continuous case, the set of DPH distributions is by definition a subset of the set of matrix geometric (MG) distributions, and they share several properties: the distribution function [3.1], the probability mass function [3.2], factorial moment expressions [3.4], rational $z$ transform [3.5], non-unique representation (section 3.5) and closure properties (section 3.6).

Although the properties of MG distributions are not completely analogous to those of ME distributions, since they are similar, we do not provide a full treatment here. We mention only that:

– the possible locations of the eigenvalues of MG distributions extend those of DPH distributions;

– MG subclasses that are non-negative by construction can be defined similarly to the ME subclasses introduced in section 4.1.1.

# 5

# Classes and Representations of Continuous Phase Type Distributions

In this section, we discuss further properties of PH distribution that have practical importance also in fitting and moment matching applications.

## 5.1. Number of parameters

The number of parameters of a rank $n$ PH distribution without probability mass at zero can be obtained in various ways. It can be computed based on the number of non-trivial coefficients of its rational Laplace transform [2.9] or based on the number of independent moments as shown in [O'CI 89, O'CI 99]. Here, we introduce a different approach to work out the number of free parameters, which we use also in the following.

THEOREM 5.1.– *The PDF of a rank $n$ PH distribution (without probability mass at zero) is determined by $2n - 1$ parameters.*

*Proof.* Let $(\boldsymbol{\alpha}, \boldsymbol{A})$ be a full rank representation of a rank $n$ PH distribution and $(-\boldsymbol{A})^{-1} = \boldsymbol{\Gamma}^{-1}\boldsymbol{E}\boldsymbol{\Gamma}$ the Jordan decomposition of $(-\boldsymbol{A})^{-1}$ normalized such that $\boldsymbol{\Gamma}\mathbb{1} = \mathbb{1}$ (which is always possible according to [TEL 07, theorem 5]). It follows that also $\boldsymbol{\Gamma}^{-1}\mathbb{1} = \mathbb{1}$. Matrix $\boldsymbol{E}$ has Jordan block structure $\boldsymbol{E} = \text{diag}\{\boldsymbol{J}_{n_j}(\lambda_j)\}$ where $\boldsymbol{J}_{n_j}(\lambda_j)$ is defined in [1.18], $\lambda_j$ is an eigenvalue of $(-\boldsymbol{A})^{-1}$ and $n_j$ is the multiplicity of $\lambda_j$.

According to [2.8], the sequence $\boldsymbol{\alpha}(-\boldsymbol{A})^{-i}\mathbb{1}, i = 0, 1, 2, ...$ determines the distribution and it can be written, based on the Jordan decomposition, as:

$$\boldsymbol{\alpha}(-\boldsymbol{A})^{-i}\mathbb{1} = \boldsymbol{\alpha}\boldsymbol{\Gamma}^{-1}\boldsymbol{E}^{i}\boldsymbol{\Gamma}\mathbb{1} = v\boldsymbol{E}^{i}\mathbb{1} \qquad [5.1]$$

where $v = \boldsymbol{\alpha}\boldsymbol{\Gamma}^{-1}$. If the distribution has no mass at zero, then $\boldsymbol{\alpha}\mathbb{1} = \boldsymbol{\alpha}(-\boldsymbol{A})^{0}\mathbb{1} = v\boldsymbol{E}^{0}\mathbb{1} = v\mathbb{1} = 1$ holds. Accordingly, $v$ is determined by $n - 1$ potentially complex elements. Matrix $E$ instead is determined by the $n$ potentially complex and potentially coincident eigenvalues of $(-\boldsymbol{A})^{-1}$. In total, there are $2n-1$ parameters. □

Based on theorem 5.1, we refer to a representation composed of $2n-1$ free parameters as compact and a representation composed of more than $2n-1$ free parameters as redundant. A redundant description of the distribution requires the definition of more than $2n-1$ parameters. For example, the $(\boldsymbol{\alpha}, \boldsymbol{A})$ vector–matrix representation of PH$(\boldsymbol{\alpha}, \boldsymbol{A})$ is redundant because it is composed of $n^2+n-1$ free parameters ($n-1$ in $\boldsymbol{\alpha}$ because we must have $\boldsymbol{\alpha}\mathbb{1}=1$, and $n^2$ in $\boldsymbol{A}$).

In the special case when $\boldsymbol{E}$ is a diagonal matrix, that is, the eigenvalues of $\boldsymbol{E}$ are distinct, the PDF provided in [2.2] simplifies to [2.11] and equation [5.1], for $i \geq 1$, to:

$$\boldsymbol{\alpha}(-\boldsymbol{A})^{-i}\mathbb{1} = \sum_{j=1}^{n} c_j \lambda_j^i, \qquad [5.2]$$

where coefficients $c_j, j=1,2,...,n$ are the entries of $\boldsymbol{\alpha}\boldsymbol{\Gamma}^{-1}$ and $\lambda_j, j=1,2,...,n$ are the eigenvalues of $\boldsymbol{E}$. Both $c_j$ and $\lambda_j$ can take complex values.

## 5.2. Representations of continuous phase type distributions

The CDF, PDF, hazard rate function and Laplace transform all properly describe a distribution by a function. This section collects further practically important ways to describe a PH distribution with a finite number of free parameters. Further representations are discussed in [TEL 07].

### 5.2.1. *Matrix representation*

The vector–matrix pair $(\boldsymbol{\gamma}, \boldsymbol{G})$ will be called a *matrix representation* of a PH$(n)$ distribution if the CDF of the distribution is:

$$F_T(t) = \mathrm{P}(T<t) = 1 - \boldsymbol{\gamma} e^{\boldsymbol{G}t}\mathbb{1}. \qquad [5.3]$$

The size of the vector–matrix pair $(\boldsymbol{\gamma}, \boldsymbol{G})$ can be larger than $n$ (see section 2.5.2). The matrix representation is not restricted by any sign constraint (each element can be negative, zero or positive).

### 5.2.2. *Markovian representation*

A subset of the matrix representations are the *Markovian representations.*

DEFINITION 5.2.– *The vector–matrix pair* $(\boldsymbol{\alpha}, \boldsymbol{A})$ *is a* Markovian representation *of a PH($n$) distribution if it satisfies the following sign constraints:*

– $\forall i : \alpha_i \geq 0$ *(the initial probabilities are non-negative);*

– $\forall i \neq j : A_{ij} \geq 0$ *(the transition rates between the transient states are non-negative);*

– $\forall i : a_i = -\sum_j A_{ij} \geq 0$ *(the transition rates to the absorbing state are non-negative).*

*A matrix representation that violates these sign constraints is referred to as a* non-Markovian representation.

Assuming that the size of the $(\boldsymbol{\alpha}, \boldsymbol{A})$ Markovian representation is $n$, it is composed of $n^2 + n$ elements and subject to the $\boldsymbol{\alpha}\mathbb{1} = 1$ constraint.

Both the matrix and the Markovian representations are real valued, but not unique and not necessarily minimal, where a representation is minimal if its size is identical with its rank.

Markovian representations form a true subset of matrix representations. All size $n$ matrix representations of a PH($n$) distribution (including the Markovian ones) can be transformed to one another by a similarity transformation of the same size (see section 2.5.1).

### 5.2.3. *Laplace representation*

The Laplace transform of a rank $n$ PH distribution is an order $n$ rational function [2.9]. The set of $2n - 1$ independent coefficients of the properly normalized rational Laplace transform (i.e. $\varphi_1, \ldots, \varphi_{n-1}$ and $\varrho_0, \ldots, \varrho_{n-1}$ in [2.9]) is referred to as the *Laplace representation.* The Laplace representation is compact, unique and real valued.

### 5.2.4. *Moment representation*

The first $2n-1$ moments, $m_1, \ldots, m_{2n-1}$, define a rank $n$ PH distribution [LIE 90]. We refer to the vector $\boldsymbol{m} = \begin{pmatrix} m_1 & \ldots & m_{2n-1} \end{pmatrix}$ as the *moment representation* of a PH distribution. The moment representation is compact, unique and positive real valued.

The advantage of working with a Markovian representation is that it ensures that the associated distribution is a PH distribution. In fact, the main drawback of all the other representations is that they do not indicate explicitly if the associated distribution is PH or not. In practice, one approach to decide if a non-Markovian matrix or a moment or a Laplace representation is associated with a PH distribution is to transform the representation into a Markovian matrix representation.

## 5.3. Transformations between continuous phase type representations

### 5.3.1. *From matrix and Markovian representations to Laplace and moment representations*

The properties of a PH distribution can be derived from its matrix and Markovian representations as:

$$f(t) = \boldsymbol{\alpha}\, e^{\boldsymbol{A}t}(-\boldsymbol{A})\mathbb{1} = \boldsymbol{\gamma}\, e^{\boldsymbol{G}t}(-\boldsymbol{G})\mathbb{1}, \qquad [5.4]$$

$$f^*(s) = \boldsymbol{\alpha}(s\boldsymbol{I}-\boldsymbol{A})^{-1}(-\boldsymbol{A})\mathbb{1} = \boldsymbol{\gamma}(s\boldsymbol{I}-\boldsymbol{G})^{-1}(-\boldsymbol{G})\mathbb{1}, \qquad [5.5]$$

$$m_n = n!\, \boldsymbol{\alpha}(-\boldsymbol{A})^{-n}\mathbb{1} = n!\, \boldsymbol{\gamma}(-\boldsymbol{G})^{-n}\mathbb{1}. \qquad [5.6]$$

### 5.3.2. *From Laplace representation to moment representation*

The $i$th moment of the PH distribution can be obtained from the Laplace representation as:

$$m_i = (-1)^i \left.\frac{d^i}{ds^i} f^*(s)\right|_{s=0}.$$

The reverse direction of this transformation (from moment representation to Laplace representation) can be obtained by a *Padé approximation* and is discussed in section 6.5.

### 5.3.3. *From moment representation to matrix representation*

This transformation is based on the procedure provided in [LIE 90], which first generates a real valued matrix $\boldsymbol{K}$ such that the reduced moments $r_i = m_i/i! = \mathbf{e}_1\boldsymbol{K}^i\mathbf{e}_1^T$ for $i = 1, 2, \ldots, 2n-1$ and $\boldsymbol{K}$ is full rank. To obtain a matrix representation in the form given in section 5.2.1, a similarity transformation is required to transform the "closing vector", $\mathbf{e}_1^T$ to $\mathbb{1}$. Let $\boldsymbol{T}$ be the square matrix whose elements are:

$$T_{ij} = \begin{cases} 1 & i \geq j, \\ 0 & \text{otherwise.} \end{cases} \qquad [5.7]$$

Then with $\boldsymbol{\gamma} = \mathbf{e}_1\boldsymbol{T}^{-1}$ and $\boldsymbol{G} = -(\boldsymbol{T}\boldsymbol{K}\boldsymbol{T}^{-1})^{-1}$, we have:

$$m_i = i!\, \mathbf{e}_1\boldsymbol{K}^i\mathbf{e}_1^T = i!\, \boldsymbol{\gamma}(-\boldsymbol{G})^{-i}\mathbb{1}$$

and, accordingly, $(\gamma, G)$ is a matrix representation of the PH distribution whose moments are $m_i$.

### 5.3.4. *From Laplace representation to matrix representation*

A matrix representation from the Laplace representation can be obtained by calculating the moments from the Laplace transform and applying the transformation given in section 5.3.3. Here, we describe a direct way provided by the so-called *companion form* [FAC 03].

Given a rational Laplace transform:

$$f^*(s) = \frac{\varphi_{n-1}s^{n-1} + \varphi_{n-2}s^{n-2} + \ldots + \varphi_1 s + \varphi_0}{s^n + \varrho_{n-1}s^{n-1} + \varrho_{n-2}s^{n-2} + \ldots + \varrho_1 s + \varrho_0},$$

the *companion form* is

$$f^*(s) = \nu(sI - C)^{-1}(-C)\mathbf{e}_1^T,$$

where

$$\nu = \begin{pmatrix} \frac{\varphi_0}{\varrho_0} & \frac{\varphi_1}{\varrho_0} & \ldots & \frac{\varphi_{n-1}}{\varrho_0} \end{pmatrix}$$

and

$$C = \begin{pmatrix} 0 & 1 & \ddots & 0 & 0 \\ \vdots & \ddots & \ddots & \ddots & \vdots \\ 0 & 0 & \ddots & 1 & 0 \\ 0 & 0 & \ldots & 0 & 1 \\ -\varrho_0 & -\varrho_1 & \ldots & -\varrho_{n-2} & -\varrho_{n-1} \end{pmatrix}.$$

Note that in the case of PH distributions without mass at zero $\varphi_0 = \varrho_0$.

For an explanation of the companion form, we refer to section 1.7. Applying the similarity transform defined in [5.7], we obtain $\gamma = \nu T^{-1}$ and $G = TCT^{-1}$, a matrix representation of the PH distribution.

### 5.3.5. *From matrix representation to Markovian representation*

The Markovian representation of most PH distributions is not unique but any size $n$ Markovian representation of a PH($n$) distribution is in similarity relation with any size $n$ non-Markovian matrix representation of the same distribution according to [2.14].

As a result, given a non-Markovian representation $(\boldsymbol{\alpha}, \boldsymbol{A})$, in order to transform it into a Markovian representation, we need to find a non-singular similarity matrix $\boldsymbol{B}$ with which $(\boldsymbol{\gamma} = \boldsymbol{\alpha}\boldsymbol{B}, \boldsymbol{G} = \boldsymbol{B}^{-1}\boldsymbol{A}\boldsymbol{B})$ is a Markovian representation.

In [TEL 07], a greedy heuristic procedure, called ToMarkovian and shown in Algorithm 5.1, was proposed to find such similarity matrix. The procedure is based on the observation that if $\boldsymbol{B}_1$ and $\boldsymbol{B}_2$ are non-singular matrices such that $\boldsymbol{B}_1\mathbb{1} = \boldsymbol{B}_2\mathbb{1} = \mathbb{1}$ (i.e. $\boldsymbol{B}_1$ and $\boldsymbol{B}_2$ are similarity matrices), then $\boldsymbol{B} = \boldsymbol{B}_1\boldsymbol{B}_2$ is a non-singular matrix that satisfies $\boldsymbol{B}\mathbb{1} = \mathbb{1}$ (i.e. $\boldsymbol{B}$ is a similarity matrix). Exploiting this, the procedure intends to modify the representation in small consecutive steps until the representation becomes Markovian. The elementary transformation matrix is:

$$\boldsymbol{B}_{i,j}(b) = \begin{pmatrix} 1 & 0 & 0 & 0 & \dots \\ 0 & 1 & 0 & 0 & \dots \\ \vdots & \vdots & \ddots & \ddots & \dots \\ 0 & b & 0 & 1-b & 0 \\ 0 & 0 & 0 & 0 & 1 \end{pmatrix}, \qquad [5.8]$$

where the diagonal entry of row $i$ is $1-b$ and the entry in position $i, j$ is $b$. A similarity transformation with $\boldsymbol{B}_{i,j}(b)$ modifies only columns $i$ and $j$ of $\boldsymbol{\alpha}$ and $\boldsymbol{A}$ and row $i$ of $\boldsymbol{A}$ with an effect proportional to $b$. The parameter $b$ can be considered as the step size of the algorithm.

The goal function that the procedure intends to minimize is:

$$\varepsilon(\boldsymbol{\gamma}, \boldsymbol{G}) = \sum_{i=1}^{n} \left( e^{-\gamma_i} + \sum_{j=1, j\neq i}^{n} e^{-G_{ij}} + e^{\sum_{j=1}^{n} G_{ij}} \right).$$

The negative elements of the representation dominates the sum in the goal function and eliminating large negative elements of the representation efficiently decreases the goal function.

## 5.4. Properties of the matrix representation of PH($n$) distributions

### 5.4.1. $2n-1$ *versus* $n^2+n-1$ *parameters*

According to theorem 5.1, a PH($n$) distribution is characterized by $2n-1$ parameters, while its matrix representation has $n^2+n-1$ parameters. This difference is due to the redundancy of the matrix representation. The number of "superfluous" parameters is $n^2+n-1-(2n-1) = n^2-n$.

**Algorithm 5.1** Transform the $(\gamma, G)$ representation to a same size Markovian one.

**procedure** TOMARKOVIAN$(\gamma, G)$
  $b \leftarrow 1, err \leftarrow \varepsilon(\gamma, G), (\alpha, A) \leftarrow (\gamma, G)$
  **for** $c = 1$ to $10$ **do** // The step size is decreased 10 times
    $\hat{\imath} \leftarrow 1$
    **while** $\hat{\imath} > 0$ **do**
      $\hat{\imath} \leftarrow 0, \hat{\jmath} \leftarrow 0$
      **for** $i = 1$ to $n$ **do**
        **for** $j = 1$ to $n, j \neq i$ **do**
          **if** $\varepsilon((\alpha B_{i,j}(b), B_{i,j}^{-1}(b) A B_{i,j}(b))) < err$ **then**
            $\hat{\imath} \leftarrow i, \hat{\jmath} \leftarrow j, err \leftarrow \varepsilon((\alpha B_{i,j}(b), B_{i,j}^{-1}(b) A B_{i,j}(b)))$
          **end if**
        **end for**
      **end for**
      **if** $\hat{\imath} > 0$ **then**
        $(\alpha, A) \leftarrow (\alpha B_{\hat{\imath},\hat{\jmath}}(b), B_{\hat{\imath},\hat{\jmath}}^{-1}(b) A B_{\hat{\imath},\hat{\jmath}}(b))$
        **if** $(\alpha, A)$ is Markovian **then**
          **return** $(\alpha, A)$ // A Markovian representation is found
        **end if**
      **end if**
    **end while**
    $b \leftarrow b/2$ // The step is halved
  **end for**
  **return** $(\gamma, G)$ // No Markovian representation is found
**end procedure**

As discussed in section 2.5.1, given an $(\alpha, A)$ matrix representation of a PH($n$) distribution, different matrix representations of the same size can be obtained by similarity transforms. The similarity matrix used to obtain a different representation has $n^2 - n$ free parameters (entries in each row have to sum up to one), equal to the number of superfluous parameters in the matrix representation.

### 5.4.2. *Different matrix representations*

It is worth noting that in some sense the majority of the matrix representations of a PH($n$) distribution is non-Markovian. Given a Markovian representation of a PH($n$) distribution, there can be (special cases apart, there are) similarity transforms that provide a different Markovian representation of the same PH($n$) distribution, as shown in example 2.12. However, a random similarity transform most probably leads to a non-Markovian representation.

EXAMPLE 5.3.– *Given:*

$$\boldsymbol{\alpha} = \begin{pmatrix} 0.1 & 0.2 & 0.7 \end{pmatrix}, \ \boldsymbol{A} = \begin{pmatrix} -5 & 2 & 1 \\ 0 & -4 & 3 \\ 0 & 0 & -2 \end{pmatrix},$$

*as in example 2.12, and using the similarity matrix:*

$$\boldsymbol{T} = \begin{pmatrix} 0 & 0 & 1 \\ -1 & 3 & -1 \\ 0 & -1 & 2 \end{pmatrix}$$

*we obtain the following non-Markovian matrix representation:*

$$\boldsymbol{\alpha T} = \begin{pmatrix} -0.2 & 1.3 & -1.5 \end{pmatrix}, \ \boldsymbol{T}^{-1}\boldsymbol{AT} = \begin{pmatrix} -14 & 38 & -31 \\ -4 & 12 & -14 \\ -2 & 7 & -9 \end{pmatrix},$$

*where many sign restrictions of Markovian representations are violated. Vector $\boldsymbol{\alpha T}$ has negative elements and matrix $\boldsymbol{T}^{-1}\boldsymbol{AT}$ has positive diagonal and negative off-diagonal elements.*

It is also interesting to note that most similarity transformations from a Markovian representation to another Markovian representation result in a new Markov chain that has no intuitive relation with the original one, as in example 2.12. On the other hand, some similarity transforms correspond to commonly used manipulations of the state space, as illustrated by the next example.

EXAMPLE 5.4.– *Given:*

$$\boldsymbol{\alpha} = \begin{pmatrix} 0.1 & 0.2 & 0.7 \end{pmatrix}, \ \boldsymbol{A} = \begin{pmatrix} -5 & 2 & 1 \\ 0 & -4 & 3 \\ 0 & 0 & -2 \end{pmatrix},$$

*by reordering the states of the underlying Markov chain a new representation of the PH distribution can be obtained with vector–matrix pair:*

$$\begin{pmatrix} 0.7 & 0.1 & 0.2 \end{pmatrix}, \ \begin{pmatrix} -2 & 0 & 0 \\ 1 & -5 & 2 \\ 3 & 0 & -4 \end{pmatrix}.$$

*This new representation can be obtained from $(\boldsymbol{\alpha}, \boldsymbol{A})$ via a similarity transformation using:*

$$\boldsymbol{T} = \begin{pmatrix} 0 & 1 & 0 \\ 0 & 0 & 1 \\ 1 & 0 & 0 \end{pmatrix}.$$

*That is*

$$\boldsymbol{\alpha T} = \begin{pmatrix} 0.7 & 0.1 & 0.2 \end{pmatrix}, \ \boldsymbol{T}^{-1}\boldsymbol{AT} = \begin{pmatrix} -2 & 0 & 0 \\ 1 & -5 & 2 \\ 3 & 0 & -4 \end{pmatrix}.$$

## 5.5. Subclasses of continuous phase type distributions

There are many subclasses of PH distributions considered in the literature. One important reason to consider subclasses of PH distributions is that restricting the structure of the underlying Markov chain results in a lower number of parameters of its matrix representation. In many cases, there is such a constraint that the matrix representation is unique and it is composed of $2n - 1$ parameters, that is, it is also a compact representation.

### 5.5.1. *Subclasses with real eigenvalues*

#### 5.5.1.1. *Hyper-exponential distributions*

One of the simplest subclass of PH distributions is the set of hyper-exponential distributions. It is a convex combination of exponential distributions with PDF:

$$f(t) = \sum_{i=1}^{n} p_i \lambda_i e^{-\lambda_i t}, \qquad [5.9]$$

where $p_i, \lambda_i \in \mathbb{R}^+$ and $\sum_{i=1}^{n} p_i = 1$.

The straightforward Markovian representation of this distribution is:

$$\boldsymbol{\alpha} = \begin{pmatrix} p_1 & p_2 & \dots & p_n \end{pmatrix}, \ \boldsymbol{A} = \begin{pmatrix} -\lambda_1 & & & \\ & -\lambda_2 & & \\ & & \ddots & \\ & & & -\lambda_n \end{pmatrix},$$

and the associated Markov chain is depicted in Figure 5.1. Assuming that $\lambda_i < \lambda_{i+1}$ for $i = 1, \dots, n-1$, the Markovian representation is unique and has $2n - 1$ free parameters ($n - 1$ parameters in $\boldsymbol{\alpha}$ because $\sum_{i=1}^{n} p_i = 1$, and $n$ parameters in $\boldsymbol{A}$). The matrix structure directly indicates that the eigenvalues are real.

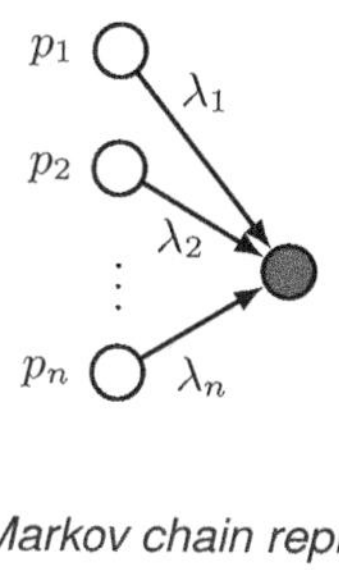


**Figure 5.1.** *Markov chain representation of a hyper-exponential distribution*

The following theorems summarize the limitations of hyper-exponential distributions.

THEOREM 5.5.– *The PDF of a hyper-exponential distribution is monotone decreasing.*

*Proof.* According to [5.9], the PDF of a hyper-exponential distribution is the sum of $n$ monotone decreasing functions $p_i\lambda_i e^{-\lambda_i t}$. □

THEOREM 5.6.– *The SCV of a hyper-exponential distribution is not less than* 1.

*Proof.* The moments of hyper-exponential distributions can be obtained from the properties of exponential distributions summarized in section 1.10. Let us consider the following interpretation of a hyper-exponential distribution:

$$Y = \begin{cases} X_1, \text{ with probability } p_1, \\ \vdots \qquad\qquad \vdots \\ X_n, \text{ with probability } p_n, \end{cases}$$

where $X_i$ is exponentially distributed with parameter $\lambda_i$. The $k$th moment of $Y$ is:

$$m_k = \mathbb{E}(Y^k) = \sum_{i=1}^{n} p_i \mathbb{E}(X_i^k) = \sum_{i=1}^{n} p_i \frac{k!}{\lambda_i^k}$$

and its SCV is:

$$SCV = \frac{m_2 - m_1^2}{m_1^2} = \frac{m_2}{m_1^2} - 1 = 2\frac{\sum_{i=1}^n p_i \left(\frac{1}{\lambda_i}\right)^2}{\left(\sum_{i=1}^n p_i \frac{1}{\lambda_i}\right)^2} - 1.$$

We introduce the discrete random variable $Z$ such that:

$$Z = \begin{cases} \frac{1}{\lambda_1}, \text{ with probability } p_1, \\ \vdots \qquad\qquad \vdots \\ \frac{1}{\lambda_n}, \text{ with probability } p_n. \end{cases}$$

The $k$th moment of $Z$ is:

$$m_k^{(Z)} = \mathbb{E}(Z^k) = \sum_{i=1}^{n} p_i \left(\frac{1}{\lambda_i}\right)^k.$$

It follows that the SCV of $Y$ can be written based on the moments of $Z$ as:

$$SCV = 2\frac{\sum_{i=1}^n p_i \left(\frac{1}{\lambda_i}\right)^2}{\left(\sum_{i=1}^n p_i \frac{1}{\lambda_i}\right)^2} - 1 = 2\frac{m_2^{(Z)}}{\left(m_1^{(Z)}\right)^2} - 1 \geq 2 - 1 = 1,$$

where we used the fact that the variance of $Z$ is non-negative, that is, $m_2^{(Z)} \geq \left(m_1^{(Z)}\right)^2$. □

THEOREM 5.7.– *The hazard rate function (see* [2.4]*) of the hyper-exponential distribution is monotone decreasing.*

*Proof.* The complementary cumulative density function (CCDF) and the derivative of the PDF of the hyper-exponential distribution are:

$$\check{F}(t) = 1 - F(t) = \sum_{i=1}^{n} p_i e^{-\lambda_i t},$$

$$f'(t) = \frac{d}{dt} f(t) = -\sum_{i=1}^{n} p_i \lambda_i^2 e^{-\lambda_i t}.$$

The derivative of the hazard rate function is:

$$\frac{d}{dt}\lambda(t) = \frac{d}{dt}\frac{f(t)}{\check{F}(t)} = \frac{\check{F}(t) f'(t) + f(t)^2}{\check{F}(t)^2} \qquad [5.10]$$

which is negative when the numerator is negative.

We introduce the discrete random variable $V$ such that:

$$V = \begin{cases} \lambda_1 & \text{with probability } q_1, \\ \vdots & \qquad \vdots \\ \lambda_n & \text{with probability } q_n, \end{cases}$$

where

$$q_i = \frac{p_i e^{-\lambda_i t}}{\sum_{j=1}^{n} p_j e^{-\lambda_j t}} = \frac{p_i e^{-\lambda_i t}}{\check{F}(t)} > 0.$$

The $k$th moment of $V$ is:

$$m_k^{(V)} = \mathbb{E}\left(V^k\right) = \sum_{i=1}^{n} q_i \lambda_i^k = \sum_{i=1}^{n} \frac{p_i e^{-\lambda_i t}}{\check{F}(t)} \lambda_i^k.$$

Using the non-negative variance of $V$, we have:

$$-m_2^{(V)} + \left(m_1^{(V)}\right)^2 < 0$$

$$-\sum_{i=1}^{n} \frac{p_i e^{-\lambda_i t}}{\check{F}(t)} \lambda_i^2 + \left(\sum_{i=1}^{n} \frac{p_i e^{-\lambda_i t}}{\check{F}(t)} \lambda_i\right)^2 < 0$$

$$\frac{f'(t)}{\check{F}(t)} + \left(\frac{f(t)}{\check{F}(t)}\right)^2 < 0$$

$$\check{F}(t) f'(t) + f(t)^2 < 0.$$

That is, the numerator of the right-hand side (RHS) of [5.10] is negative. □

We note that theorem 5.6 also follows from theorem 5.7 because any distribution with decreasing hazard rate has a SCV not less than 1 according to [BAR 65, p. 33, corollary 4.9].

### 5.5.1.2. *Hypo-exponential distribution*

The hypo-exponential distribution is obtained as a sum of independent exponentially distributed random variables with rates $\lambda_1, \ldots, \lambda_n$, where $\lambda_i \in \mathbb{R}^+$. Its PDF can be obtained by the convolution of the PDFs of the exponential phases, which simplifies to a closed form expression when the rates are different [ROS 19, p. 310]:

$$f(t) = \lambda_1 e^{-\lambda_1 t} * \ldots * \lambda_n e^{-\lambda_n t} = \sum_{i=1}^{n} a_i \lambda_i e^{-\lambda_i t},$$

where $*$ denotes the convolution operator and $a_i = \prod_{j=1, j\neq i}^{n} \frac{\lambda_j}{\lambda_j - \lambda_i}$. It is important to note that in contrast to [5.9], where coefficients $p_i$ are positive, coefficients $a_i$ might be negative. In Laplace domain, the convolution is a product and the Laplace transform of the hypo-exponential distribution can be expressed as:

$$f^*(s) = \prod_{i=1}^{n} \frac{\lambda_i}{s + \lambda_i},$$

which is applicable in the presence of equal rates as well.

The straightforward Markovian representation of the hypo-exponential distribution is:

$$\alpha = \begin{pmatrix} 1 & 0 & \ldots & 0 \end{pmatrix}, \; \boldsymbol{A} = \begin{pmatrix} -\lambda_1 & \lambda_1 & & \\ & -\lambda_2 & \lambda_2 & \\ & & \ddots & \ddots \\ & & & -\lambda_n \end{pmatrix},$$

and the associated Markov chain is depicted in Figure 5.2. Assuming $\lambda_i \leq \lambda_{i+1}$ for $i = 1, \ldots, n-1$, the Markovian representation is unique and has $n$ free parameters.

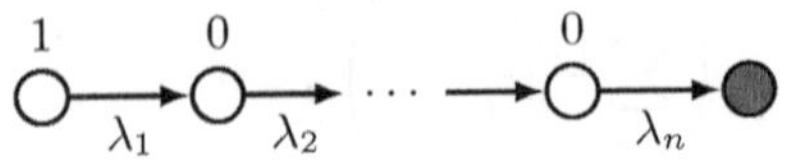


**Figure 5.2.** *Markov chain representation of a hypo-exponential distribution*

The eigenvalues of $\boldsymbol{A}$ are real and the PDF starts from zero, that is, $f(0) = 0$ if $n \geq 2$, and it is flat around zero because $\frac{d^i}{dt^i} f(t)|_{t=0} = 0$ for $i = 1, \ldots, n-2$. The PDF is monotone increasing until its unique peak and is monotone decreasing after that.

THEOREM 5.8.– *The SCV of a hypo-exponential distribution is less than* 1 *for* $n \geq 2$.

*Proof.* A hypo-exponentially distributed random variable $Y$ is a sum of independent, exponentially distributed random variables $X_1, \ldots, X_n$ with parameter $\lambda_1, \ldots, \lambda_n$. That is, $Y = \sum_{i=1}^n X_i$. For the mean and the variance of $Y$, we have:

$$\mathbb{E}(Y) = \sum_{i=1}^{n} \mathbb{E}(X_i) = \sum_{i=1}^{n} \frac{1}{\lambda_i},$$

$$D_Y^2 = \sum_{i=1}^{n} D_{X_i}^2 = \sum_{i=1}^{n} \frac{1}{\lambda_i^2},$$

from which

$$SCV = \frac{D_Y^2}{\mathbb{E}(Y)^2} = \frac{\sum_{i=1}^n \frac{1}{\lambda_i^2}}{\left(\sum_{i=1}^n \frac{1}{\lambda_i}\right)^2} = \frac{\sum_{i=1}^n \frac{1}{\lambda_i^2}}{\sum_{i=1}^n \frac{1}{\lambda_i^2} + \sum_{i=1}^n \sum_{j=i+1}^n \frac{2}{\lambda_i \lambda_j}} < 1.$$

□

THEOREM 5.9.– *The hazard rate function of the hypo-exponential distribution is monotone increasing.*

*Proof.* According to [BAR 65, p. 36, theorem 5.1], if $X$ and $Y$ are independent random variables with monotone increasing hazard rate functions, then $X + Y$ has a monotone increasing hazard rate function as well.

The sum $Y = \sum_{i=1}^n X_i$ can also be written as $Y = Y_n$ with $Y_1 = X_1$ and $Y_i = X_i + Y_{i-1}$ for $i = 2, \ldots, n$. Noting that the exponential distribution is with monotone increasing hazard rate function (it is the extreme case because its hazard rate function is constant) and applying the result on the sum of independent random variables with monotone increasing hazard rate functions $n - 1$ times result in the theorem. □

Again, theorem 5.8 also follows from theorem 5.9, because any distribution with increasing hazard rate has an SCV not greater than 1 [BAR 65, p. 33, corollary 4.9].

#### 5.5.1.3. *Erlang distribution*

A special subset of hypo-exponential distributions is the Erlang distribution in which the summed up independent exponential random variables have the same rate

$(\lambda_1 = \ldots = \lambda_n = \lambda)$. The Markovian representation of the Erlang$(n, \lambda)$ distribution is:

$$\boldsymbol{\alpha} = \begin{pmatrix} 1 & 0 & \ldots & 0 \end{pmatrix}, \; \boldsymbol{A} = \begin{pmatrix} -\lambda & \lambda & & \\ & -\lambda & \lambda & \\ & & \ddots & \ddots \\ & & & -\lambda \end{pmatrix},$$

and the associated Markov chain is depicted in Figure 5.3. Matrix $\boldsymbol{A}$ has a single eigenvalue $(-\lambda)$ whose multiplicity is $n$. Having fixed the size, $n$, the Erlang distribution has a single parameter, $\lambda$.

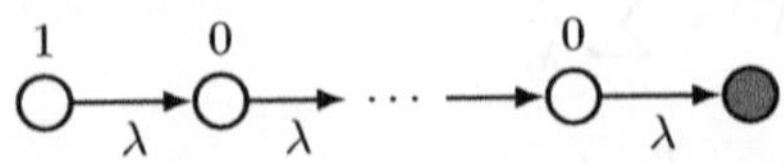


**Figure 5.3.** *Markov chain representation of an Erlang distribution*

Erlang distributions have several important properties that motivate their use in many practical situations. The simple closed form expression for its PDF [1.23] results in simple closed form expressions for its moments:

$$m_k = \frac{(n+k-1)!}{\lambda^k (n-1)!}.$$

A further important extremal property of Erlang distributions is discussed in section 6.1.

#### 5.5.1.4. *Hyper-Erlang distribution*

The hypo- and hyper-composition (i.e. the summation of random variables and the random selection from a number of random variables according to a discrete distribution, respectively) can be applied to all subclasses of PH distributions in a similar way as to exponential distributions. As an example, we define hyper-Erlang distributions here, a class of distributions we are going to use later.

DEFINITION 5.10.– *An RV X has order-N* hyper-Erlang distribution *if its PDF is:*

$$f(x) = \sum_{i=1}^{N} \alpha_i \frac{(\lambda_i x)^{r_i - 1}}{(r_i - 1)!} \lambda_i e^{-\lambda_i x}, \qquad [5.11]$$

*with* $\alpha_i, \lambda_i > 0, r_i \in \{1, 2, 3, ...\}$ *for* $i = 1, \ldots, N$ *and* $\sum_{i=1}^{N} \alpha_i = 1$.

That is, $X$ is Erlang$(r_i, \lambda_i)$ distributed with probability $\alpha_i$ for $i = 1, \ldots, N$. An example for a hyper-Erlang distribution is depicted in Figure 5.4. The associated initial vector and transient generator matrix are structured according to the $N$ Erlang distributions composing the hyper-Erlang distribution as:

$$\boldsymbol{\alpha} = (\alpha_1 \underbrace{0 \ldots 0}_{r_1-1} \; \alpha_2 \underbrace{0 \ldots 0}_{r_2-1} \; \ldots \; \alpha_N \underbrace{0 \ldots 0}_{r_N-1}), \qquad [5.12]$$

$$\boldsymbol{A} = \left(\begin{array}{ccc|ccc|c|ccc}
-\lambda_1 & \lambda_1 & & & & & & & & \\
 & \ddots & \ddots & & & & & & & \\
 & & -\lambda_1 & & & & & & & \\
\hline
 & & & -\lambda_2 & \lambda_2 & & & & & \\
 & & & & \ddots & \ddots & & & & \\
 & & & & & -\lambda_2 & & & & \\
\hline
 & & & & & & \ddots & & & \\
\hline
 & & & & & & & -\lambda_N & \lambda_N & \\
 & & & & & & & & \ddots & \ddots \\
 & & & & & & & & & -\lambda_N
\end{array}\right)
\begin{array}{l} \left.\vphantom{\begin{array}{c}1\\1\\1\end{array}}\right\} r_1 \\ \left.\vphantom{\begin{array}{c}1\\1\\1\end{array}}\right\} r_2 \\ \\ \left.\vphantom{\begin{array}{c}1\\1\\1\end{array}}\right\} r_N \end{array}$$

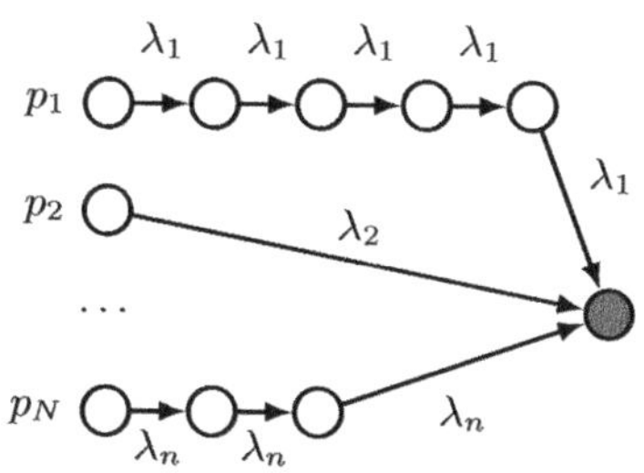


**Figure 5.4.** *Hyper-Erlang distribution with* $r_1 = 5, r_2 = 1, \ldots, r_N = 3$

The Laplace transform of $f(x)$ is:

$$f^*(s) = \sum_{i=1}^{N} \alpha_i \left(\frac{\lambda_i}{s+\lambda_i}\right)^{r_i} \qquad [5.13]$$

whose rank, computed based on definition 2.6, is $R = \sum_{i=1}^{N} r_i$.

The $k$th moment of hyper-Erlang distributions is calculated as:

$$m_k = \int_0^\infty x^k f(x)dx = \sum_{i=1}^{N} \alpha_i \frac{(k+r_i-1)!}{(r_i-1)!} \frac{1}{\lambda_i^k}, \qquad [5.14]$$

where $m_0 = 1$ is guaranteed by $\sum_{i=1}^{N} \alpha_i = 1$.

A Hyper-Erlang distribution is a combination of hyper- and hypo-structures because it is a mixture (hyper structure) of hypo-exponential distributions (Erlang distributions). Its special case when $r_1 = r_2 = \ldots = r_N = 1$ leads to a hyper-exponential distribution with SCV not less than 1. On the other hand, with $N = 1$ it is a size $r_1$ Erlang distribution with SCV equal to $1/r_1 \leq 1$.

#### 5.5.1.5. *Acyclic phase type distributions*

Any PH distribution that can be represented by a Markov chain without loop is referred to as an acyclic phase type (APH) distribution. If a Markov chain does not contain a loop, then with a proper reordering of the states its generator matrix becomes an upper triangular matrix. A general upper triangular matrix of size $n$ has $(n^2 + n)/2$ parameters which, for what concerns a PH($n$) distribution, is a redundant representation. The practical importance of the APH class is due to the following essential result by Cumani [CUM 82].

THEOREM 5.11.– *Any APH($n$) distribution can be represented by a Markovian representation in the form:*

$$\boldsymbol{\alpha} = \begin{pmatrix} p_1 & p_2 & \ldots & p_n \end{pmatrix}, \; \boldsymbol{A} = \begin{pmatrix} -\lambda_1 & \lambda_1 & & \\ & -\lambda_2 & \lambda_2 & \\ & & \ddots & \ddots \\ & & & -\lambda_n \end{pmatrix}, \qquad [5.15]$$

*where $\lambda_i \leq \lambda_{i+1}$ for $i = 1, \ldots, n-1$. The associated Markov chain is depicted in Figure 5.5.*

The major advantage of APH distributions is that the Markov representation in Figure 5.5 has $2n - 1$ parameters ($n - 1$ in $\boldsymbol{\alpha}$ because $\sum_{i=1}^{n} p_i = 1$, and $n$ in $\boldsymbol{A}$), and some easy to check constraints, namely:

- $p_i \geq 0$ for $i = 1, \ldots, n$;
- $0 < \lambda_1 \leq \lambda_2 \leq \lambda_3 \leq \cdots \leq \lambda_n$.

It follows that the Markovian representation in [5.15] with the associated constraints is a compact and unique representation.

The eigenvalues of $\boldsymbol{A}$ are $-\lambda_1, \ldots, -\lambda_n$. Similar to section 5.5.1.2, utilizing the relation of the PDF, the Laplace transform and the $(\boldsymbol{\alpha}, \boldsymbol{A})$ matrix representation in [5.15], for $f(t) = -\boldsymbol{\alpha} e^{\boldsymbol{A}t} \boldsymbol{A} \mathbb{1}$ we have:

$$f(t) = \sum_{i=1}^{n} p_i f_i(t) \text{ with } f_i(t) = \lambda_i e^{-\lambda_i t} * \lambda_{i+1} e^{-\lambda_{i+1} t} * \ldots * \lambda_n e^{-\lambda_n t} \qquad [5.16]$$

and

$$f^*(s) = \int_0^\infty e^{-st} f(t)dt = \sum_{i=1}^{n} p_i f_i^*(s) \text{ with } f_i^*(s) = \prod_{j=i}^{n} \frac{\lambda_j}{s+\lambda_j}. \qquad [5.17]$$

All the subclasses of PH distributions discussed before (sections 5.5.1.1-5.5.1.4) are APH distributions. This implies that the APH class is flexible both for what concerns the hazard rate (it can be increasing, constant or decreasing) and the SCV (it can be lower than 1, 1 or larger than 1).

The Markov representation in Figure 5.5 is convenient when a Markov chain structure with many potential initial states and a single final state is beneficial in the application, where final state refers to any state from which the absorbing state is directly reachable. There are several other Markovian representations of APH distributions with simple structure. If the application of interest is such that having a single initial state and many potential final states is beneficial, then the Markovian representation in Figure 5.7 is useful. If both the number of initial and final states need to be low, then the Markovian representation in Figure 5.6 is a favorable option. In [CUM 82], the structures in Figures 5.5–5.7 are referred to as CF1, CF2 and CF3, respectively, where CF stands for canonical form. Based on a paper with an essentially different subject [COX 55], the CF3 structure is also known as Coxian distribution in many references (e.g. [TRI 01, BUC 14, ROS 19]).

The identity of the CF forms can be seen by comparing the possible trajectories of the Markov chains. In CF1, the process starts at state $i$ with probability $p_i$, spends an exponentially distributed time with parameter $\lambda_i$ in state $i$, an exponentially distributed time with parameter $\lambda_{i+1}$ in state $i+1$, and so on, and finally it spends an exponentially distributed time with parameter $\lambda_n$ in state $n$.

The Markov chain of CF3 reverses this trajectory. The process starts in state 1, it spends an exponentially distributed time with parameter $\lambda_n$ in state 1, an exponentially distributed time with parameter $\lambda_{n-1}$ in state 2, and so on. The overall probability of jumping to the absorbing state after leaving state $n+1-i$ (the state with parameter $\lambda_i$) is $\eta_i \prod_{j=i+1}^{n}(1-\eta_j)$ and probabilities $\eta_i$ are such that $\eta_i \prod_{j=i+1}^{n}(1-\eta_j) = p_i$.

In CF2, the process starts at state 1, spends an exponentially distributed time with parameter $\lambda_n$ in state 1, after that with probability $p_i$ it jumps to state $i+1$ where it spends an exponentially distributed time with parameter $\lambda_i$, an exponentially distributed time with parameter $\lambda_{i+1}$ in state $i+2$, and so on. Finally, it stays in state $n$ for an exponentially distributed time with parameter $\lambda_{n-1}$.

### 5.5.2. *Subclasses with complex eigenvalues*

All the above discussed PH subclasses have only real eigenvalues. In order to have complex eigenvalues, the Markov chain has to contain loops.

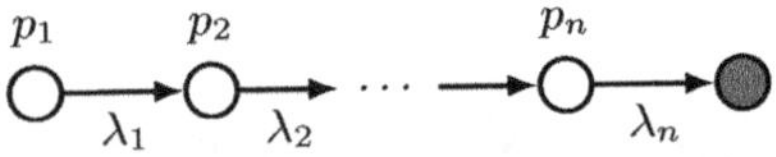


**Figure 5.5.** *Markov chain representation of an APH distribution, where $\lambda_i \le \lambda_{i+1}$ for $i = 1, \ldots, n-1$ (CF1 structure in [CUM 82])*

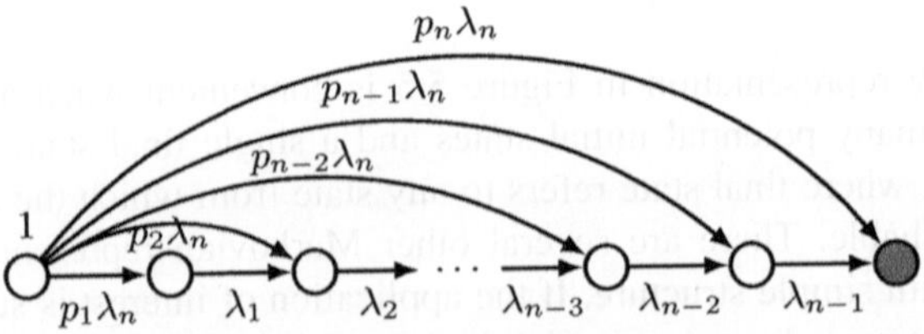


**Figure 5.6.** *CF2 structure in [CUM 82]*

$$\eta_n = p_n$$

$$\eta_i = \frac{p_i}{\prod_{j=i+1}^{n} 1 - \eta_j}$$

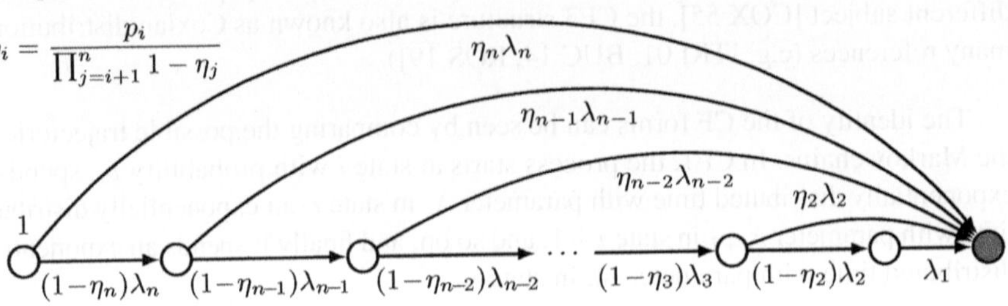


**Figure 5.7.** *CF3 structure in [CUM 82], also referred to as Coxian distribution*

### 5.5.2.1. *Feedback Erlang distribution*

The size of a *feedback Erlang distribution* is $n \ge 3$ and its Markovian representation is:

$$\boldsymbol{\alpha} = \begin{pmatrix} p_1 & p_2 & \cdots & p_n \end{pmatrix}, \; \boldsymbol{A} = \begin{pmatrix} -\lambda & \lambda & & \\ & -\lambda & \lambda & \\ & & \ddots & \ddots \\ z\lambda & & & -\lambda \end{pmatrix},$$

depicted in Figure 5.8.

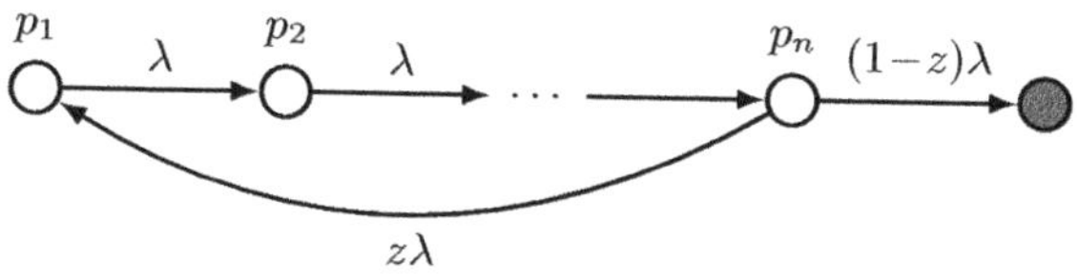


**Figure 5.8.** *Feedback Erlang distribution*

In the following, we will use complex distributions composed of such cyclic structures as a building block, that we refer to as a feedback Erlang block (FEB).

DEFINITION 5.12.– A feedback Erlang block *is a cyclic PH structure of at least three phases whose generator matrix is in the form:*

$$\boldsymbol{A} = \begin{pmatrix} -\lambda & \lambda & & \\ & \ddots & \ddots & \\ & & -\lambda & \lambda \\ z\lambda & & & -\lambda \end{pmatrix}. \qquad [5.18]$$

*The parameters of a FEB are the transition rate $\lambda$, the feedback probability $z$ and the size (the number of phases) $k$.*

COMMENT 5.13.– *For notational convenience, we extend this definition to the degenerate FEB with parameters $\lambda$, $z = 0$, $k = 1$ which corresponds to a single exponential phase and has a single eigenvalue equal to $-\lambda$.*

The eigenvalue structure of a FEB is already discussed in example 2.18 and depicted in the right side of Figure 2.8. The eigenvalues of $\boldsymbol{A}$ are $\lambda(\sqrt[k]{z}e^{\Im i2\pi/k} - 1)$ for $i = 1, \ldots, k$, and they are located at regular distances on a circle around $-\lambda$ whose radius is $\lambda\sqrt[k]{z}$. The eigenvalue with largest real part corresponds to the right most point of the circle ($i = k$). Accordingly, it is real and equal to $\lambda(\sqrt[k]{z} - 1)$. The rightmost complex eigenvalue pair is $-a \pm \Im b = \lambda(\sqrt[k]{z}e^{\pm\Im 2\pi/k} - 1)$, which is obtained at $i = 1$ and $i = k - 1$. This particular eigenvalue structure of FEBs is the main reason for their widespread application [MOC 99]. In many applications, the following reverse problem needs to be solved: find $k$, $\lambda$ and $z$ such that the rightmost complex eigenvalue pair is $-a \pm \Im b$.

THEOREM 5.14.– *The parameters of the smallest FEB whose right most complex eigenvalue pair is $-a \pm \Im b$ can be obtained with:*

$$k = \left\lceil \frac{\pi}{\arctan\left(\frac{a}{b}\right)} \right\rceil, \qquad [5.19]$$

$$\lambda = \frac{1}{2}\left(2a - b\tan\frac{\pi}{k} + b\cot\frac{\pi}{k}\right), \qquad [5.20]$$

$$z = \left(1 - \left(a - b\tan\frac{\pi}{k}\right)\right)^k, \qquad [5.21]$$

*where $\lceil x \rceil$ denotes the smallest integer not smaller than $x$.*

*Proof.* Substituting $k$, $\lambda$, $z$ into $\lambda(\sqrt[k]{z}e^{\pm\Im 2\pi/k} - 1)$ gives $-a \pm \Im b$. □

### 5.5.2.2. *Unicyclic PH distribution*

The simplest Markovian representation of a unicyclic PH distribution is:

$$\boldsymbol{\alpha} = (p_1\ p_2\ \ldots\ p_n)\,,\ \boldsymbol{A} = \begin{pmatrix} -\lambda_1 & \lambda_1 & & \\ \gamma_2 & -\lambda_2 - \gamma_2 & \lambda_2 & \\ \vdots & & \ddots & \ddots \\ \gamma_n & & & -\lambda_n - \gamma_n \end{pmatrix},$$

as depicted in Figure 5.9.

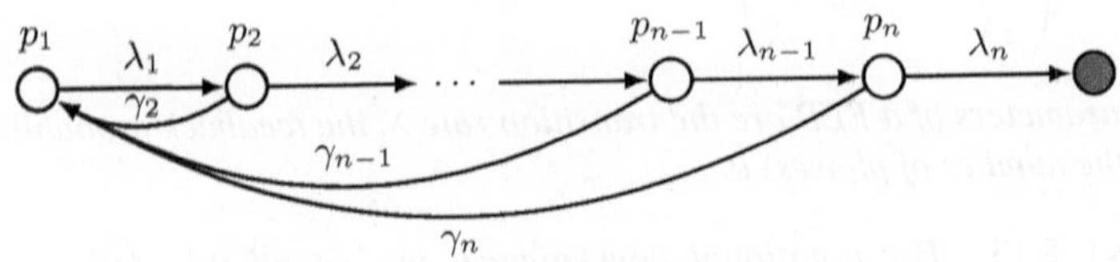


**Figure 5.9.** *Unicyclic PH distribution*

The Markovian representation of a unicyclic PH distribution has $3n - 2$ parameters and $\boldsymbol{A}$ can have complex eigenvalues. The importance of this class comes from a long lasting conjecture of O'Cinneide [O'CI 99], which assumed that any rank $n$ PH distribution can be represented as a unicyclic distribution of $n$ phases. This conjecture was proved to be false in [HE 05].

## 5.6. Canonical Markovian representation

A canonical Markovian representation of a PH($n$) distribution is a unique, size $n$ Markovian vector–matrix representation with $2n - 1$ parameters. For what concerns APH($n$) distributions, CF1 (Figure 5.5) is one such representation.

For $n = 1$, the PH($n$) class contains only the exponential distribution and $\boldsymbol{\alpha} = (1), \boldsymbol{A} = (-\lambda)$ is the trivial canonical representation. For $n = 2$, the APH($n$) class is identical to the PH($n$) class, that is, with $n = 2$ having a loop does not extend the set of distributions that one can obtain. Accordingly, CF1 is a canonical representation for the PH(2) class.

For $n \geq 3$, the set of APH($n$) distributions is a real subset of the set of PH($n$) distributions. The canonical Markovian representation of a PH(3) distribution has the structure [HOR 09]:

$$\boldsymbol{\alpha} = \begin{pmatrix} p_1 & p_2 & p_3 \end{pmatrix}, \; \boldsymbol{A} = \begin{pmatrix} -x_1 & 0 & x_{13} \\ x_2 & -x_2 & 0 \\ 0 & x_3 & -x_3 \end{pmatrix},$$

where $p_1 + p_2 + p_3 = 1$, $x_1 \geq x_2 \geq x_3 \geq 0$ and $x_1 > x_{13}$. The associated Markov chain is depicted in Figure 5.10 and it has six parameters ($x_1, x_2, x_3, x_{13}$ and any two of $p_1, p_2$ and $p_3$), which is more than $2n - 1 = 2 \cdot 3 - 1 = 5$. In order to bring the number of parameters to 5, an additional constraint is needed. This additional constraint depends on the considered PH(3) distribution and it is one of the following:

– $x_{13} = 0$,

– $x_1 = x_2$,

– $p_2 = 0$.

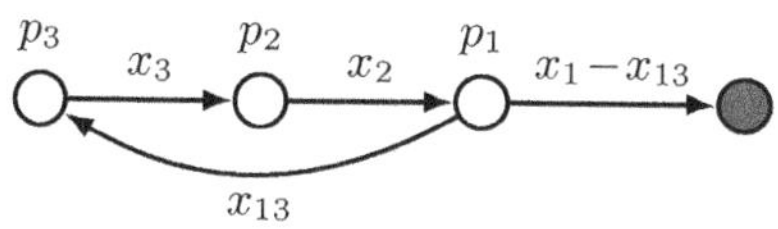


**Figure 5.10.** *The Canonical representation of PH(3) distributions*

The procedure to transform any Markovian or non-Markovian $(\boldsymbol{v}, \boldsymbol{H})$ representation of size 3 into a PH(3) canonical form is provided in Algorithm 5.2. If the algorithm returns **error**, then $(\boldsymbol{v}, \boldsymbol{H})$ is not a PH(3) distribution.

For $n \geq 4$, a canonical representation is not known for a general PH($n$) distribution.

The canonical representation of the DPH($n$) class is more complex than the one of the PH($n$) class because the entries of the transition probability matrix must not be greater than 1 and this results in additional constraints. The canonical representation of the DPH($n$) class is unknown for $n \geq 4$ and it is studied for $n = 2$ and $n = 3$ in [PAP 13].

## 5.7. Analysis of a non-Markovian representation

A size $n$, full rank, non-Markovian vector-matrix pair, $(\boldsymbol{\gamma}, \boldsymbol{G})$, falls into one of the following cases:

– Case 1: a non-Markovian representation of a PH($n$) distribution.

**Algorithm 5.2** Transformation to Canonical PH(3) representation.

**procedure** TOCANONICALPH3($\boldsymbol{v}, \boldsymbol{H}$)<br>
// **input**: $\boldsymbol{v}, \boldsymbol{H}$, any size 3 matrix representation<br>
// **output**: $\boldsymbol{\pi}, \boldsymbol{A}$, canonical representation if $\boldsymbol{v}, \boldsymbol{H}$ is a PH(3)<br>
    **if** $v_1 + v_2 + v_3 \neq 1$ **then**<br>
        **error** "Probability mass at 0"<br>
    **end if**<br>
    $\lambda_1, \lambda_2, \lambda_3$ = decreasingly ordered eigenvalues of $-\boldsymbol{H}$<br>
    **if** $\lambda_3 < 0$ **or** $\lambda_3 \in \mathbb{C}$ **then**<br>
        **error** "Invalid eigenvalues or negative density at 0"<br>
    **end if**<br>
    $a_0 = \lambda_1 \lambda_2 \lambda_3,\ a_1 = \lambda_1 \lambda_2 + \lambda_1 \lambda_3 + \lambda_2 \lambda_3,\ a_2 = \lambda_1 + \lambda_2 + \lambda_3$<br>
    **if** $a_2^2 - 3\,a_1 < 0$ **then**<br>
        **error** "Invalid characteristic polynomial"<br>
    **end if**<br>
    $\gamma_u = \frac{1}{3}\,(a_2 + 2\sqrt{a_2^2 - 3\,a_1}),\quad \gamma_0 = \frac{1}{3}\,(a_2 + \sqrt{a_2^2 - 3\,a_1}),$<br>
    $\gamma_\ell = \begin{cases} \lambda_1 & \textbf{if } \lambda_1 \in \text{real} \\ \gamma_0 & \textbf{if } \lambda_1 \in \text{complex} \end{cases}$<br>
    **if** $\boldsymbol{v}\,\boldsymbol{H}\,\mathbb{1} > 0$ **or** $(\boldsymbol{v}\,\boldsymbol{H}\,\mathbb{1} = 0$ **and** $\boldsymbol{v}\,\boldsymbol{H}^2\,\mathbb{1} > 0)$ **then**<br>
        **error** "Negative density around 0"<br>
    **end if**<br>
    $\gamma_2 = \begin{cases} -\boldsymbol{v}\,\boldsymbol{H}^2\,\mathbb{1}\,/\,\boldsymbol{v}\,\boldsymbol{H}\,\mathbb{1} & \textbf{if } \boldsymbol{v}\,\boldsymbol{H}\,\mathbb{1} < 0 \\ 0 & \textbf{if } \boldsymbol{v}\,\boldsymbol{H}\,\mathbb{1} = 0 \end{cases}$<br>
    **if** $\gamma_2 > \gamma_u$ **then**<br>
        **error** "$\pi_2$ is negative"<br>
    **end if**<br>
    $x_1 = \max\{\gamma_2, \gamma_\ell\}$<br>
    $x_{13} = x_1 - a_0\,/\,(x_1^2 - a_2\,x_1 + a_1)$<br>
    $x_2 = \frac{1}{2}\left(a_2 - x_1 + \sqrt{(a_2 - x_1)^2 - 4\,(x_1^2 - a_2\,x_1 + a_1)}\right)$<br>
    $x_3 = \frac{1}{2}\left(a_2 - x_1 - \sqrt{(a_2 - x_1)^2 - 4\,(x_1^2 - a_2\,x_1 + a_1)}\right)$<br>
    $\pi_1 = \boldsymbol{v}\,\boldsymbol{H}\,\mathbb{1}\,/\,(x_{13} - x_1)$<br>
    $\pi_2 = \boldsymbol{v}\,(x_1\,\boldsymbol{I} + \boldsymbol{H})\,\boldsymbol{H}\,\mathbb{1}\,/\,(x_{13} - x_1)\,x_2$<br>
    $\pi_3 = \boldsymbol{v}\,(x_2\,\boldsymbol{I} + \boldsymbol{H})\,(x_1\,\boldsymbol{I} + \boldsymbol{H})\,\boldsymbol{H}\,\mathbb{1}\,/\,(x_{13} - x_1)\,x_2\,x_3$<br>
    **if** $\pi_3 < 0$ **then**<br>
        **error** "$\pi_3$ is negative"<br>
    **end if**<br>
    **return** $\boldsymbol{\pi} = \begin{pmatrix} \pi_1 & \pi_2 & \pi_3 \end{pmatrix},\ \boldsymbol{A} = \begin{pmatrix} -x_1 & 0 & x_{13} \\ x_2 & -x_2 & 0 \\ 0 & x_3 & -x_3 \end{pmatrix}$<br>
**end procedure**

– Case 2: a non-Markovian representation of a PH($m$) distribution with $m > n$.

– Case 3: a non-Markovian representation of a ME($n$) distribution which does not belong to any finite size PH class.

– Case 4: it is such that $\gamma\, e^{Gt}(-G)\mathbb{1} < 0$ for some $t > 0$, that is, it does not correspond to a distribution.

The problem of categorizing a given vector–matrix pair $(\gamma, G)$ according to the above list is considered in many papers for $n = 3$. This is due to the fact that the roots of the characteristic polynomial are available in analytic form, based on which several complicated analytic results can be obtained, for example, in [O'CI 93, FAC 03, ÉLT 08, KOL 11]. For $n > 3$, this classification problem is not solved yet. The size of the minimal Markovian representation of $(\gamma, G)$ is referred to as *PH-order* [PUL 11].

EXAMPLE 5.15.– *This example is from [O'CI 93] and used also in [FAC 03, pp. 109–110]. The vector–matrix pair $(\alpha, A)$ with:*

$$\alpha = \begin{pmatrix} p_1 & p_2 & p_3 \end{pmatrix}, \; A = \begin{pmatrix} -1 & 1 & 0 \\ 0 & -1 & 1 \\ 0 & 0 & -1 \end{pmatrix}$$

*defines the PDF*

$$f(t) = \left( p_1 \frac{t^2}{2} + p_2 t + p_3 \right) e^{-t}.$$

*Using $p_2 = 1 - p_1 - p_3$, two parameters characterize the distribution. Its PDF belongs to a PH(3) distribution if $p_1 > 0, p_3 > 0$ and $p_2 = 1 - p_1 - p_3 > 0$. If $p_1 > 0, p_3 > 0$, $p_2 < 0$ and $p_2^2 = (1-p_1-p_3)^2 \leq \frac{2p_1p_3(n-3)}{n-2}$, then the PDF is non-negative, and it has a Markovian representation of size $n > 3$.*

*Figure 5.11 depicts those regions of parameters $p_1, p_3$ for which $f(t)$ belongs to different PH($n$) classes. PDFs from different PH($n$) classes are shown in Figure 5.12. From $p_2 = 1 - p_1 - p_3$, we have that $(\alpha, A)$ is Markovian when $p_1 > 0, p_3 > 0$ and $p_1 + p_3 < 1$. The corresponding region is represented by the triangle labeled as PH(3) in Figure 5.11 (see Case 1). An internal point of the triangle is $p_1 = p_3 = 1/3$, which means $\alpha = \begin{pmatrix} 1/3 & 1/3 & 1/3 \end{pmatrix}$, and the associated PDF is plotted in solid line in Figure 5.12. Outside the triangle $(\alpha, A)$ is non-Markovian due to a negative value in $\alpha$. The white area represents the cases when $f(t)$ is not a PDF because it has negative values for some $t > 0$, as in Figure 5.12 with $\alpha = \begin{pmatrix} 3 & -3 & 1 \end{pmatrix}$ (see Case 4). The boundary of the ME(3) region is the limit where $f(t) = 0$ for some $t > 0$ (see Case 3). It is the case in Figure 5.12 when $\alpha = \begin{pmatrix} 2 & -2 & 1 \end{pmatrix}$. Inside the ME(3) region, for any $p_1, p_3$ point there is an $n$ for which the $(1 - p_1 - p_3)^2 \leq \frac{2p_1p_3(n-3)}{n-2}$ condition holds and the* *associated PDF has a Markovian representation of size* $n$ *(see Case 2). For example, with* $\boldsymbol{\alpha} = \begin{pmatrix} 1 & -1 & 1 \end{pmatrix}$ *we have the following size 4 Markovian representation of* $f(t)$*:*

$$\boldsymbol{\beta} = \begin{pmatrix} 1/2 & 0 & 0 & 1/2 \end{pmatrix}, \ \boldsymbol{B} = \begin{pmatrix} -1 & 1 & 0 & 0 \\ 0 & -1 & 1 & 0 \\ 0 & 0 & -1 & 1 \\ 0 & 0 & 0 & -2 \end{pmatrix}.$$

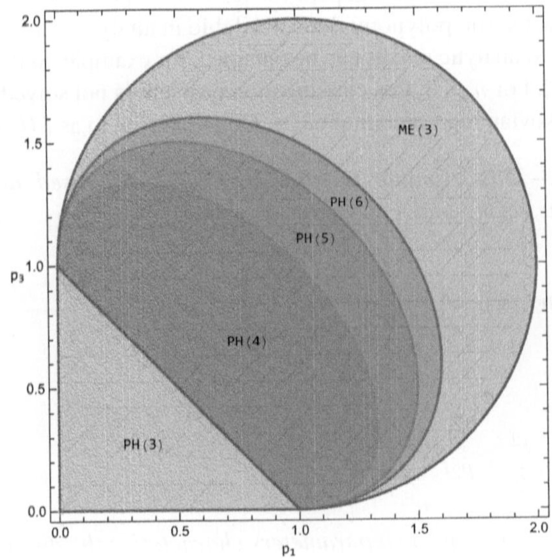


**Figure 5.11.** *Regions of* $p_1, p_3$ *parameters (with* $p_2 = 1 - p_1 - p_3$*) belonging to different PH(*$n$*) classes. For a color version of this figure, see www.iste.co.uk/horvath/phase.zip*

*The size* 5 *PH distribution with representation:*

$$\boldsymbol{\tau} = \begin{pmatrix} 1/2 & 0 & 0 & 1/4 & 1/4 \end{pmatrix}, \ \boldsymbol{T} = \begin{pmatrix} -1 & 1 & 0 & 0 & 0 \\ 0 & -1 & 1 & 0 & 0 \\ 0 & 0 & -1 & 1/2 & 1/2 \\ 0 & 0 & 0 & -2 & 2 \\ 0 & 0 & 0 & 0 & -4 \end{pmatrix}$$

*is such that its rank is* 3*, because* $(\boldsymbol{\alpha} = \begin{pmatrix} 1 & -1 & 1 \end{pmatrix}, \boldsymbol{A})$ *is its full rank non-Markovian representation and its PH order is* 4*, because* $(\boldsymbol{\beta}, \boldsymbol{B})$ *is its size* 4 *Markovian representation and there is no size 3 Markovian representation of this distribution [FAC 03].*

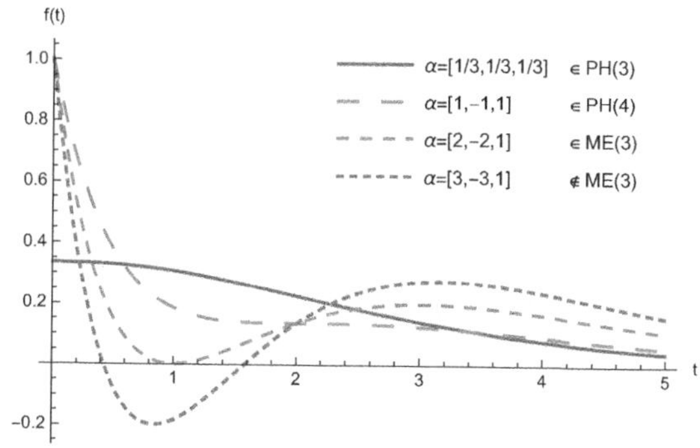


**Figure 5.12.** *Examples of density functions from different PH($n$) classes. For a color version of this figure, see www.iste.co.uk/horvath/phase.zip*

Whether a non-Markovian vector-matrix pair $(\gamma, \boldsymbol{G})$ represents a PH($n$) distribution or not is an open question for $n > 3$. Only numerical procedures are available to investigate such cases:

– If there is a similarity matrix $\boldsymbol{T}$ with which $(\gamma\boldsymbol{T}, \boldsymbol{T}^{-1}\boldsymbol{G}\boldsymbol{T})$ is a Markovian representation, then $(\gamma, \boldsymbol{G})$ represents a PH($n$) distribution. Algorithm 5.1 presents a numerical procedure to find such a similarity transform and provides the resulting Markovian representation.

– The numerical procedure in [REI 14] intends to transform $(\gamma, \boldsymbol{G})$ into a Markovian representation $(\boldsymbol{\alpha}, \boldsymbol{A})$ of size $m$ where $m > n$. According to [O'CI 90, HOR 15], this is possible if the real part of the complex eigenvalues of $\boldsymbol{G}$ are less than its dominant eigenvalue (real eigenvalue with maximal real part) and $\gamma\, e^{\boldsymbol{G}t}(-\boldsymbol{G})\mathbb{1}$ is strictly positive for $t \in [0, \infty)$. Since the procedure in [REI 14] is numerical, these conditions can be checked up to the applied numerical precision. The procedure can terminate with three different results:

- it finds a Markovian representation $(\boldsymbol{\alpha}, \boldsymbol{A})$ of size $m > n$ meaning that $(\gamma, \boldsymbol{G})$ represents a PH distribution of PH-order at most $m$ (see Case 2);

- it finds a $t$ for which $\gamma\, e^{\boldsymbol{G}t}(-\boldsymbol{G})\mathbb{1} < 0$, that is, $(\gamma, \boldsymbol{G})$ does not represent any distribution (see Case 4);

- it stops after a finite number of steps without a decisive result.

### 5.7.1. *Monocyclic representation*

In [MOC 99], a special representation called *monocyclic* is defined.

DEFINITION 5.16.– *A monocyclic PH distribution is composed of a sequence of FEBs (including degenerate FEBs of size* 1*).*

An example is depicted in Figure 5.13. Denoting the parameters of FEB$_i$ by $k_i, \nu_i, z_i$, the example is such that $k_1 = 1, k_2 = 4, k_3 = 1$ and, consequently, the first and the third FEBs are degenerate.

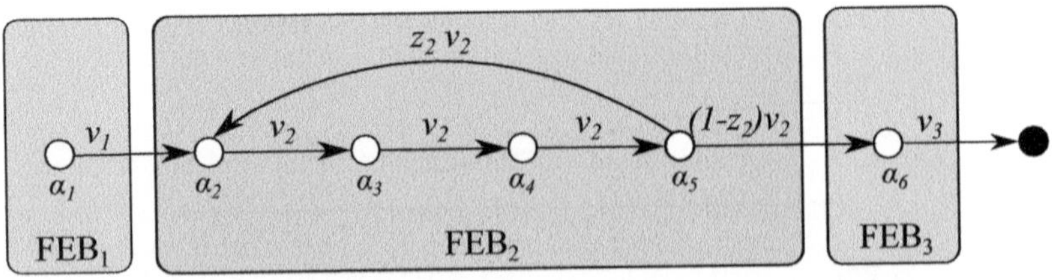


**Figure 5.13.** *Monocyclic representation of a PH distribution*

The monocyclic representation is important because of the following theorem.

THEOREM 5.17.– *[MOC 99] Every rank $n$ PH distribution has a Markovian monocyclic representation whose size is possibly larger than $n$.*

In [MOC 99], a constructive algorithm is provided to obtain the Markovian monocyclic representation starting from any, potentially non-Markovian representation. This way the algorithm in [MOC 99] can be used also for obtaining a Markovian representation of a PH distribution from a non-Markovian one.

### 5.7.2. *Transformation to a Markovian representation*

The transformation of a non-Markovian representation $(\boldsymbol{\alpha}, \boldsymbol{A})$ to a Markovian (monocyclic) representation $(\boldsymbol{\beta}, \boldsymbol{B})$ consists of the following steps [MOC 99].

1) In the first step, matrix $\boldsymbol{B}$ is constructed from a sequence of FEBs. Each FEB implements a real eigenvalue or a conjugate complex eigenvalue pair of $\boldsymbol{A}$:

- if the $j$th eigenvalue, $-\lambda_j$, is real, the parameters of the $j$th FEB are $\nu_j = \lambda_j, k_j = 1, z_j = 0$;

- if it is complex, denoting by $-\lambda_j = a_j + b_j\Im$ and $-\bar{\lambda}_j = a_j - b_j\Im$ the complex conjugate pair, the parameters $k_j, \nu_j, z_j$ of the corresponding FEB are determined based on $a_j$ and $b_j$ according to theorem 5.14.

With these parameters, $\boldsymbol{B}$ is Markovian by construction and contains all eigenvalues of $\boldsymbol{A}$ with the proper multiplicities. However, the size of $\boldsymbol{B}$ can be larger than the size of $\boldsymbol{A}$, meaning that additional eigenvalues are introduced. These extra eigenvalues do not play a role in the PDF, as $\boldsymbol{\beta}$ will eliminate their effect.

2) The second step of the procedure is calculating the initial vector $\boldsymbol{\beta}$. To this end, we need to obtain the transformation matrix $\boldsymbol{W}_1$ that transforms matrix $\boldsymbol{A}$ to

matrix $\boldsymbol{B}$. According to theorem 1.4 and corollary 1.7, matrix $\boldsymbol{W_1}$ can be computed as the solution of the linear system $\boldsymbol{A}\boldsymbol{W_1} = \boldsymbol{W_1}\boldsymbol{B}$, $\boldsymbol{W_1}\mathbb{1} = \mathbb{1}$. The initial vector is given then by $\boldsymbol{\beta} = \boldsymbol{\alpha}\boldsymbol{W_1}$, which may or may not be a non-negative probability vector.

3) The third step is necessary only if $\boldsymbol{\beta}$ contains negative elements after the second step. In this case, an Erlang block is appended to the sequence of FEBs. An Erlang block is composed of a number of consecutive exponential transitions with the same transition rate. After adding the Erlang block to $\boldsymbol{B}$, denoting by $\hat{\boldsymbol{B}}$ the resulting matrix, the corresponding transformation matrix, $\boldsymbol{W_2}$, is the solution of the linear system $\boldsymbol{B}\boldsymbol{W_2} = \boldsymbol{W_2}\hat{\boldsymbol{B}}$, $\boldsymbol{W_2}\mathbb{1} = \mathbb{1}$, and the related initial vector is $\hat{\boldsymbol{\beta}} = \boldsymbol{\beta}\boldsymbol{W_1}$. It is proven in [MOC 99] that there exists an Erlang block that makes the representation Markovian if $(\boldsymbol{\alpha}, \boldsymbol{A})$ defines a PH distribution. Unfortunately, there is no explicit way to obtain the size and the rate parameter of the Erlang block, denoted as $n_{\mathrm{Erl}}$ and $\lambda_{\mathrm{Erl}}$, respectively. A simple heuristic for finding $n_{\mathrm{Erl}}$ and $\lambda_{\mathrm{Erl}}$ is to increase $n_{\mathrm{Erl}}$ successively and to apply the secant method to try to find $\lambda_{\mathrm{Erl}}$ such that $\hat{\boldsymbol{\beta}}$ is a probability vector [REI 14].

Algorithm 5.3 provides a formal description of this procedure, whose implementation is available in the BuTools package [BUT 15].

**Algorithm 5.3** Transform the $(\boldsymbol{\alpha}, \boldsymbol{A})$ representation to a Markovian monocyclic representation.

**procedure** TOMONOCYCLIC($\boldsymbol{\alpha}, \boldsymbol{A}$)<br>
    $n \leftarrow$ size of $\boldsymbol{A}$, $\lambda_j \leftarrow$ eigenvalues of $\boldsymbol{A}$, $\boldsymbol{B} \leftarrow ()$<br>
    **for** $j = 1$ to $n$ **do**     // Step 1<br>
        **if** $\mathcal{I}(\lambda_j) = 0$ **then**<br>
            add a FEB with $\nu_j = \lambda_j, k_j = 1, z_j = 0$ to $\boldsymbol{B}$<br>
        **else if** $\mathcal{I}(\lambda_j) > 0$ **then**<br>
            add a FEB with $\nu_j, k_j, z_j$ according to theorem 5.14 to $\boldsymbol{B}$<br>
        **end if**<br>
    **end for**<br>
    $\boldsymbol{W_1} \leftarrow$ solution of $\boldsymbol{A}\boldsymbol{W_1} = \boldsymbol{W_1}\boldsymbol{B}, \boldsymbol{W_1}\mathbb{1} = \mathbb{1}$     // Step 2<br>
    $\boldsymbol{\beta} \leftarrow \boldsymbol{\alpha}\boldsymbol{W_1}$<br>
    **if** $\boldsymbol{\beta} \geq 0$ **then**<br>
        **return** $(\boldsymbol{\beta}, \boldsymbol{B})$<br>
    **end if**<br>
    find appropriate Erlang block with size $n_{\mathrm{Erl}}$ and rate $\lambda_{\mathrm{Erl}}$     // Step 3<br>
    $\hat{\boldsymbol{B}} \leftarrow$ add an Erlang block with $n_{\mathrm{Erl}}, \lambda_{\mathrm{Erl}}$ to $\boldsymbol{B}$<br>
    $\boldsymbol{W_2} \leftarrow$ solution of $\boldsymbol{B}\boldsymbol{W_2} = \boldsymbol{W_2}\hat{\boldsymbol{B}}, \boldsymbol{W_2}\mathbb{1} = \mathbb{1}$<br>
    **return** $(\boldsymbol{\beta}\boldsymbol{W_2}, \hat{\boldsymbol{B}})$<br>
**end procedure**

EXAMPLE 5.18.– *The algorithm can also be applied to Markovian representations. For example, the PH distribution with representation:*

$$\boldsymbol{\alpha} = \begin{pmatrix} 0 & 1 & 0 & 0 & 0 \end{pmatrix}, \; \boldsymbol{A} = \begin{pmatrix} -1 & 1 & & & \\ & -1 & 1 & & \\ 0.5 & & -1 & 0.5 & \\ & & & -1 & 1 \\ 0.5 & & & & -1 \end{pmatrix}$$

*can be transformed as follows.*

– *Step 1: The eigenvalues of* $\boldsymbol{A}$ *are* $-0.07477$, $-1.564 \pm 0.6184\Im$, $-0.8979 \pm 0.6123\Im$. *The FEBs realizing these eigenvalues are:*

- $k_1 = 1$, $\nu_1 = 0.07477$, $z_1 = 0$ *for eigenvalue* $0.07477$;

- $k_2 = 4$, $\nu_2 = 0.8979$, $z_2 = 0.216204$ *for eigenvalues* $-1.56467 \pm 0.618366\Im$;

- $k_3 = 3$, $\nu_3 = 1.208$, $z_3 = 0.2067$ *for eigenvalues* $-0.8979 \pm 0.6123\Im$;

*and the resulting generator matrix is:*

$$\boldsymbol{B} = \left(\begin{array}{c|cccc|ccc} -0.07477 & 0.07477 & 0 & 0 & 0 & 0 & 0 & 0 \\ \hline 0 & -0.8979 & 0.8979 & 0 & 0 & 0 & 0 & 0 \\ 0 & 0 & -0.8979 & 0.8979 & 0 & 0 & 0 & 0 \\ 0 & 0 & 0 & -0.8979 & 0.8979 & 0 & 0 & 0 \\ 0 & 0.1941 & 0 & 0 & -0.8979 & 0.7038 & 0 & 0 \\ \hline 0 & 0 & 0 & 0 & 0 & -1.207 & 1.207 & 0 \\ 0 & 0 & 0 & 0 & 0 & 0 & -1.207 & 1.207 \\ 0 & 0 & 0 & 0 & 0 & 0.2496 & 0 & -1.207 \end{array}\right).$$

– *Step 2: By the similarity matrix satisfying* $\boldsymbol{AW} = \boldsymbol{WB}$ *and* $\boldsymbol{W}\mathbb{1} = \mathbb{1}$, *the initial vector related to* $\boldsymbol{B}$ *is obtained as:*

$$\boldsymbol{\beta} = \boldsymbol{\alpha W} = \begin{pmatrix} 0.5508 & 0.04352 & 0.003787 & 0.1473 & 0.2546 & 0 & 0 & 0 \end{pmatrix}.$$

– *Step 3:* $\boldsymbol{\beta}$ *is a non-negative probability vector, there is no need for further transformation.*

EXAMPLE 5.19.– *An example where Step 3 of the algorithm is necessary is:*

$$\boldsymbol{\alpha} = \begin{pmatrix} -2 & 3 & 0 \end{pmatrix}, \; \boldsymbol{A} = \begin{pmatrix} -1 & 0.5 & 0.5 \\ & -1 & 1 \\ 0.5 & & -1 \end{pmatrix}.$$

– *Step 1: The eigenvalues of* $\boldsymbol{A}$ *are* $-0.2393$, $-1.3803 \pm 0.4289\Im$. *The FEBs realizing these eigenvalues are:*

- $k_1 = 1$, $\nu_1 = 0.2393$, $z_1 = 0$ *for eigenvalue* $0.2393$;

- $k_2 = 3$, $\nu_2 = 1$, $z_2 = 0.5$ *for eigenvalues* $-1.3803 \pm 0.4289\Im$;

*and the resulting generator matrix is:*

$$B = \left(\begin{array}{c|ccc} -0.2393 & 0.2393 & 0 & 0 \\ \hline 0 & -1.1327 & 1.1327 & 0 \\ 0 & 0 & -1.1327 & 1.1327 \\ 0 & 0.0947 & 0 & -1.1327 \end{array}\right).$$

– *Step 2: The initial vector for matrix* $\boldsymbol{B}$ *is* $\boldsymbol{\beta} = (0.32563\ -0.176161\ 0.850531\ 0)$.

– *Step 3:* $\boldsymbol{B}$ *needs to be extended with an Erlang block because* $\boldsymbol{\beta}$ *contains a negative element. An appropriate Erlang block is with size* $n_{\text{Erl}} = 2$ *and rate* $\lambda_{\text{Erl}} = 1.329$ *resulting in:*

$$\hat{B} = \left(\begin{array}{c|ccc||cc} -0.23931 & 0.23931 & 0 & 0 & 0 & 0 \\ \hline 0 & -1.1327 & 1.1327 & 0 & 0 & 0 \\ 0 & 0 & -1.1327 & 1.1327 & 0 & 0 \\ 0 & 0.0947 & 0 & -1.1327 & 1.038 & 0 \\ \hline\hline 0 & 0 & 0 & 0 & -1.329 & 1.329 \\ 0 & 0 & 0 & 0 & 0 & -1.329 \end{array}\right).$$

*From the solution of* $\boldsymbol{B}\hat{\boldsymbol{W}} = \hat{\boldsymbol{W}}\hat{\boldsymbol{B}}$ *and* $\hat{\boldsymbol{W}}\mathbb{1} = \mathbb{1}$, *the initial vector related to* $\hat{\boldsymbol{B}}$ *is:*

$$\hat{\boldsymbol{\beta}} = \boldsymbol{\beta}\hat{\boldsymbol{W}} = \begin{pmatrix}0.2189\ 0.1045\ 0.02415\ 0.08649\ 0.5659\ 0\end{pmatrix}.$$

*In this example, among the three representations of the same PH distribution with different sizes,* $(\boldsymbol{\alpha}, \boldsymbol{A})$, $(\boldsymbol{\beta}, \boldsymbol{B})$ *and* $(\hat{\boldsymbol{\beta}}, \hat{\boldsymbol{B}})$, *the first two are non-Markovian, while the last one is Markovian and monocyclic.*

## 5.8. Representation minimization

In definition 2.6, we defined the rank of PH$(\boldsymbol{\alpha}, \boldsymbol{A})$ based on the Laplace transform. This rank definition allows us to obtain a non-Markovian full rank representation of PH$(\boldsymbol{\alpha}, \boldsymbol{A})$ in two steps. The first step is to compute the rational Laplace transform of PH$(\boldsymbol{\alpha}, \boldsymbol{A})$ such that the redundant terms are eliminated by simplifying the numerator and the denominator by all common terms. The second step is to apply the transformation *from Laplace transform to matrix representation* presented in section 5.3.4.

There are many other ways to define the rank of PH$(\boldsymbol{\alpha}, \boldsymbol{A})$. For example, if $\boldsymbol{A}$ is diagonalizable, then the number of non-zero $c_i$ coefficients in [2.11] defines the rank of the representation. However, in all cases the implementation of numerical procedures to compute the rank of PH$(\boldsymbol{\alpha}, \boldsymbol{A})$ raises intricate numerical issues.

### 5.8.1. *Numerical issues*

In many rank computations, the rank depends on whether the difference of certain computed parameters is zero or not.

For example, if the rank is checked based on definition 2.6, then the factorization of the polynomials in the numerator and in the denominator should be compared and the identical terms can be eliminated.

If the rank is checked based on [2.11] or [2.12] according to definition 2.8, then the zero/non-zero property of the $c_i$ or $c_{ij}$ parameters decide the rank.

Using an exact arithmetic, the zero/non-zero property of a computed parameter is obtained exactly, but the use of exact arithmetic is limited to simple cases with limited sizes. When using floating point arithmetic, the zero/non-zero property of a computed parameter is subject to numerical errors.

A computationally stable procedure to obtain a full rank representation is the *Staircase method* from linear system theory [BUC 13]. While many other rank computations require the solution of the polynomial equation whose order is the size of the representation (e.g. to compute the eigenvalues, or to simplify the rational Laplace transform), the *Staircase method* is based on the singular value decomposition of the matrix of the representation, which is known to be a computationally more stable operation.

## 5.9. Markovian representation minimization

In some cases, we are interested in finding a small Markovian representation based on a full rank non-Markovian representation of size $n$. Such case can arise, for example, when the non-Markovian representation is obtained from moment matching and we need a Markovian representation of the obtained distribution.

As discussed in section 5.7, starting from a full rank non-Markovian representation of size $n$, there are only numerical procedures available for converting the representation into a Markovian one for $n > 3$. Algorithm 5.1 can be applied to obtain a Markovian representation of size $n$. If it fails, the numerical procedure in [REI 14] provides a Markovian representation if the full rank non-Markovian representation, $(\gamma, G)$, is such that a similar Markovian matrix representation of size $m$ with $m > n$ exists. The size of the obtained representation is larger than $n$ because the procedure in [REI 14] represents each complex conjugate eigenvalue pair of $(\gamma, G)$ by a FEB, which adds extra eigenvalues to the representation. This redundancy can be reduced by the heuristic procedure proposed in [HOR 14], which intends to represent more than one complex conjugate pairs by an unicyclic structure.

EXAMPLE 5.20.– *Example 5.18 from* $(\boldsymbol{\beta}, \mathbf{B})$ *to* $(\boldsymbol{\alpha}, \mathbf{A})$ *is an example for the special case when the eigenvalues of a monocyclic PH distribution with representation* $(\boldsymbol{\beta}, \mathbf{B})$ *(and the eigenvalues are represented by different FEBs) can be obtained by a smaller generator matrix with unicyclic structure (matrix* $\mathbf{A}$*). In this case, the size of the unicyclic representation is smaller than the size of the monocyclic representation.*

# 6

# Moment Matching

In this chapter, we present both moment bounds, that is, constraints that the moments must satisfy for any PH or ME distribution, and methods to obtain PH and ME distributions with given moments.

In section 6.1, we discuss a feature of practical importance, namely, the ability of PH distributions to capture low variability, for which the DPH and the PH cases differ significantly. The same feature for what concerns ME distributions is considered in section 6.3 and we show that ME distributions outperform PH distributions in this respect.

We discuss algebraic characteristics of the moments of PH and ME distributions in section 6.4 together with some of their practical consequences.

Moment matching algorithms, that is, methods that generate PH and ME distributions with given moments, are provided in sections 6.5–6.7. An extension to also match the derivative of the CDF at zero is described in section 6.8.

Explicit moment bounds for PH(2) and ME(2) distributions are reported in section 6.9 while numerical investigations of the moment bounds for PH(3) and ME(3) distributions are delineated in section 6.10.

## 6.1. Continuous phase type distributions with minimal squared coefficient of variation

Modeling real systems often involves the necessity of dealing with actions or activities whose duration has a distribution with low variability. Typical examples are deterministic timeouts in telecommunication protocols and activities that last for an (almost) deterministic period in supply chains. This raises the question: how well can a

PH distribution with a given rank capture low variability? This question has practical implications for the following reason. When PH distributions are used to approximate non-memoryless distributions to transform a non-Markovian process into a Markovian one, the size of the resulting state space depends directly on the size of the involved PH distributions [HOR 16a]. Large PH distributions easily lead to an infeasibly large state space. Thus, our aim is to characterize the lowest variability that can be realized with PH and DPH distributions of a given rank.

In probability theory and statistics, a standard measure to quantify variability of a random variable is the SCV, defined in [1.1]. The SCV of a deterministic distribution is equal to 0, and any other distribution has strictly positive SCV.

A well known result for the family of PH distributions [ALD 87] is stated by the following theorem.

THEOREM 6.1.– *[ALD 87] The minimal SCV for a full rank PH($n$) distributed random variable, $X$, is $1/n$ and it is realized by the Erlang($n$) distribution.*

An implicit but important implication of this theorem is that the minimal SCV of a full rank PH($n$) distribution does not depend on the mean of the distribution. For any $\lambda$, the SCV of the Erlang($n, \lambda$) distribution is $1/n$. However, it is trivial to show that this property for PHs does not hold for DPH's. in the case of DPH distribution, the characterization of the minimal SCV is more complex, which we illustrate with two simple examples.

EXAMPLE 6.2.– *It is easy to see that for any size $n$, the DPH with representation:*

$$\boldsymbol{\beta} = (1\ 0\ \dots\ 0) \qquad \boldsymbol{B} = \begin{pmatrix} 0 & 1 & 0 & 0 & \dots & 0 \\ 0 & 0 & 1 & 0 & \dots & 0 \\ & & & \ddots & & \\ 0 & \dots & 0 & 0 & 1 & 0 \\ 0 & \dots & 0 & 0 & 0 & 1 \\ 0 & \dots & 0 & 0 & 0 & 0 \end{pmatrix} \qquad [6.1]$$

*leads to deterministic time to absorption, with SCV equal to 0. Accordingly, for any size $n$, there exists a DPH distribution with SCV equal to 0. Vice versa, for any positive integer $n$, there exists a DPH distribution whose mean time to absorption is $n$ with SCV equal to 0.*

The above example suggests that there is a dependence between the mean and the minimal SCV of DPH distributions. The following examples regarding a PH(2) and a DPH(2) distribution investigate this fact further.

EXAMPLE 6.3.– *Let $T_C$ and $T_D$ be PH and DPH random variables with representations $(\boldsymbol{\alpha}, \boldsymbol{A})$ and $(\boldsymbol{\beta}, \boldsymbol{B})$, respectively, as:*

$$\boldsymbol{\alpha} = (1\ 0) \qquad \boldsymbol{A} = \begin{pmatrix} -\lambda_1 & \lambda_1 \\ 0 & -\lambda_2 \end{pmatrix} \qquad [6.2]$$

$$\boldsymbol{\beta} = (1\ 0) \qquad \boldsymbol{B} = \begin{pmatrix} 1-\beta_1 & \beta_1 \\ 0 & 1-\beta_2 \end{pmatrix} \qquad [6.3]$$

*with the corresponding graphical representation depicted in Figure 6.1.*

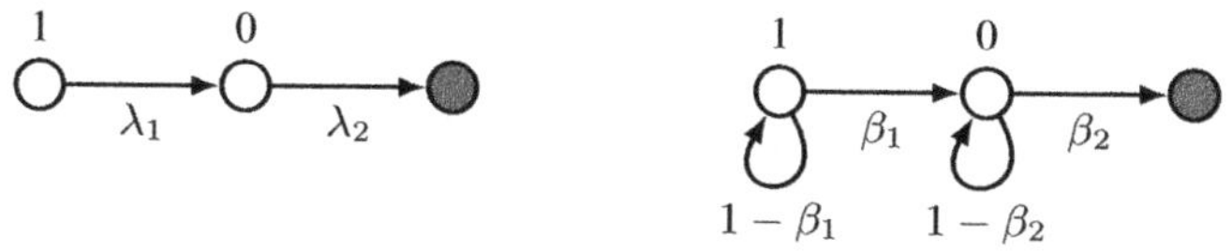


**Figure 6.1.** *Two-phase PH and DPH structures*

*The mean $m$, the variance $\sigma^2$ and the SCV of $T_C$ and $T_D$ are:*

$$m_C = \frac{1}{\lambda_1} + \frac{1}{\lambda_2}, \qquad m_D = \frac{1}{\beta_1} + \frac{1}{\beta_2},$$

$$\sigma_C^2 = \frac{1}{\lambda_1^2} + \frac{1}{\lambda_2^2}, \qquad \sigma_D^2 = \frac{1}{\beta_1^2} - \frac{1}{\beta_1} + \frac{1}{\beta_2^2} - \frac{1}{\beta_2},$$

$$SCV_C = \frac{\sigma_C^2}{m_C^2} = \frac{\lambda_1^2 + \lambda_2^2}{(\lambda_1 + \lambda_2)^2}, \quad SCV_D = \frac{\sigma_D^2}{m_D^2} = \frac{\beta_1^2 - \beta_1^2\beta_2 + \beta_2^2 - \beta_1\beta_2^2}{(\beta_1 + \beta_2)^2}. \qquad [6.4]$$

*Both distributions are characterized by two free parameters: the PH distribution by $\lambda_1$ and $\lambda_2$, and the DPH distribution by $\beta_1$ and $\beta_2$. First, we suppose to fix the value of $\lambda_1$ and $\beta_1$, and to find such $\lambda_2^{min}$ and $\beta_2^{min}$ that minimize the SCV in [6.4]. The values $\lambda_2^{min}$ and $\beta_2^{min}$ are obtained by setting the derivative of the SCV with respect to $\lambda_2$ and $\beta_2$ equal to 0, and are given by:*

$$\lambda_2^{min} = \lambda_1, \qquad \beta_2^{min} = \min\left(\frac{\beta_1(2+\beta_1)}{2-\beta_1}, 1\right),$$

*where the* min *function is used because we must have $0 \leq \beta_2 \leq 1$. The minimal SCV of the PH structure is obtained when the parameters $\lambda_1$ and $\lambda_2$ are equal (in accordance with theorem 6.1), while the DPH structure exhibits the minimal SCV when, in general, $\beta_1$ differs from $\beta_2$.*

*In order to investigate the dependence of the minimal coefficient of variation with respect to the mean, we assume now that the two pairs of free parameters of the considered structures are $\lambda_2$ and $m_C$, and $\beta_2$ and $m_D$, respectively. Rearranging* [6.4], *we have:*

$$\lambda_1 = \frac{\lambda_2}{m_C\lambda_2 - 1}, \qquad \beta_1 = \frac{\beta_2}{m_D\beta_2 - 1},$$

$$SCV_C = \frac{2 - 2m_C\lambda_2 + m_C^2\lambda_2^2}{m_C^2\lambda_2^2}, \; SCV_D = \frac{2 - 2m_D\beta_2 - m_D\beta_2^2 + m_D^2\beta_2^2}{m_D^2\beta_2^2}. \qquad [6.5]$$

*For a given mean ($m_C$ and $m_D$), the minimal SCV is obtained by equating to 0 the derivative of SCV with respect to $\lambda_2$ and $\beta_2$, respectively. The parameters that minimize SCV are:*

$$\lambda_2^{min} = \frac{2}{m_C}, \qquad \beta_2^{min} = \frac{2}{m_D},$$

*where as a result of the initial probability vector $(1 \;\; 0)$, we have $m_D \geq 2$ and, consequently, $\beta_2^{min} \leq 1$. Substituting these values into* [6.5], *we obtain:*

$$\lambda_1 = \frac{2}{m_C}, \quad \beta_1 = \frac{2}{m_D},$$

$$SCV_C = \frac{1}{2}, \; SCV_D = \frac{1}{2} - \frac{1}{m_D}.$$

*For the PH distribution, the minimal SCV is obtained (in accordance again with theorem 6.1) with $\lambda_1 = \lambda_2$ and it is independent of the mean $m_C$. For the DPH distribution, differently from the previous case, the minimum is attained with $\beta_1 = \beta_2$ (i.e. we have a negative binomial distribution which is the discrete counterpart of the Erlang distribution), but the value of the minimum depends on the mean $m_D$.*

The following theorem characterizes the minimal SCV that can be realized by a DPH($n$) distribution.

THEOREM 6.4.– *[TEL 00] The minimal SCV for a DPH($n$) random variable $X$ with mean $m$ is:*

$$\min SCV_X = \frac{\{m\}(1 - \{m\})}{m^2}, \quad \text{if } m \leq n,$$
$$\min SCV_X = \frac{1}{n} - \frac{1}{m}, \quad \text{if } m > n, \qquad [6.6]$$

*where $\{m\}$ denotes the fractional part of $m$, that is, $\{m\} = m - \lfloor m \rfloor$. The structures of the DPH distributions that realize the minimal SCV are depicted in Figures 6.2 and 6.3, respectively.*

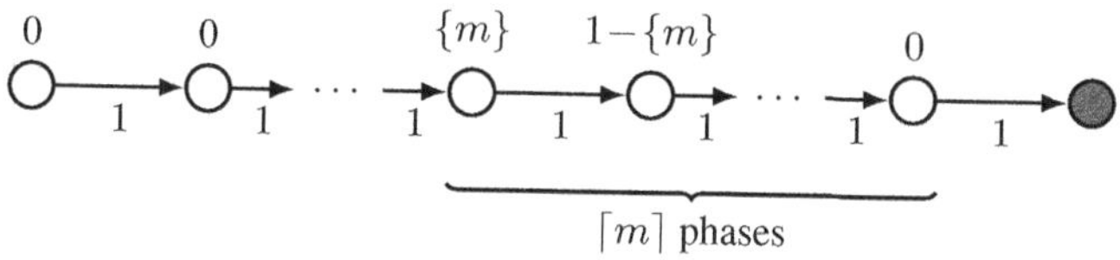


**Figure 6.2.** *DPH distribution that realizes minimal SCV in the case of $m \leq n$*

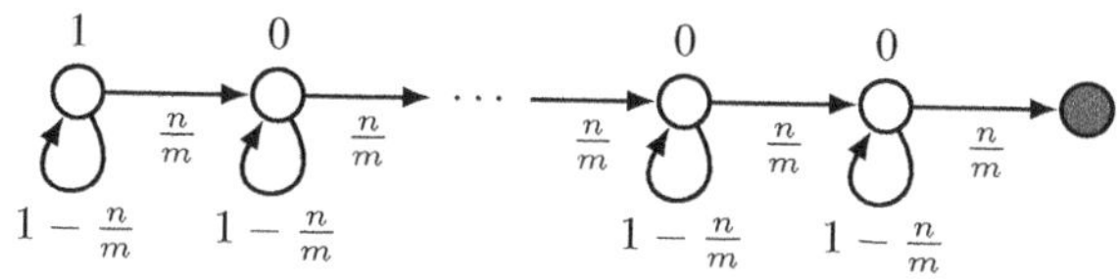


**Figure 6.3.** *DPH distribution that realizes minimal SCV in the case of $m > n$*

Accordingly, in the case of $m \leq n$, the minimal SCV is realized by a structure in which only $\lceil m \rceil$ phases are really used. As a result, the remaining $n - \lceil m \rceil$ phases, which have initial probability equal to 0, can be eliminated without affecting the distribution. On the other hand, if $m > n$, then the negative binomial distribution (the discrete counterpart of the Erlang distribution) exhibits the minimal SCV.

As pointed out also by the previous examples, an important difference between DPH and PH distributions is that for DPH distributions the minimal SCV depends on the mean while in the case of PH distributions it does not. Another difference is that in the discrete case, in contrast with the continuous case, increasing the number of phases beyond $m$ does not have any effect on the minimal SCV. Instead, in the case of $m > n$, PH and DPH distributions are more similar and converge to identical limits as $m \to \infty$.

In Figure 6.4, we plot the minimal SCV for DPH and PH distributions first as a function of the mean for a fixed number of phases (on the left), and then as a function of the number of phases for a fixed value of the mean (on the right). When the number of phases is equal to $n$, then the minimal SCV of the PH family is $1/n$ independently of the mean. Instead, for the DPH family, the mean plays an important rule. In the range $1 \leq m \leq n$, the minimal SCV depends on the fractional part of $m$ and can even be 0. If $m > n$, then the minimal SCV increases as $m$ increases and tends to $1/n$, that is, to the minimal SCV of the PH class. When the mean is fixed, for what concerns the DPH family, increasing the number of phases lowers the minimal SCV only up to $n = \lceil m \rceil$, then it remains constant. For the PH family, the minimal SCV decreases continuously with the increasing number of phases.

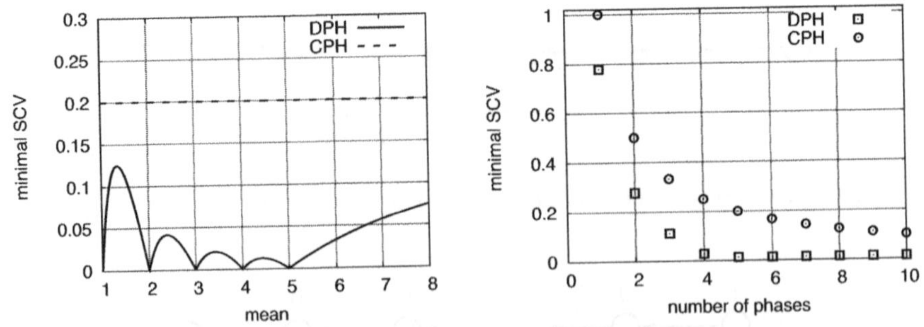


**Figure 6.4.** *Minimal SCV of DPH and PH distributions as a function of the mean with $n = 5$ phases (left) and as a function of the number of phases with mean $m = 4.5$ (right)*

Theorems 6.1 and 6.4 suggest that the DPH family is more powerful in exhibiting low SCV. In order to better quantify the difference, we consider the following calculations. Given the mean, we determine the lowest possible SCV for the DPH class, which is:

$$\frac{\{m\}(1 - \{m\})}{m^2} \qquad [6.7]$$

and can be realized using $\lceil m \rceil$ phases. Then the reciprocal of [6.7] rounded up to the nearest integer gives the necessary number of phases for a PH distribution to realize the same SCV. Figure 6.5 depicts the results of the calculation as a function of the mean. As an example, for $m = 5.5$, the minimal SCV for the DPH family is 0.00826 obtained with six phases and we need a PH distribution with 121 phases to realize such a low SCV. With $m = 5.1$, the calculation results in 290. Note also, that as the mean approaches an integer value, the minimal SCV of the DPH family tends to 0 and, consequently, the necessary number of phases for a PH distribution to exhibit that SCV tends to infinity.

We assumed so far that one step in the DTMC associated with a given DPH distribution are defined on a discrete time scale, that is the model time is a natural number. When DPH distributions are used to model (continuous) timed activities using equidistant intervals, then the length of that interval is a parameter of the discrete time model. We will refer to this parameter, denoted by $\delta$, as the scale factor and call the resulting distribution scaled DPH distribution. For example, if an "unscaled" DPH distribution is with support $\{1, 2, 3, ...\}$, then with $\delta = 0.5$ the support becomes $\{0.5, 1, 1.5, ...\}$ and with $\delta = 2$ the support is $\{2, 4, 6, ...\}$. The scale factor largely affects the shape and the properties of a DPH distribution over the continuous time axis and provides an additional degree of freedom in the application of DPH distributions.

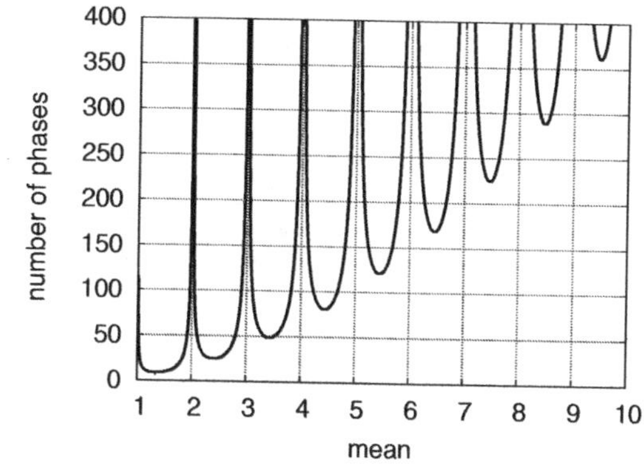


**Figure 6.5.** *Necessary number of phases for a PH distribution to realize the lowest possible SCV of the DPH class as a function of the mean $m$. Note that the number of phases to realize the lowest possible SCV in the case of DPH distributions is equal to $\lceil m \rceil$*

For scaled DPH distributions, theorem 6.4 can be restated as follows.

THEOREM 6.5.– *[BOB 04] The minimal SCV for a scaled DPH random variable X with rank $n$, mean $m$ and scale factor $\delta$ is:*

$$\min SCV_X = \frac{\{m/\delta\}\,(1 - \{m/\delta\})}{(m/\delta)^2}, \quad \textit{if } m/\delta \le n,$$
$$\min SCV_X = \frac{1}{n} - \frac{\delta}{m}, \qquad \textit{if } m/\delta > n. \qquad [6.8]$$

*The structures of the DPH distributions that realize the minimal SCV are depicted in Figures 6.6 and 6.7, respectively.*

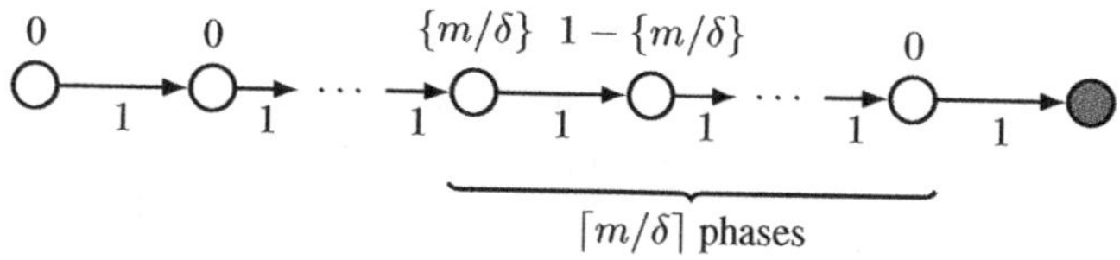


**Figure 6.6.** *DPH distribution that realizes minimal SCV in the case of $m/\delta \le n$*

Figure 6.8 illustrates the effect of the scale factor on the minimal SCV. The figure on the left side provides the minimal SCV as a function of the mean for rank $n = 5$. In accordance with [6.8], the minimal SCV curves alternate from $\delta$ to $\delta n$ such that the minimal SCV is 0 for all integer multiple of $\delta$ in between $\delta$ and $\delta n$. From $\delta n$ on, the minimal SCV curves increase to $1/n$, which is the minimal SCV of the PH class. As $\delta$ decreases the minimal SCV curve converges earlier to $1/n$.

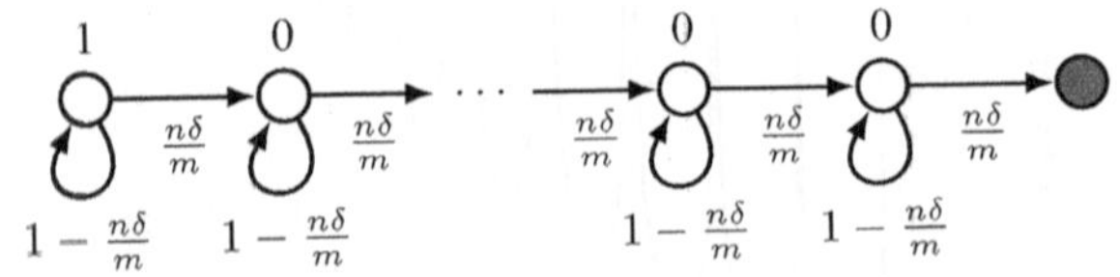

**Figure 6.7.** *Scaled DPH distribution that realizes minimal SCV in the case of $m/\delta > n$*

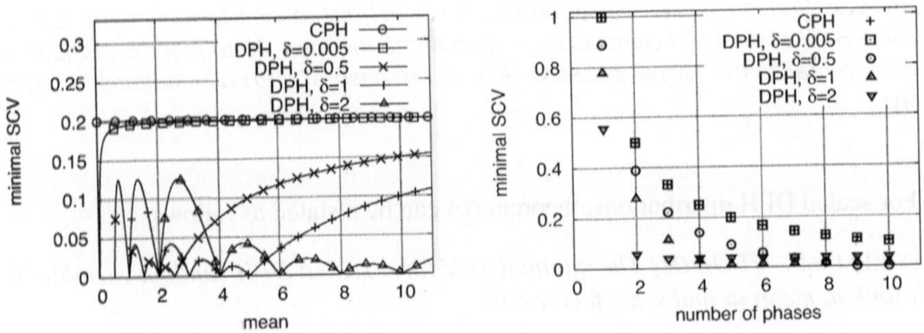


**Figure 6.8.** *Minimal SCV of DPH and PH distributions as a function of the mean with $n = 5$ phases (left) and as a function of the number of phases with mean $m = 4.5$ (right) with different scale factors*

The curve with $\delta = 0.005$ suggests that as $\delta$ tends to 0 the minimal SCV of the scaled DPH class tends to $1/n$, which can be shown in a more formal manner [BOB 04]. The right side of Figure 6.8 provides the minimal SCV as a function of the number of phases with mean $m = 4.5$. The minimal SCV values of the PH class and the DPH class with $\delta = 0.005$ are indistinguishable in the figure. With $n < m/\delta$ the larger the scale factor the smaller the minimal SCV. With $n \geq m/\delta$ increasing $n$ does not decrease the minimal SCV of the DPH class and the minimal SCV depends on $\{m/\delta\}(1 - \{m/\delta\})$. For example, for $m = 4.5$, $\delta = 0.5$ and $n \geq 9$ the minimal SCV is 0, while for $m = 4.5$, $\delta = 2.0$ and $n \geq 4$, it is 0.037.

## 6.2. Moment bounds of continuous phase type distributions based on the steepest increase property

The extremal behavior of the Erlang distribution among the PH distributions and lemma 2.21 establish further relations between the moments of PH distributions [HE 19].

COROLLARY 6.6.– *The kth moment of the PH($n$) distributed random variable $X$ with dominant eigenvalue $-\lambda$ is upper bounded by:*

$$\mathbb{E}(X^k) \leq \frac{(n+k-1)!}{(n-1)!\lambda^k}, \qquad \text{for } k \geq 1. \qquad [6.9]$$

*Proof.* Let $Y$ be Erlang$(n, \lambda)$ distributed. According to lemma 2.23, the CDFs of $X$ and $Y$ satisfy $F(t) \geq F_{(n,\lambda)}(t)$ for $t > 0$. Using this, we have:

$$\mathbb{E}(X^k) = \int_{t=0}^{\infty} t^k f_X(t)dt = \int_{t=0}^{\infty} (1 - F(t))kt^{k-1}dt$$
$$\leq \int_{t=0}^{\infty} (1 - F_{(n,\lambda)}(t))kt^{k-1}dt = \mathbb{E}(Y^k) = \frac{(n+k-1)!}{(n-1)!\lambda^k},$$

where we used [1.25] in the last step. □

COROLLARY 6.7.– *The kth moment of the PH($n$) distributed random variable $X$ is lower bounded by:*

$$\frac{(n+k-1)!}{(n-1)!}\left(\frac{\mathbb{E}(X)}{n}\right)^k \leq \mathbb{E}(X^k) \qquad \text{for } k \geq 1. \qquad [6.10]$$

*Proof.* For $k = 1$, LHS of [6.10] is equal to its RHS. A random variable $V$ is smaller in mean residual life than random variable $X$ if $\mathbb{E}(\max\{0, V - x\}) \leq \mathbb{E}(\max\{0, X - x\})$ holds for all real $x$ [STO 83]. According to O'Cinneide [O'CI 91a], if $V$ is Erlang$(n, n/\mathbb{E}(X))$ distributed and $X$ is PH$(n)$ distributed, then $V$ is smaller in mean residual life than $X$. That is, for any $g(t) > 0$, we have:

$$\int_{x=0}^{\infty} g(x)\mathbb{E}(\max\{0, V - x\})\, dx \leq \int_{x=0}^{\infty} g(x)\mathbb{E}(\max\{0, X - x\})\, dx. \quad [6.11]$$

For $k > 1$, let $g(x) = k(k-1)x^{k-2}$, the LHS of [6.11] is:

$$\int_{x=0}^{\infty} g(x)\mathbb{E}(\max\{0, V - x\})\, dx = \int_{x=0}^{\infty} g(x) \int_{t=x}^{\infty} (t - x) f_V(t)dtdx =$$
$$\int_{x=0}^{\infty} g(x) \int_{t=x}^{\infty} (1 - F_V(t))dtdx = \int_{t=0}^{\infty} (1 - F_V(t)) \int_{x=0}^{t} g(x)dxdt =$$
$$\int_{t=0}^{\infty} (1 - F_V(t))kt^{k-1}dt = \mathbb{E}(V^k) = \frac{(n+k-1)!}{(n-1)!}\left(\frac{\mathbb{E}(X)}{n}\right)^k.$$

Applying the same derivation also for the RHS of [6.11] gives the corollary. □

Note that [6.9] upper bounds the $k$the moment of $X$ based on its dominant eigenvalue, while [6.10] lower bounds it based on its mean. For a PH$(n)$ distributed random variable $X$ with dominant eigenvalue $-\lambda$, a by-product of [6.9] and [6.10] is that $\mathbb{E}(X) = n/\lambda$ if and only if $X$ is Erlang$(n, \lambda)$ distributed.

Using lemma 2.21, the next lemma refines the upper bound for the $k+1$-st moment based on the $k$-th one, which gives corollary 6.6 an alternative proof.

LEMMA 6.8.– *For $k = 0, 1, \ldots$, the $k+1$-st moment of PH($n$) distributed random variable $X$ with dominant eigenvalue $-\lambda$ is bounded by:*

$$\mathbb{E}(X^{k+1}) \leq \frac{n+k}{\lambda}\mathbb{E}(X^k), \qquad [6.12]$$

*and the equality holds only when $X$ is Erlang$(n, \lambda)$ distributed.*

*Proof.* Multiplying both sides of [2.23] by $t^k$ and integrating from 0 to $\infty$ gives the following identities for the LHS and the RHS:

$$LHS = \int_{t=0}^{\infty} t^k d(tf(t)) = [t^{k+1}f(t)]_0^{\infty} - \int_{t=0}^{\infty} tf(t)dt^k$$

$$= -k\int_{t=0}^{\infty} tf(t)t^{k-1}dt = -k\mathbb{E}(X^k)\,;$$

$$RHS = \int_{t=0}^{\infty} t^k(n-\lambda t)f(t)dt = n\int_{t=0}^{\infty} t^k f(t)dt - \lambda\int_{t=0}^{\infty} t^{k+1}f(t)dt$$

$$= m\mathbb{E}(X^k) - \lambda\mathbb{E}(X^{k+1}),$$

from which we have

$$-k\mathbb{E}(X^k) \leq n\mathbb{E}(X^k) - \lambda\mathbb{E}(X^{k+1})\,.$$

When $X$ is Erlang$(n, \lambda)$ distributed, the equality in [6.12] comes from the fact that lemma 2.21 gives equality for Erlang distribution for all $t > 0$. □

Applying [6.12] for $k = 0$ and $k = 1$ enables us to derive the following upper bounds on the mean and the SCV of $X$ using $SCV_X = \frac{\mathbb{E}(X^2)}{\mathbb{E}(X)^2} - 1$:

$$\mathbb{E}(X) \leq \frac{n}{\lambda}, \qquad [6.13]$$

$$SCV_X \leq \frac{n+1-\lambda\mathbb{E}(X)}{\lambda\mathbb{E}(X)}. \qquad [6.14]$$

Interestingly, equation [6.14] gives an upper bound for $SCV_X$, while the lower bound for $SCV_X$ is much more widely known as $SCV_X \geq 1/n$. Hence we have (see Figure 6.9):

$$\frac{1}{n} \leq SCV_X \leq \frac{n+1}{\lambda \mathbb{E}(X)} - 1.$$

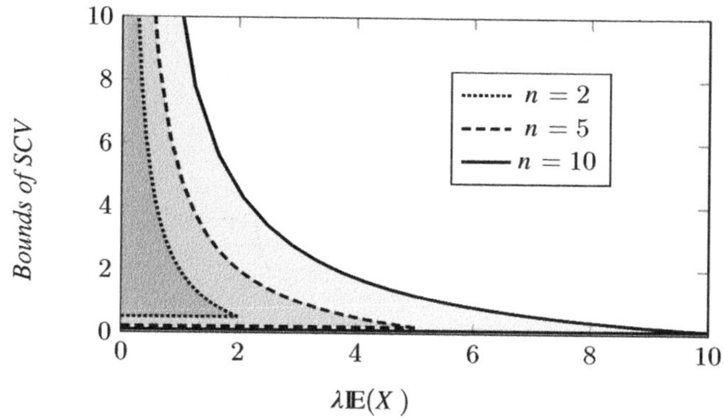


**Figure 6.9.** *Bounds of SCV for ordinary PH distributions. The lower bound is the well known $1/n$ bound provided in [ALD 87, O'CI 91a], while the upper bound is provided based on the steepest increase property*

## 6.3. Matrix exponential distributions with minimal squared coefficient of variation

The set of ME(1) distributions is identical with the set of PH(1) distributions, that is, with the set of exponential distributions. It was shown in [LIE 90] that the sets of PH(2) and ME(2) distributions are identical. Consequently, the ME(1) and ME(2) distributions have the same characteristics as their PH counterpart.

For $n > 2$, the PH($n$) distributions constitute a proper subset of the set of ME($n$) distributions and the SCV of ME($n$) distributions can be smaller than the minimal SCV of PH($n$) distributions, which is $1/n$. For the time being, only conjectures exist about the minimal SCV of ME($n$) distributions for $n > 2$. Moreover, numerical optimization of the involved parameters is required to find the ME distribution that exhibits minimal SCV. In the following we provide the conjecture and show corresponding ME distributions that are believed to exhibit minimal SCV.

### 6.3.1. *Matrix exponential distributions with complex eigenvalues*

The following conjecture provides a parametric family of functions that is believed to contain the PDF of the ME distribution with minimal SCV.

CONJECTURE 1.– *For $n$ larger than 2, the PDF of the ME($n$) distribution that exhibits minimal SCV has the form:*

$$f(t) = ce^{-\lambda t} \prod_{i=1}^{\frac{n-1}{2}} \sin^2(\omega\lambda t - \phi_i) \quad \text{if } n \text{ is odd,} \qquad [6.15]$$

*and:*

$$f(t) = c\left(e^{-\lambda t}\left(a_2 + \prod_{i=1}^{\frac{n-2}{2}} \sin^2(\omega\lambda t - \phi_i)\right) - a_1 e^{-\lambda_1 t}\right) \quad \text{if } n \text{ is even.} \qquad [6.16]$$

*In both cases, the parameters are real and non-negative.*

We note that the PDF in [6.15] is non-negative by construction. The parameters that minimize the SCV of these PDFs are determined by optimization. Details of this optimization process are provided in [HOR 16c, HOR 20]. For low values of $n$, the optimal parameters are given in Table 6.1 for odd values of $n$ and in Table 6.2 for even values of $n$. The PDF of some of the ME distributions exhibiting minimal SCV are plotted in Figure 6.10.

| $n$ | $\omega$ | $\phi_1, \phi_2, \ldots$ |
|---|---|---|
| 3 | 0.517965 | 0.168519 |
| 5 | 0.474055 | 0.106184, 0.532534 |
| 7 | 0.442459 | 0.0755237, 0.381414, 0.860304 |
| 9 | 0.418775 | 0.0576237, 0.292994, 0.664694, 1.12488 |
| 11 | 0.400272 | 0.0460437, 0.235534, 0.537594, 0.911894, 1.33678 |

**Table 6.1.** *The parameters of ME($n$) distributions with minimal SCV based on conjecture 1 for odd $n$, where $\lambda = 1$ and $c$ is such that $\int_0^\infty f(t)dt = 1$*

Table 6.3 lists the SCV of the ME distributions defined by the parameters in Tables 6.1 and 6.2. The last column provides the necessary number of phases for a PH distribution to exhibit the same SCV. It is apparent that ME distributions are advantageous when low variability must be captured and keeping the size low is of concern. Another conclusion from Table 6.3 is that even values of $n$ are not efficient in decreasing the SCV. As a result, we focus on the odd values of $n$ in the rest of this section.

The density function defined in [6.15] has an exponential decay according to $e^{-\lambda t}$ and it has got zeros in a periodic manner: $f(t) = 0$ for $t = \phi_i/(\omega\lambda) + k\pi$, where $k \in \mathbb{N}$. Considering this periodic behavior, we assume that $0 \le \phi_i < \omega\lambda\pi$ and the zeros in $(0, \pi)$ belong to the main cycle (when $k = 0$).

| $n$ | $\lambda_1$ | $a_1$ | $a_2$ | $\omega$ | $\phi$ |
|---|---|---|---|---|---|
| 4 | 1.94907 | 0.5896 | 0.2246 | 0.5197 | 0.6487 |
| 6 | 8.34112 | 0.01551 | 0.00074 | 0.4779 | $0.2183, 0.5957$ |
| 8 | 13.917 | 0.001936 | 0.000045 | 0.444 | $0.1383, 0.4148,$ $0.8983$ |
| 10 | 20.0137 | 0.000315 | $4.535 \cdot 10^{-6}$ | 0.4196 | $0.09831, 0.3145,$ $0.6887, 1.15$ |
| 12 | 26.64 | $5.899 \cdot 10^{-5}$ | $5.894 \cdot 10^{-7}$ | 0.4008 | $0.07473, 0.2509,$ $0.5545, 0.9296,$ $1.355$ |

**Table 6.2.** *Parameters of ME($n$) distributions with minimal SCV based on conjecture 1 for even $n$, where $\lambda = 1$ and $c$ is such that $\int_0^\infty f(t)dt = 1$*

| $n$ | SCV | $\lceil 1/\text{SCV} \rceil$ |
|---|---|---|
| 3 | 0.2009 | 5 |
| 4 | 0.14981 | 7 |
| 5 | 0.08126 | 13 |
| 6 | 0.07552 | 14 |
| 7 | 0.04288 | 24 |
| 8 | 0.04135 | 25 |
| 9 | 0.02616 | 39 |
| 10 | 0.02559 | 40 |
| 11 | 0.01749 | 58 |
| 12 | 0.01724 | 59 |

**Table 6.3.** *Minimal SCV of ME($n$) distributions based on conjecture 1. The last column provides the necessary number of phases for a PH distribution to exhibit the same SCV*

EXAMPLE 6.9.– *Functions with a single high peak and negligible values far from the main peak exhibit low SCV. The density function defined in* [6.15] *equals zero at the points $t = \frac{\phi_i}{\omega\lambda} \pm k\pi$ and it has local peaks between consecutive zeros. The height of the peaks depends on the distance between the neighbouring zeros. The more distant the zeros are the larger the peaks. To the left of this main peak, densely located zeros ensure negligible values, while to the right of the main peak, the exponential decay and further zeros suppress the function.*

*To demonstrate these effects, Figure 6.11 shows* [6.15] *with both linear and logarithmic $y$-axis in the case of $c = 1$, $\omega = 1$, $\lambda = 1$, $n = 7$, $\phi_1 = 0.1$, $\phi_2 = 0.5$, $\phi_2 = 1$. The graph with logarithmic $y$-axis also plots the exponential decay ($e^{-t}$), which is a straight line due to the logarithmic scaling.*

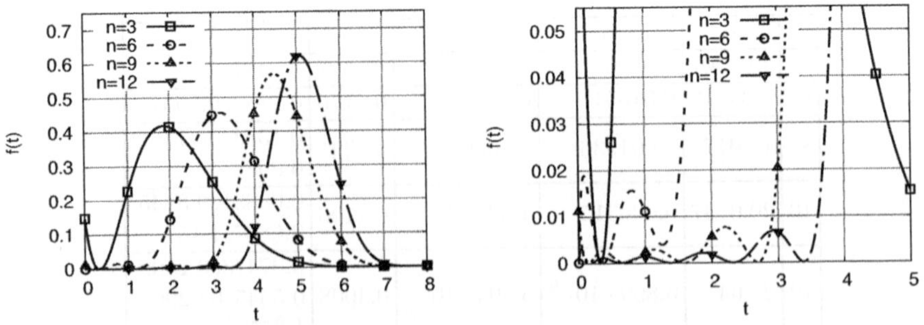


**Figure 6.10.** *Probability density function of ME distributions with minimal SCV. On the right side, we zoom in on the values close to zero to illustrate the oscillating nature of the PDF*

*The main peak of the curve is around t* $\approx 2$. *On the left of this peak, one of the* $sin^2(\omega t - \phi_i)$ *terms is close to its zero* $(0.1, 0.5, 1)$ *and consequently takes a small value, which keeps* $f(t)$ *small. The* $f(t)$ *function has no zero between* $1$ *and* $\pi + 0.1$ *(which is the first zero of the second period). This large gap of zeros allows* $f(t)$ *to have a large peak in the middle of the interval. Right to the main peak* $(t > 2)$*, the function decays due to two reasons: the exponential decay* $(e^{-t})$ *and the zeros of the next period.*

*The logarithmic scaling of the* $y$*-axis makes the location of the zeros visible. While the zeros at* $\pi + 0.1, \pi + 0.5, \pi + 1$ *keep the function low between* $3, 4$ *and* $5$*, the gap in the zeros between* $\pi + 1$ *and* $2\pi + 0.1$ *allows a peak at around t* $\approx 2 + \pi$. *This peak is suppressed only by the exponential decay and remains visible also in the graph with linear* $y$*-axis.*

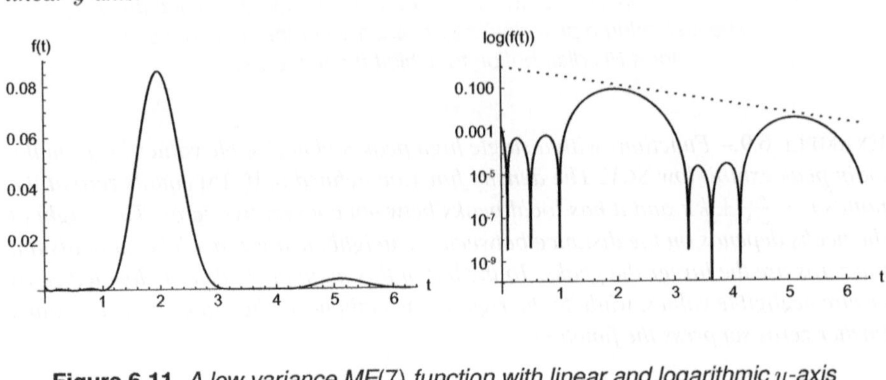


**Figure 6.11.** *A low variance ME(7) function with linear and logarithmic* $y$*-axis*

In spite of the availability of conjecture 1 for more than a decade [ÉLT 06], it has not been proven or disproven yet. Independent of the fact weather conjecture 1 holds true, the non-negative set of functions characterized by [6.15] define ME distributions with very low SCV.

Since the lowest possible SCV values were obtained by numerical optimization, approximation of the decay of the SCV as a function of $n$ is also a numerical problem. In [HOR 16c], where optimization results are available up to $n \approx 100$, it was obtained that the decay rate is approximately $2/n^2$, which is clearly much faster than that of the PH class (which is $1/n$). In [HOR 20], this approximation is refined up to $n = 2,000$ according to Figure 6.12. The same decay rate is reported in [ALM 21] up to $n = 40,000$.

The matrix representation of [6.15] is based on the trigonometric-exponential relation:

$$f(t) = ce^{-\lambda t} \prod_{j=1}^{N} \cos^2\left(\frac{\omega\lambda t - \phi_j}{2}\right) = \sum_{k=0}^{n-1} \eta_k e^{-\beta_k t}, \qquad [6.17]$$

whose details are provided in [HOR 20]. In [6.17], $n$ is odd, $N = (n-1)/2$, $\beta_0$ is real, $\beta_k$ has positive imaginary part for $1 \le k \le N$ and complex conjugate negative imaginary part for $N+1 \le k \le 2N$. Based on the exponential form, the following matrix representation can be obtained:

$$f(t) = \sum_{k=0}^{n-1} \eta_k e^{-\beta_k t} = -\boldsymbol{\alpha} e^{\boldsymbol{A}t} \boldsymbol{A} \mathbb{1}, \qquad [6.18]$$

where $\boldsymbol{\alpha} = \begin{pmatrix} \frac{\eta_0}{\beta_0} & \frac{\eta_1}{\beta_1} & \cdots & \frac{\eta_{n-1}}{\beta_{n-1}} \end{pmatrix}$ and $\boldsymbol{A} = \text{diag}\left\langle\begin{pmatrix} -\beta_0 & -\beta_1 & \ldots & -\beta_{n-1} \end{pmatrix}\right\rangle$.

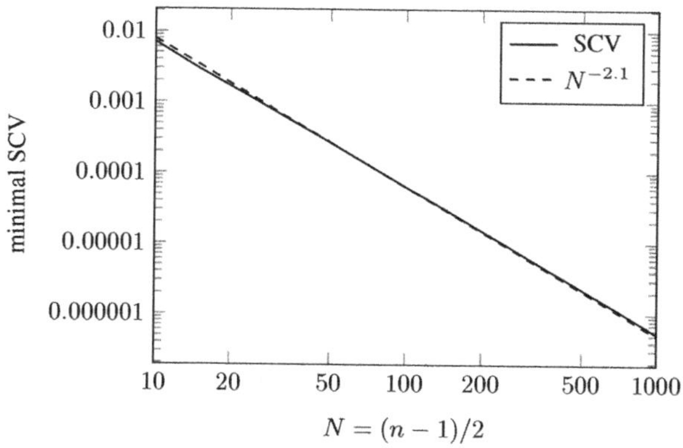


**Figure 6.12.** *The minimal SCV as a function of $N = (n-1)/2$ in log-log scale*

### 6.3.2. *Matrix exponential distributions with real eigenvalues*

Both PH and ME distributions can have complex eigenvalues, while some special subclasses of PH and ME distributions are known to have only real eigenvalues (see section 5.5). As in the general case with complex eigenvalues, there is a conjecture for ME distributions with only real eigenvalues exhibiting minimal SCV.

CONJECTURE 2.– *For $n$ larger than 2, the PDF of the ME distribution with only real eigenvalues that exhibits minimal SCV has the form:*

$$f(t) = ce^{-\lambda t} \prod_{j=1}^{(n-1)/2} (\lambda t - \tau_j)^2 \quad \textit{if } n \textit{ is odd,} \qquad [6.19]$$

*and:*

$$f(t) = cte^{-\lambda t} \prod_{j=1}^{(n-2)/2} (\lambda t - \tau_j)^2 \quad \textit{if } n \textit{ is even.} \qquad [6.20]$$

*In both cases, the parameters are real and non-negative.*

The density functions defined in [6.19] and [6.20] have exponential decay according to $e^{-\lambda t}$ and they have $N = (n-1)/2$ and $N = (n-2)/2$ zeros, respectively, at $\tau_1, \ldots, \tau_N$.

EXAMPLE 6.10.– *Low variance density functions according to* [6.19] *and* [6.20] *have one main peak and the functions left to the main peak are suppressed by the zeros (left to the main peak) and right to the main peak they are suppressed by the exponential decay and by the zeros right to the main peak (if any). Figure 6.13 demonstrates these effects when all zeros are left to the main peak with linear and logarithmic $y$-axis in the case of $c = 1$, $\lambda = 1$, $n = 7$, $\tau_1 = 0.1$, $\tau_2 = 0.5$, $\tau_3 = 1$. The graph with logarithmic $y$-axis also plots the exponential decay $e^{-t}$.*

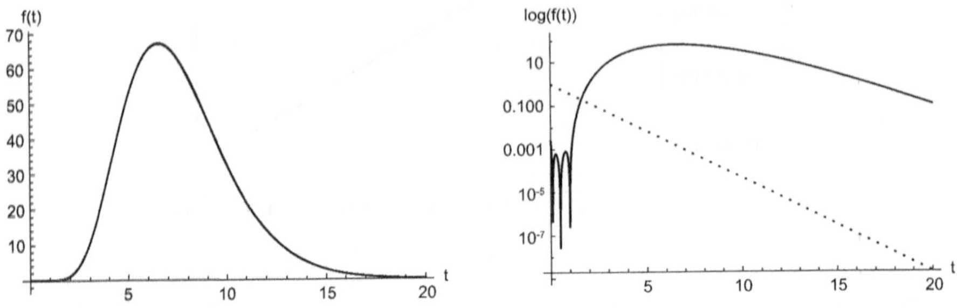


**Figure 6.13.** *A low variance ME(7) function with real eigenvalues and with linear and logarithmic $y$-axis*

*The main peak of the curve is around* $t \approx 7$. *Left of this peak, one of the* $(t - \tau_i)$ *terms is close to its zero* $(0.1, 0.5, 1)$ *and consequently takes a small value, which keeps* $f(t)$ *small. The function* $f(t)$ *has no zero above* $1$ *and increases from this point on up to the main peak where the exponential decay starts to dominate the polynomial increase in the quadratic terms.*

*Compared to Figure 6.11, the right-hand plot in Figure 6.13 indicates that the decay of the function after the peak is much slower. It is due to the fact that in Figure 6.11 the exponential decay is multiplied by a bounded (periodic) function, while in Figure 6.13 the exponential decay competes with a degree* $n - 1$ *polynomial.*

The parameters that minimize the SCV of the density function defined in conjecture 2 are determined by optimization [MÉS 22]. Minimal SCV results for low-rank ME distributions are presented in Table 6.4. For higher ranks, we only report that the decay of the lowest possible SCV as a function of the size is around $N^{-1.85}$, where $N = (n-1)/2$ is the number of the zeros of $f(t)$. This empirical decay is based on Figure 6.14, which plots the minimal SCV as a function of the size according to [MÉS 22].

| n | SCV | $\lceil 1/\mathrm{SCV} \rceil$ |
|---|---|---|
| 3 | 0.276583 | 4 |
| 4 | 0.19333 | 6 |
| 5 | 0.138453 | 8 |
| 6 | 0.108623 | 10 |
| 7 | 0.0861277 | 12 |
| 8 | 0.0717026 | 14 |
| 9 | 0.0600486 | 17 |
| 10 | 0.0518365 | 20 |
| 11 | 0.0449173 | 23 |
| 12 | 0.0397335 | 26 |
| 13 | 0.0352403 | 29 |
| 14 | 0.031726 | 32 |
| 15 | 0.0286172 | 35 |

**Table 6.4.** *Minimal SCV of ME($n$) distributions with real eigenvalues based on conjecture 2. The last column provides the necessary number of phases for a PH distribution to exhibit the same SCV*

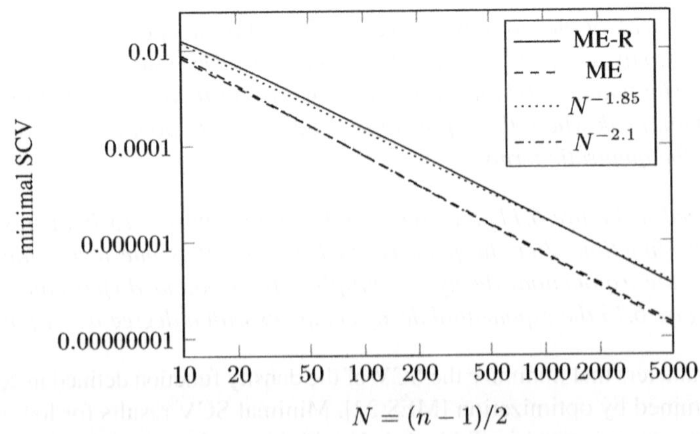


**Figure 6.14.** *The minimal SCV of ME($n$) distributions and ME($n$) distributions with real eigenvalues (ME-R) for odd $n$ as a function of $N = (n-1)/2$ in log-log scale. To indicate the decay rates, the functions $N^{-1.85}$ and $N^{-2.1}$ are plotted as well*

EXAMPLE 6.11.– *Figure 6.15 plots the PDF of ME(7) with complex eigenvalues, ME(7) with real eigenvalues and PH(7) distributions with minimal SCV, when the mean of the distributions is* 1. *The figure indicates how much the main peak of the ME(7) with complex eigenvalues is larger than the main peaks of the other two distributions. The sharp breakdowns on the plot with logarithmic $y$-axis refer to the points where the PDFs are equal to zero. According to theorem 2.15, the Erlang distribution is positive on* $[0, \infty)$. *The ME(7) with real eigenvalues have zero density points only left of the main peak, while the ME(7) with complex eigenvalues have zero density points periodically on* $[0, \infty)$.

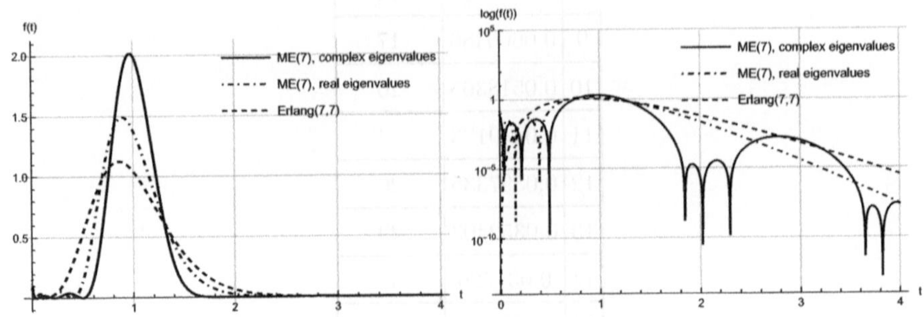


**Figure 6.15.** *PDF of ME(7) with complex eigenvalues, ME(7) with real eigenvalues and PH(7) distributions with minimal SCV whose means equal to* 1

*It is interesting to note, that the PDF at zero and the asymptotic decay rate of the ME(7), ME(7) with real eigenvalues and PH(7) distributions are* 0.08224, 0.1585, 0 *and* 3.937, 9.08117, 7, *respectively. It means that different sets of functions find their minimal SCV with different features. Neither the density at zero nor the asymptotic decay determines the SCV itself.*

## 6.4. Characteristics of moments of continuous phase type and matrix exponential distributions

Throughout this section, we consider only full rank PH and ME distributions (Definition 2.14), that is, where the size of the vector–matrix representation is equal to the degree of the denominator of the corresponding Laplace transform.

Given the full rank vector–matrix representation $(\boldsymbol{\alpha}, \boldsymbol{A})$ of a PH($n$) or a ME($n$) distribution, we construct the doubly infinite Hankel matrix as:

$$\boldsymbol{R} = \begin{pmatrix} \ddots & \vdots & \vdots & \vdots & \vdots & \vdots & ⋰ \\ \cdots & r_{-4} & r_{-3} & r_{-2} & r_{-1} & r_0 & \cdots \\ \cdots & r_{-3} & r_{-2} & r_{-1} & r_0 & r_1 & \cdots \\ \cdots & r_{-2} & r_{-1} & r_0 & r_1 & r_2 & \cdots \\ \cdots & r_{-1} & r_0 & r_1 & r_2 & r_3 & \cdots \\ \cdots & r_0 & r_1 & r_2 & r_3 & r_4 & \cdots \\ ⋰ & \vdots & \vdots & \vdots & \vdots & \vdots & \ddots \end{pmatrix}$$

where

$$r_i = \boldsymbol{\alpha}(-\boldsymbol{A})^{-i}\mathbb{1}.$$

For $i \geq 0$, $r_i = m_i/i!$ is the $i$th reduced moment of the distribution according to [2.7]. For $i < 0$, $r_i$ is closely related with the $-i$th derivative of the CDF at 0, because for $k \geq 1$, the $k$th derivative of the CDF at 0 is:

$$\left.\frac{d^k F(t)}{dt^k}\right|_{t=0} = -\boldsymbol{\alpha} \left.\frac{d^k}{dt^k} e^{\boldsymbol{A}t}\right|_{t=0} \mathbb{1} = -\boldsymbol{\alpha}\boldsymbol{A}^k\mathbb{1} = (-1)^{k+1}\boldsymbol{\alpha}(-\boldsymbol{A})^k\mathbb{1}. \quad [6.21]$$

Let $c_1$ and $c_2$ be two sets of indices. $\boldsymbol{R}_{c_1,c_2}$ denotes the submatrix of $\boldsymbol{R}$ which consists of rows according to $c_1$ and columns according to $c_2$. For example:

$$\boldsymbol{R}_{\{1,3,4\},\{0,2\}} = \begin{pmatrix} r_1 & r_3 \\ r_3 & r_5 \\ r_4 & r_6 \end{pmatrix}.$$

THEOREM 6.12.– *The rank of the doubly infinite Hankel matrix, $\boldsymbol{R}$, composed of the reduced moments of a rank $n$ PH or ME distribution, is $n$.*

*Proof.* $\boldsymbol{R}$ can be expressed as:

$$\boldsymbol{R} = \begin{pmatrix} \vdots \\ \boldsymbol{\alpha}(-\boldsymbol{A})^2 \\ \boldsymbol{\alpha}(-\boldsymbol{A})^1 \\ \boldsymbol{\alpha} \\ \boldsymbol{\alpha}(-\boldsymbol{A})^{-1} \\ \boldsymbol{\alpha}(-\boldsymbol{A})^{-2} \\ \vdots \end{pmatrix} \left(\cdots\ (-\boldsymbol{A})^2\mathbb{1}\ (-\boldsymbol{A})^1\mathbb{1}\ \mathbb{1}\ (-\boldsymbol{A})^{-1}\mathbb{1}\ (-\boldsymbol{A})^{-2}\mathbb{1}\ \cdots\right)$$

where the size of the first matrix is $\infty \times n$ and the size of the second matrix is $n \times \infty$. As a result, the rank of $\boldsymbol{R}$ is at most $n$. On the other hand, the $\boldsymbol{R}_{\{0,1,\ldots,n-1\},\{0,1,\ldots,n-1\}}$ sub-matrix of $\boldsymbol{R}$ whose size is $n \times n$ is non-singular when $(\boldsymbol{\alpha}, \boldsymbol{A})$ is full rank [LIE 90]. Accordingly, the row vectors $\boldsymbol{\alpha}(-\boldsymbol{A})^0, \boldsymbol{\alpha}(-\boldsymbol{A})^{-1}, \ldots, \boldsymbol{\alpha}(-\boldsymbol{A})^{-n+1}$, as well as the column vectors $(-\boldsymbol{A})^0\mathbb{1}, (-\boldsymbol{A})^{-1}\mathbb{1}, \ldots, (-\boldsymbol{A})^{-n+1}\mathbb{1}$ are linearly independent and the rank of $\boldsymbol{R}$ is $n$. □

COROLLARY 6.13.– *A direct consequence of theorem 6.12 is that the determinant of any $(n+1) \times (n+1)$ submatrix of $\boldsymbol{R}$ is 0. That is, for any $i_0, \ldots, i_n, j_0, \ldots, j_n \in \mathbb{Z}$:*

$$\det\left(\boldsymbol{R}_{\{i_0,\ldots,i_n\},\{j_0,\ldots,j_n\}}\right) = 0. \qquad [6.22]$$

Corollary 6.13 has multiple important consequences. It can be used to estimate the rank of a PH distribution based on its reduced moments.

COROLLARY 6.14.– *If the reduced moments of PH($\boldsymbol{\alpha}, \boldsymbol{A}$) are $\ldots, r_{-2}, r_{-1}, r_0, r_1, r_2, \ldots$ and there exists $i_0, \ldots, i_{n-1}, j_0, \ldots, j_{n-1} \in \mathbb{Z}$ such that:*

$$\det\left(\boldsymbol{R}_{\{i_0,\ldots,i_{n-1}\},\{j_0,\ldots,j_{n-1}\}}\right) \neq 0, \qquad [6.23]$$

*then the rank of PH($\boldsymbol{\alpha}, \boldsymbol{A}$) is at least $n$.*

EXAMPLE 6.15.– *According to corollary 6.14, the rank of PH($\boldsymbol{\gamma}, \boldsymbol{G}$) from example 2.7 is at least one, because for $i \in \mathbb{Z}$, the $i$th reduced moments of PH($\boldsymbol{\gamma}, \boldsymbol{G}$) are:*

$$r_i = \boldsymbol{\gamma}(-\boldsymbol{G})^{-i}\mathbb{1} = \lambda^{-i}$$

*and with these reduced moments:*

$$\det\left(\boldsymbol{R}_{\{0\},\{0\}}\right) = \det(1) = 1.$$

*Corollary 6.14 does not provide any upper bound on the rank directly. In this example, the determinant of any $2 \times 2$ block of $\boldsymbol{R}$ can be computed in closed form, since for any $i, j \in \mathbb{Z}$:*

$$\det\left(\boldsymbol{R}_{\{i_0,i_1\},\{j_0,j_1\}}\right) = \det\begin{pmatrix} \lambda^{-i_0-j_0} & \lambda^{-i_1-j_0} \\ \lambda^{-i_1-j_0} & \lambda^{-i_1-j_1} \end{pmatrix} = 0,$$

*which indicates that none of the* $2 \times 2$ *blocks of* $\boldsymbol{R}$ *have non-zero determinant, which implies that the rank of PH(*$\gamma, \boldsymbol{G}$*) is one.*

EXAMPLE 6.16.– *This example demonstrates why corollary 6.14 provides only a lower bound on the rank. Consider PH(*$\gamma, \boldsymbol{G}$*) with:*

$$\gamma = \begin{pmatrix} \frac{1}{2} & \frac{1}{4} & \frac{1}{4} \end{pmatrix} \quad \text{and } \boldsymbol{G} = \begin{pmatrix} -1 & 1 & 0 \\ 0 & -2 & 1 \\ 1 & 0 & -3 \end{pmatrix}$$

*whose reduced moments are:*

$$r_{-4} = 37/4, r_{-3} = 11/4, r_{-2} = 1, r_{-1} = 3/4,$$
$$r_0 = 1, r_1 = 3/2, r_2 = 9/4, r_3 = 67/20, r_4 = 497/100.$$

*With these reduced moments, we have* $\det \begin{pmatrix} r_0 & r_1 \\ r_1 & r_2 \end{pmatrix} = 0$, *while* $\det \begin{pmatrix} r_{-1} & r_0 \\ r_0 & r_1 \end{pmatrix} = -\frac{1}{8}$ *and* $\det \begin{pmatrix} r_1 & r_2 \\ r_2 & r_3 \end{pmatrix} = -\frac{3}{80}$. *That is, the fact that the determinant of a* $2 \times 2$ *block of* $\boldsymbol{R}$ *is zero does not imply that the determinants of all* $2 \times 2$ *blocks of* $\boldsymbol{R}$ *are zero. In theory, it means that an exhaustive search is needed for* $2 \times 2$ *blocks of* $\boldsymbol{R}$ *with non-zero determinant to characterize the rank. In practice, however, it is enough to check the determinant of some* $2 \times 2$ *blocks of* $\boldsymbol{R}$ *and if all of the determinants are zero, then the rank is* 1 *in the vast majority of the cases.*

On the other hand, corollary 6.13 can be used to establish relations among the entries of the sequence $\ldots, r_{-2}, r_{-1}, r_0, r_1, r_2, \ldots$ and, consequently, among the moments of a full rank PH($n$) distribution and the derivatives of its CDF at zero. Specifically, for a full rank PH($n$) distribution and any $i \in \mathbb{Z}$, we have:

$$\det\left(\boldsymbol{R}_{\{0,1,2,\ldots,n\},\{i,i+1,i+2,\ldots,i+n\}}\right) = 0, \qquad [6.24]$$

which is an equation that the reduced moments satisfy.

The following example illustrates the relations among the reduced moments based on [6.24].

EXAMPLE 6.17.– *Based on* [6.24], *one can compute the* $r_i$ *reduced moments of a full rank PH(*$n$*) and ME(*$n$*) distribution for* $i = 2n, 2n+1, 2n+2, \ldots$ *if* $r_0, r_1, ..., r_{2n-1}$ *are known. Assuming* $r_0 = 1, r_1, ..., r_{2n-1}$ *are known:*

$$\det\left(\boldsymbol{R}_{\{0,1,\ldots,n\},\{0,1,\ldots,n\}}\right) = \det \begin{pmatrix} r_0 & r_1 & \cdots & r_n \\ r_1 & r_2 & \cdots & r_{n+1} \\ \vdots & \vdots & \iddots & \vdots \\ r_n & r_{n+1} & \cdots & r_{2n} \end{pmatrix} = 0$$

*is a linear equation for $r_{2n}$, from which $r_{2n}$ can be computed. Next, assuming that $r_1, ..., r_{2n-1}, r_{2n}$ are known:*

$$\det\left(\boldsymbol{R}_{\{0,1,...,n\},\{1,2,...,n+1\}}\right) = \det\begin{pmatrix} r_1 & r_2 & \cdots & r_{n+1} \\ r_2 & r_3 & \cdots & r_{n+2} \\ \vdots & \vdots & \iddots & \vdots \\ r_{n+1} & r_{n+2} & \cdots & r_{2n+1} \end{pmatrix} = 0$$

*is a linear equation for $r_{2n+1}$, and all subsequent reduced moments can be computed iteratively in a similar manner.*

*The $r_{-1}, r_{-2}, \ldots$ reduced moments of full rank PH(n) and ME(n) distributions can be computed from $r_0, r_1, ..., r_{2n-1}$ as follows. If $r_0 = 1, r_1, ..., r_{2n-1}$ are known:*

$$\det\left(\boldsymbol{R}_{\{0,1,...,n\},\{-1,0,...,n-1\}}\right) = \det\begin{pmatrix} r_{-1} & r_0 & \cdots & r_{n-1} \\ r_0 & r_1 & \cdots & r_n \\ \vdots & \vdots & \iddots & \vdots \\ r_{n-1} & r_n & \cdots & r_{2n-1} \end{pmatrix} = 0$$

*is a linear equation for $r_{-1}$, and subsequent negative reduced moments can be computed similarly.*

EXAMPLE 6.18.– *Different PH(3) distributions sharing five moments. Assuming that:*

$$r_0 = 1,\ r_{10} = 0.000295999,\ r_{20} = 6.9987 \cdot 10^{-13},\ r_{30} = 1.01758 \cdot 10^{-23},$$

$$r_{40} = 5.24414 \cdot 10^{-36},\ r_{50} = 2.23015 \cdot 10^{-49} \qquad [6.25]$$

*using [6.24] with $n = 3$ and $i = 0, 1, ..., 44$, we obtain 45 equations with 45 unknowns. These equations are not linear and they have different solutions that result in reduced moments that can be realized with a PH(3) distribution. For example, the reduced moments of:*

$$\boldsymbol{\alpha}_1 = (0.1\ 0.2\ 0.7),\ \boldsymbol{A}_1 = \begin{pmatrix} -1 & 1 & 0 \\ 0 & -2 & 2 \\ 1 & 0 & -3 \end{pmatrix}$$

*and:*

$$\boldsymbol{\alpha}_2 = (0.5861\ 0.3095\ 0.1044),\ \boldsymbol{A}_2 = \begin{pmatrix} -2.514 & 0 & 0.7969 \\ 2.514 & -2.514 & 0 \\ 0 & 0.9272 & -0.9272 \end{pmatrix}$$

*are according to [6.25]. Table 6.5 lists the first reduced moments of the two PH distributions. The differences vanish for large moments, but the numerical value of the* *relative differences, $(r_i^{(1)} - r_i^{(2)})/r_i^{(1)}$, in the last row of the table indicate that the $r_{10}$ reduced moments are identical while the $r_9$ and $r_{11}$ reduced moments are different. This periodicity of the moments remain valid for higher order moments, that is, every tenth moment of the two distributions remain identical. It is interesting to note that the PDF of the two distributions exhibit a similar behavior to that of the moments (see Figure 6.16). The two PDFs cross each others infinitely many times. This also shows that two PH distributions whose PDFs are identical at infinitely many points are not necessarily identical.*

| $r_i$ | 0. | 1. | 2. | 3. | 4. | 5. |
|---|---|---|---|---|---|---|
| $(\alpha_1, A_1)$ | 1. | 1.5 | 1.48 | 1.03 | 0.537 | 0.224 |
| $(\alpha_2, A_2)$ | 1. | 1.54 | 1.5 | 1.032 | 0.538 | 0.224 |
| $r_i$ | 6. | 7. | 8. | 9. | 10. | 11. |
| $(\alpha_1, A_1)$ | 0.0782 | 0.0233 | 0.0061 | 0.00141 | 0.000296 | 0.0000562 |
| $(\alpha_2, A_2)$ | 0.0782 | 0.0233 | 0.0061 | 0.00141 | 0.000296 | 0.0000562 |
| rel. diff. | $-1.1 \cdot 10^{-5}$ | $-1.52 \cdot 10^{-6}$ | $-1.78 \cdot 10^{-7}$ | $-1.51 \cdot 10^{-8}$ | $-3.66 \cdot 10^{-16}$ | $3.94 \cdot 10^{-10}$ |

**Table 6.5.** *The reduced moments of the two PH(3) distributions of example 6.18*

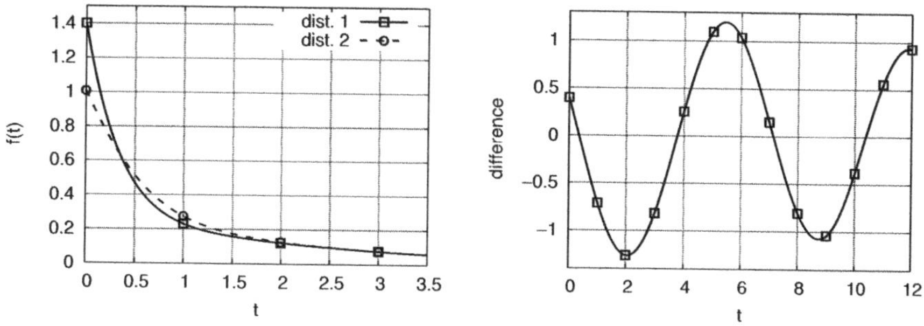


**Figure 6.16.** *Density function of the two PH(3) distributions of example 6.18, and their difference multiplied $e^{2.8t}$*

EXAMPLE 6.19.– *Fitting three moments out of five. We assume that we are given the first five reduced moments of a random variable:*

$$r_1 = 1,\ r_2 = 1.25,\ r_3 = 4.1\dot{6},\ r_4 = 104.1\dot{6},\ r_5 = 8333.\dot{3}, \qquad [6.26]$$

*and our aim is to construct a PH distribution with two phases to approximate it. A PH distribution with two phases is determined by its first three reduced moments. However, we are not limited to using $r_1, r_2$ and $r_3$. For example, we can choose to fit reduced moments $r_1, r_3$ and $r_5$. In order to obtain PH distribution with two phases and reduced moments $r_1, r_3$ and $r_5$, we do the following. We assume that $r_1, r_3$ and $r_5$ are reduced moments of a PH distribution with two phases and we compute the first three reduced* *moments of this distribution based on* [6.24]. *Then the fitting is performed based on these first three reduced moments according to section 6.5.*

*Table 6.6 reports fittings that are different in the choice of the utilized three reduced moments. In all the cases, the mean is maintained and hence we have to choose two out of the four remaining reduced moments. The reduced moments that are set according to* [6.26] *are indicated in bold.*

| $r_1$ | $r_2$ | $r_3$ | $r_4$ | $r_5$ |
|---|---|---|---|---|
| **1** | **1.25** | $\mathbf{4.1\dot{6}}$ | 35.5903 | 371.817 |
| **1** | **1.25** | 6.31469 | $\mathbf{104.1\dot{6}}$ | 1990.25 |
| **1** | **1.25** | 9.3729 | 265.489 | $\mathbf{8333.\dot{3}}$ |
| **1** | 1.09182 | $\mathbf{4.1\dot{6}}$ | $\mathbf{104.1\dot{6}}$ | 3353.48 |
| **1** | 1.05945 | $\mathbf{4.1\dot{6}}$ | 163.533 | $\mathbf{8333.\dot{3}}$ |
| **1** | 1.01543 | 2.27679 | $\mathbf{104.1\dot{6}}$ | $\mathbf{8333.\dot{3}}$ |

**Table 6.6.** *Matching different three reduced moments of example 6.19 with PH(2) distributions*

*In Figure 6.17, the PDF of the different fitted distributions are depicted. The legend indicates the reduced moments that are fitted. The figure depicts the fitting with three phases as well which matches all five reduced moments. The logarithmic plot suggests that a better fitting of the lower moments results in a better body fitting, while the better fitting of the higher moments results in better tail fitting.*

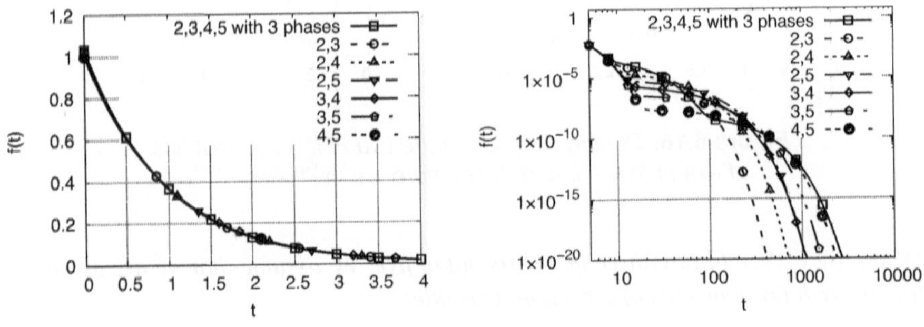


**Figure 6.17.** *The PDF of different matching PH(2) distributions of example 6.19. The body of the distribution is on the left with the linear $y$-axis and its tail is on the right with the logarithmic $y$-axis*

EXAMPLE 6.20.– *PH matching based on the derivatives of the CDF at zero. PH($n$) distributions are uniquely defined by their first $2n - 1$ reduced moments and a method is available to compute an $(\alpha, \mathbf{A})$ vector–matrix representation based on them (see* *section 6.5). This method can also be extended to "derivative matching" utilizing the relation of moments and derivatives at zero.*

*Given the first three derivatives of the CDF of a PH(2) distribution at* $t = 0$ *as* $r_{-1} = 4,\ r_{-2} = 17,\ r_{-3} = 76$, *we can use:*

$$\det\begin{pmatrix} r_{-3} & r_{-2} & r_{-1} \\ r_{-2} & r_{-1} & r_0 \\ r_{-1} & r_0 & r_1 \end{pmatrix} = 0, \quad \det\begin{pmatrix} r_{-2} & r_{-1} & r_0 \\ r_{-1} & r_0 & r_1 \\ r_0 & r_1 & r_2 \end{pmatrix} = 0,$$

$$\det\begin{pmatrix} r_{-1} & r_0 & r_1 \\ r_0 & r_1 & r_2 \\ r_1 & r_2 & r_3 \end{pmatrix} = 0$$

*with* $r_0 = 1$ *to obtain* $r_1, r_2, r_3$. *Solving the equations from left to right, the number of unknowns is always 1, and the equations are linear. The solution is:*

$$r_1 = \frac{4}{15}, \quad r_2 = \frac{17}{225}, \quad r_3 = \frac{76}{3375}.$$

*From* $r_1, r_2, r_3$, *we obtain:*

$$\boldsymbol{\alpha} = \begin{pmatrix} 0.2 & 0.8 \end{pmatrix}, \quad \boldsymbol{A} = \begin{pmatrix} -3 & 3 \\ 0 & -5 \end{pmatrix}.$$

EXAMPLE 6.21.– *PH matching based on derivatives and moments. Given* $r_{-3}, r_{-2}, \dots, r_2$, *a ME(3) distribution can be constructed by the following steps:*

– *compute* $r_0, r_1, \dots, r_5$ *using the determinants of the* $4 \times 4$ *submatrices of the Hankel matrix* $\boldsymbol{R}$;

– *calculate vector* $\boldsymbol{\alpha}$ *and matrix* $\boldsymbol{A}$ *based on* $r_0, \dots, r_{2n-1}$ *such that* $r_i = \boldsymbol{\alpha}(-\boldsymbol{A})^{-i}\mathbb{1}$ *for* $i = 0, 1, \dots, 2n-1$ *using the procedure described in section 6.5;*

– *check the non-negativity of* $f(t) = \boldsymbol{\alpha} e^{\boldsymbol{A}t}(-\boldsymbol{A})\mathbb{1}$.

*For example, starting from:*

$$r_{-3} = \frac{24}{5}, \quad r_{-2} = 0, \quad r_{-1} = \frac{6}{5}, \quad r_0 = 1, \quad r_1 = \frac{3}{5}, \quad r_2 = \frac{7}{20},$$

*we obtain* $r_3 = 5/24$ *using* $\det\left(\boldsymbol{R}_{\{-3,-2,-1,0\},\{0,1,2,3\}}\right) = 0$, $r_4 = 1/8$ *using* $\det\left(\boldsymbol{R}_{\{-2,-1,0,1\},\{0,1,2,3\}}\right) = 0$ *and* $r_5 = 3/40$ *using* $\det\left(\boldsymbol{R}_{\{-1,0,1,2\},\{0,1,2,3\}}\right) = 0$. *Using* $r_0, \dots, r_5$, *we compute* $\boldsymbol{\alpha}$ *and* $\boldsymbol{A}$ *which are:*

$$\boldsymbol{\alpha} = \begin{pmatrix} \frac{1}{3} & \frac{1}{3} & \frac{1}{3} \end{pmatrix}, \quad \boldsymbol{A} = \begin{pmatrix} -\frac{152}{9} & \frac{7702}{225} & -\frac{3962}{225} \\ -\frac{395}{81} & \frac{631}{81} & -\frac{341}{81} \\ \frac{305}{81} & -\frac{721}{81} & \frac{251}{81} \end{pmatrix}.$$

*According to the test provided in section 5.7, the above $\boldsymbol{\alpha}$ and $\boldsymbol{A}$ provide an appropriate density function.*

## 6.5. Matching $2n-1$ moments with size $n$ matrix representations

In this section, we present a procedure to compute an $(\boldsymbol{\alpha}, \boldsymbol{A})$ matrix representation of size $n$ based on the first $2n-1$ moments, $m_1, \ldots, m_{2n-1}$, such that:

$$m_k = k!\boldsymbol{\alpha}(-\boldsymbol{A})^{-k}\mathbb{1},$$

for $k \in \{1, \ldots, 2n-1\}$.

### 6.5.1. *Padé approximation for continuous distributions*

A full rank PH($n$) distribution is determined by $2n-1$ moments. A representation composed of $2n-1$ parameters is given by the coefficients of the Laplace transform of the distribution in the continuous case and by the coefficients of the $z$-transform of the distribution in the discrete case. These parameters can be found by applying the so called Padé approximation. We first discuss the continuous case.

According to [1.10], the following relation holds between the Laplace transform of the PDF of any distribution and its moments:

$$f^*(s) = \sum_{i=0}^{\infty}(-1)^i \frac{m_i}{i!} s^i. \qquad [6.27]$$

According to [2.9], $f^*(s)$ is a rational function of $s$ for a PH($n$) distribution. The Padé approximation intends to approximate the parameters of the rational function such that:

$$\frac{\varphi_{n-1}s^{n-1} + \varphi_{n-2}s^{n-2} + \ldots + \varphi_1 s + \varphi_0}{s^n + \varrho_{n-1}s^{n-1} + \varrho_{n-2}s^{n-2} + \ldots + \varrho_1 s + \varphi_0} \approx \sum_{i=0}^{2n-1}(-1)^i \frac{m_i}{i!} s^i, \qquad [6.28]$$

where we used $\varrho_0 = \varphi_0$.

More precisely, based on the moments $m_1, \ldots, m_{2n-1}$, the Padé approximation sets the parameters of the rational function such that the first $2n$ coefficient of the polynomial:

$$\varphi_0 + \varphi_1 s + \varphi_2 s^2 + \cdots + \varphi_{n-1}s^{n-1} + \varphi_n s^n + \ldots + \varphi_{2n-1}s^{2n-1}, \qquad [6.29]$$

where $\varphi_n = \varphi_{n+1} = \ldots = \varphi_{2n-1} = 0$ are equal to the first $2n$ coefficient of the polynomial:

$$\left(\varphi_0 + \varrho_1 s + \varrho_2 s^2 + \cdots + \varrho_{n-1}s^{n-1} + s^n\right) \sum_{i=0}^{2n-1}(-1)^i \frac{m_i}{i!} s^i. \qquad [6.30]$$

Due to $m_0 = 1$, the equality holds for the constant term of the polynomial $\varphi_0 = \varrho_0$. The $2n-1$ unknown coefficients can be obtained by solving a system of $2n-1$ linear equations given by:

$$\varphi_i = \sum_{j=0}^{i} (-1)^j \frac{m_j}{j!} \varrho_{i-j} \qquad \text{for } i = 1, ..., 2n-1, \qquad [6.31]$$

with $\varrho_0 = \varphi_0$, $\varrho_n = 1$ and $\varrho_i = 0$ for $i > n-1$.

Having obtained the coefficients of the Laplace transform, other representations can be calculated based on the relations provided in section 5.3.

EXAMPLE 6.22.– *As an example, consider $n = 3$ which results in the following set of linear equations:*

$$\begin{aligned} \varphi_1 &= \varrho_1 - \varphi_0 m_1 \\ \varphi_2 &= \varrho_2 - \varrho_1 m_1 + \tfrac{1}{2}\varphi_0 m_2 \\ 0 &= -\varrho_2 m_1 + \tfrac{1}{2}\varrho_1 m_2 - \tfrac{1}{6}\varphi_0 m_3 \\ 0 &= -m_1 + \tfrac{1}{2}\varrho_2 m_2 - \tfrac{1}{6}\varrho_1 m_3 + \tfrac{1}{24}\varphi_0 m_4 \\ 0 &= \tfrac{m_2}{2} - \tfrac{1}{6}\varrho_2 m_3 + \tfrac{1}{24}\varrho_1 m_4 - \tfrac{1}{120}\varphi_0 m_5 \end{aligned} \qquad [6.32]$$

It is not guaranteed that the obtained Laplace transform corresponds to an appropriate distribution. Moreover, small changes in the moments can cause large changes in the result. This is illustrated by the next example.

EXAMPLE 6.23.– *Consider:*

$$\boldsymbol{\alpha} = \left( \frac{1}{10} \ \frac{2}{5} \ \frac{1}{2} \right), \quad \boldsymbol{A} = \begin{pmatrix} -10 & 5 & 2 \\ 2 & -8 & 3 \\ 1 & 0 & -2 \end{pmatrix}$$

*for which the first five moments are:*

$$\frac{354}{545}, \frac{54012}{59405}, \frac{12577854}{6475145}, \frac{784074864}{141158161}, \frac{305740641264}{15386239549}. \qquad [6.33]$$

*Applying Padé approximation with the above moments yields Laplace transform:*

$$\frac{2s^2 + \frac{166}{5}s + 109}{s^3 + 20s^2 + 104s + 109} \qquad [6.34]$$

*from which by inverse Laplace transform we obtain:*

$$0.0305201e^{-12.2183t} + 0.736932e^{-6.38433t} + 1.23255e^{-1.39733t},$$

*which is a valid PDF. By repeating the same procedure with moments given as:*

$$\frac{354}{545}, \frac{54012}{59405}, \frac{12577854}{6475145}, \frac{784074864}{141158161} + \frac{1}{1000}, \frac{305740641264}{15386239549}, \qquad [6.35]$$

*we obtain:*

$$0.759824e^{-6.43651t} + 1.23229e^{-1.39724t} - 2.15183 \cdot 10^{-6}e^{0.653635t}$$

*which, due to the presence of the positive exponent, is not a valid PDF.*

As pointed out already and illustrated in example 6.23, the above moment matching approach based on Padé approximation does not guarantee a proper distribution. Given $2n - 1$ moments, $m_1, ..., m_{2n-1}$, four types of outcome are possible:

Case 1) The Laplace transform of a valid distribution corresponding to a ME($n$) distribution, which can be realized by a PH($n$) distribution; in this case, a vector–matrix ME representation can be found based on section 5.3.4 and a vector–matrix PH representation based on section 5.3.5.

Case 2) The Laplace transform of a valid distribution corresponding to a ME($n$) distribution which cannot be realized by a PH($n$) distribution; in this case, a PH($k$) representation with $k > n$ can be found based on section 5.7.

Case 3) The Laplace transform of a valid distribution corresponding to a ME($k$) distribution with $k < n$; in this case, the ME($k$) distribution realizes only $2k - 1$ of the given $2n - 1$ moments and the $2n - 1$ moments cannot be realized with a ME($n$) distribution.

Case 4) The Laplace transform does not provide a valid distribution; this implies also in this case that the given $2n - 1$ moments cannot be realized by a ME($n$) distribution.

The following example illustrates each case.

EXAMPLE 6.24.–

*Case 1) The moments in* [6.33] *belong to the first case. The Laplace transform is determined by Padé approximation as given in* [6.34]. *A vector–matrix ME representation is (section 5.3.2):*

$$\begin{pmatrix} \frac{171}{5} & -\frac{156}{5} & -2 \end{pmatrix}, \begin{pmatrix} 109 & -109 & 0 \\ 109 & -110 & 1 \\ 214 & -194 & -19 \end{pmatrix}$$

*and a PH representation is given as (section 5.3.5):*

$$\begin{pmatrix} 0.163688 & 0.226094 & 0.610218 \end{pmatrix}, \begin{pmatrix} -12.2183 & 0 & 0 \\ 6.38433 & -6.38433 & 0 \\ 0 & 1.39733 & -1.39733 \end{pmatrix}.$$

*Case 2) For the second case, we use the first five moments:*

$1, 1.6, 3.52, 12.32, 221.76,$

*which can be realized by a ME(3) distribution whose one possible vector–matrix realization is:*

$$\begin{pmatrix} \frac{1}{3} & \frac{1}{3} & \frac{1}{3} \end{pmatrix}, \begin{pmatrix} -2.96158 & 2.33486 & 1.59791 \\ -0.366378 & -0.337173 & -0.271666 \\ -0.300289 & -0.329494 & -0.395001 \end{pmatrix}$$

*and cannot be realized by a PH(3) distribution. It is possible however to construct a PH(18) distribution based on the numerical procedure in [REI 14] (section 5.9) with vector–matrix description:*

$$(1.936 \cdot 10^{-6} \;\; 8.502 \cdot 10^{-8} \;\; 8.875 \cdot 10^{-8} \;\; 9.265 \cdot 10^{-8} \;\; 8.605 \cdot 10^{-8} \;\; 8.941 \cdot 10^{-8}$$

$$9.298 \cdot 10^{-8} \;\; 9.774 \cdot 10^{-8} \;\; 1.161 \cdot 10^{-7} \;\; 2.930 \cdot 10^{-7} \;\; 2.213 \cdot 10^{-6} \;\; 0.0000222$$

$$0.0002183 \;\; 0.002013 \;\; 0.01702 \;\; 0.1253 \;\; 0.6834 \;\; 0.1719),$$

$$\begin{pmatrix} -\lambda_1 & \lambda_1 & 0 & 0 & 0 & 0 & 0 & \dots & 0 & 0 \\ 0 & -\lambda_2 & \lambda_2 & 0 & 0 & 0 & 0 & \dots & 0 & 0 \\ 0 & 0 & -\lambda_2 & \lambda_2 & 0 & 0 & 0 & \dots & 0 & 0 \\ 0 & \lambda_3 & 0 & -\lambda_2 & \lambda_2 - \lambda_3 & 0 & 0 & \dots & 0 & 0 \\ 0 & 0 & 0 & 0 & -\lambda_4 & \lambda_4 & 0 & \dots & 0 & 0 \\ & & & & \vdots & & & & & \\ 0 & 0 & 0 & \dots & 0 & 0 & 0 & -\lambda_4 & \lambda_4 & 0 \\ 0 & 0 & 0 & \dots & 0 & 0 & 0 & 0 & -\lambda_4 & \lambda_4 \\ 0 & 0 & 0 & \dots & 0 & 0 & 0 & 0 & 0 & -\lambda_4 \end{pmatrix},$$

*with:*

$$\lambda_1 = 0.0749856, \lambda_2 = 1.78286, \lambda_3 = 0.000046943, \lambda_4 = 1.99458.$$

*That is, the real eigenvalue is $\lambda_1$, the complex eigenvalue pair is represented by a FEB of size 3 with parameters $\lambda_2$ and $\lambda_3$ and an Erlang block of size 14 with parameters $\lambda_4$ is needed for a non-negative initial vector.*

*Case 3) For the third case, consider the first five moments:*

$$1, 2, 6, x, y \qquad [6.36]$$

*for which the Padé approximation, independently of $x$ and $y$, results in the Laplace transform:*

$$\frac{1}{s+1},$$

*which is the Laplace transform of the exponential distribution, that is a PH(1) distribution. This happens because for $m_1 = 1, m_2 = 2, m_3 = 6$ the value of $m_4$*

*and $m_5$ does not have any impact in* [6.32]. *If $x \neq 24$ and $y \neq 120$, then the Padé approximation provides a distribution with different moments from those given in input. This result is correct in the sense that the moments in* [6.36] *cannot be realized by a ME(3) distribution if either $x \neq 24$ or $y \neq 120$. Adding two more moments and setting $x$ and $y$, for example, as in:*

$$1, 2, 6, 25, 125, 6!, 7!,$$

*the Padé approximation provides a size 4 result, but this does not correspond to a proper distribution.*

*Case 4) The moments in* [6.35], *as already shown in example 6.23, belong to the fourth case.*

### 6.5.2. Padé approximation for discrete distributions

If $X$ is a DPH($n$) distributed random variable, then the z-transform of its PMF is (see [3.5]):

$$T(z) = \frac{a_1 z + a_2 z^2 + \cdots + a_n z^n}{b_0 + b_1 z + \cdots + b_{n-1} z^{n-1} + z^n}, \qquad [6.37]$$

where:

$$\sum_{i=1}^{n} a_i = 1 + \sum_{i=0}^{n-1} b_i$$

is the normalizing condition of the PMF to ensure $T(1) = 1$. According to [1.13], the factorial moments are determined by the derivatives of $T(z)$ around $z = 1$. To apply a similar approach as in the continuous time case, where the moments are determined by the derivatives of the LT around $s = 0$, we introduce a change of variable $\hat{z} = z - 1$, which results in:

$$\hat{T}(\hat{z}) = \frac{\hat{a}_0 + \hat{a}_1 \hat{z} + \cdots + \hat{a}_n \hat{z}^n}{\hat{b}_0 + \hat{b}_1 \hat{z} + \cdots + \hat{b}_{n-1} \hat{z}^{n-1} + \hat{z}^n}, \qquad [6.38]$$

where the additional constraint:

$$\sum_{i=0}^{n} (-1)^i \hat{a}_i = 0 \qquad [6.39]$$

comes from the fact that there is no constant term in the numerator of $T(z)$. Additionally, $T(z)|_{z=1} = \hat{T}(\hat{z})|_{\hat{z}=0} = 1$ implies $\hat{a}_0 = \hat{b}_0$. That is, there are $2n + 1$ coefficients in [6.38], and we have two constraints for them.

With this change of variable, we can apply the Padé approximation at $\hat{z} = 0$ to perform the factorial moment matching based on:

$$\hat{T}(\hat{z}) = \sum_{i=0}^{\infty} \frac{f_i}{i!} \hat{z}^i \approx \sum_{i=0}^{2n-1} \frac{f_i}{i!} \hat{z}^i.$$

The coefficients in [6.38] can be determined by a system of linear equations that is obtained by setting the first $2n$ coefficients of:

$$\hat{a}_0 + \hat{a}_1 \hat{z} + \cdots + \hat{a}_n \hat{z}^n$$

equal to those of:

$$\left(\hat{a}_0 + \hat{b}_1 \hat{z} + \cdots + \hat{b}_{n-1} \hat{z}^{n-1} + \hat{z}^n\right) \sum_{i=0}^{2n-1} \frac{f_i}{i!} \hat{z}^i$$

together with considering [6.39].

There is no guarantee that the $z$ transform expression produced by Padé approximation corresponds to a proper PMF for an arbitrary set of factorial moments. There are four types of outcome, which are the same as in the continuous time case and are discussed above. We illustrate the discrete time case by a simple example for which the outcome is a DPH distribution.

EXAMPLE 6.25.– *Consider the first five moments:*

$$4.425, 38.3, 512, 9200, 207340$$

*from which the factorial moments* $f_1, ..., f_5$ *are:*

$$4.425, 33.875, 405.95, 6522.75, 131451.2.$$

*The parameters of* $\hat{T}(\hat{z})$ *are the solution of the linear system:*

$$\hat{a}_0 = \hat{b}_0,$$

$$\hat{a}_1 = \hat{b}_1 + \hat{b}_0 f_1,$$

$$\hat{a}_2 = \hat{b}_2 + \hat{b}_1 f_1 + \frac{1}{2} \hat{b}_0 f_2,$$

$$\hat{a}_3 = \hat{b}_2 f_1 + \frac{1}{2} \hat{b}_1 f_2 + \frac{1}{6} \hat{b}_0 f_3 + 1,$$

$$0 = \frac{1}{2} \hat{b}_2 f_2 + \frac{1}{6} \hat{b}_1 f_3 + \frac{1}{24} \hat{b}_0 f_4 + f_1,$$

$$0 = \frac{1}{6} \hat{b}_2 f_3 + \frac{1}{24} \hat{b}_1 f_4 + \frac{1}{120} \hat{b}_0 f_5 + \frac{f_2}{2},$$

$$\hat{a}_0 - \hat{a}_1 + \hat{a}_2 - \hat{a}_3 = 0.$$

*By using the solution of the above system in* [6.38] *and inverting the change of variable, we obtain the z transform:*

$$T(z) = \frac{-0.63261z^3 + 2.00092z^2 - 1.45015z}{z^3 - 6.00657z^2 + 10.0291z - 5.10441},$$

*from which a vector–matrix representation can be obtained (using the discrete counterpart of sections 5.3.2 and 5.3.3):*

$$\boldsymbol{\beta} = \begin{pmatrix}0.392 & 0.6078 & 0.0001314\end{pmatrix}, \quad \boldsymbol{B} = \begin{pmatrix} 0.2753 & 0 & 0 \\ 0.1997 & 0.8003 & 0 \\ 0 & 0.1108 & 0.8892 \end{pmatrix}.$$

A different approach to provide an ME($n$) distribution given $2n-1$ moments was proposed in [LIE 90]. It is based on the partial realization algorithm [GRA 83] and provides a vector–matrix representation directly. Its discrete time counterpart, which is easy to derive from the original algorithm, provides MG distributions.

## 6.6. Matching any three moments with minimal acyclic phase type distributions

Based on Bobbio et al. [BOB 05], this chapter presents an explicit method to compose APH distributions of minimal size with any given first three moments. One of the main difficulties here is that the PH family is limited with respect to its moments and the moment bounds in general are hard to obtain. This problem however is somewhat easier for the APH distributions and there exist bounds for the first three moments of APH($n$) distributions. The characterization of these bounds are based on a basic transformation, which extends the APH($n-1$) class with an additional phase in order to describe the APH($n$) class. We refer to the class of size $n$ APH distributions with possible mass at 0 as APHM($n$).

The Erlang($n, \mu$) distribution plays a special role in this chapter. Hence, we recall that its first three (non-normalized) moments are:

$$m_1 = \frac{n}{\mu}, \quad m_2 = \frac{n(n+1)}{\mu^2}, \quad m_3 = \frac{n(n+1)(n+2)}{\mu^3}$$

and its normalized moments (defined in [1.5]) are:

$$n_2 = \frac{n+1}{n}, \quad n_3 = \frac{n+2}{n},$$

which are independent of $\mu$.

### 6.6.1. *Moment bounds*

Before presenting the moment bounds of the APH($n$) and APHM($n$) classes, we present some important properties of APH distributions.

COROLLARY 6.26.– *Any APH($n$) distribution can be realized as the composition of an APH($n-1$) distribution with representation $(\boldsymbol{\alpha}, \boldsymbol{A})$ and an additional phase with transition rate $\lambda$ according to Figure 6.18 (gray circles represent the absorbing states). The initial distribution of the composed PH distribution is defined by the initial distribution of the APH($n-1$) distribution $\boldsymbol{\alpha}$ and an additional parameter $0 \leq p \leq 1$.*

*Proof.* The corollary is a consequence of the result of Cumani [CUM 82], according to which any APH($n$) distribution can be transformed into the canonical form depicted in Figure 5.5. □

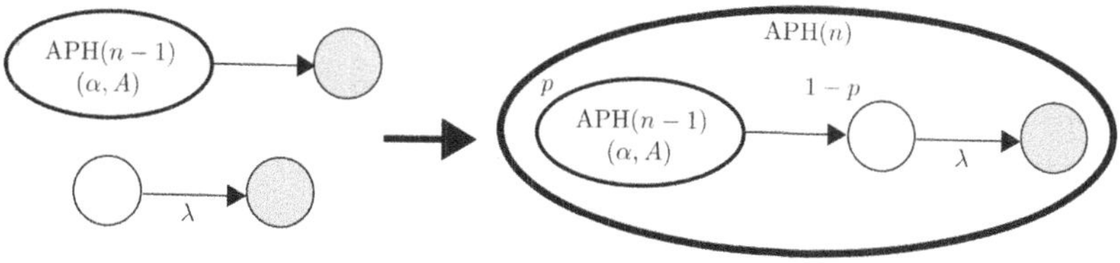


**Figure 6.18.** *Realization of an APH($n$) distribution by the composition of an APH($n-1$) distribution and an additional phase*

COROLLARY 6.27.– *Let $m_1$, $n_2$ and $n_3$ be the first moment, the second and third normalized moments of an APH($n-1$) distribution, respectively. The extension of this APH($n-1$) distribution with an additional phase with transition rate $\lambda$ and parameter $0 \leq p \leq 1$ according to Figure 6.18 results in an APH($n$) distribution with the normalized moments:*

$$\hat{n}_2 = \frac{2(1+pa) + a^2 p n_2}{(1+pa)^2}, \qquad \hat{n}_3 = \frac{6(1+pa) + a^2 p n_2 (n_3 a + 3)}{2(1+pa)^2 + a^2 p n_2 (1+pa)}, \qquad [6.40]$$

*where $a = \lambda m_1$.*

*Proof.* Let $m_1$, $m_2$ and $m_3$ denote the first three moments of the APH($n-1$) distribution. The first three moments of the APH($n$) distribution resulting from the composition can be written as:

$$\hat{m}_1 = p\left(m_1 + \frac{1}{\lambda}\right) + (1-p)\frac{1}{\lambda}, \quad \hat{m}_2 = p\left(m_2 + \frac{2m_1}{\lambda} + \frac{2}{\lambda^2}\right) + (1-p)\frac{2}{\lambda^2},$$

$$\hat{m}_3 = p\left(m_3 + \frac{3m_2}{\lambda} + \frac{6m_1}{\lambda^2} + \frac{6}{\lambda^3}\right) + (1-p)\frac{6}{\lambda^3}.$$

Using the definition of the normalized moments in [1.5] and $a = \lambda m_1$ results in the corollary. □

We will use the basic transformation provided in corollary 6.27 to connect the bounds of the APH($n-1$) class to the bounds of the APH($n-1$) class. Corollary 6.26 shows that this transformation is sufficient to describe the whole APH($n$) class. The second and third normalized moments resulting from a given transformation will be denoted as $\{\hat{n}_2, \hat{n}_3\} = \mathcal{T}(\{n_2, n_3\}, a, p)$.

The main results about the moment bounds of the APH($n$) and APHM($n$) class are as follows.

THEOREM 6.28.– *$n_2$ and $n_3$ can be the second and third normalized moments of an APH($n$) distribution if they satisfy the following constraints: $n_2 \geq \frac{n+1}{n}$ and:*

$$n_3 \text{ lower bound: } \begin{cases} l_n \leq n_3 & \text{if } \dfrac{n+1}{n} \leq n_2 \leq \dfrac{n+4}{n+1}, & a) \\ \dfrac{n+1}{n} n_2 < n_3 & \text{if } \dfrac{n+4}{n+1} \leq n_2, & b) \end{cases}$$

$$n_3 \text{ upper bound: } \begin{cases} n_3 \leq u_n & \text{if } \dfrac{n+1}{n} \leq n_2 \leq \dfrac{n}{n-1}, & c) \\ n_3 < \infty & \text{if } \dfrac{n}{n-1} < n_2, & d) \end{cases} \qquad [6.41]$$

*where $l_n$ and $u_n$ are:*

$$l_n = \frac{(3+a_n)(n-1)+2a_n}{(n-1)(1+a_n p_n)} - \frac{2a_n(n+1)}{2(n-1)+a_n p_n(n a_n + 2n - 2)}, \qquad [6.42]$$

$$u_n = \frac{1}{n^2 n_2}\left(2(n-2)(n n_2 - n - 1)\sqrt{1+\frac{n(n_2-2)}{n-1}} + (n+2)(3nn_2 - 2n - 2)\right) \qquad [6.43]$$

*and:*

$$p_n = \frac{(n+1)(n_2-2)}{3n_2(n-1)}\left(\frac{-2\sqrt{n+1}}{\sqrt{4(n+1)-3nn_2}} - 1\right),$$

$$a_n = \frac{n_2 - 2}{p_n(1-n_2) + \sqrt{p_n^2 + \dfrac{p_n n(n_2-2)}{n-1}}}.$$

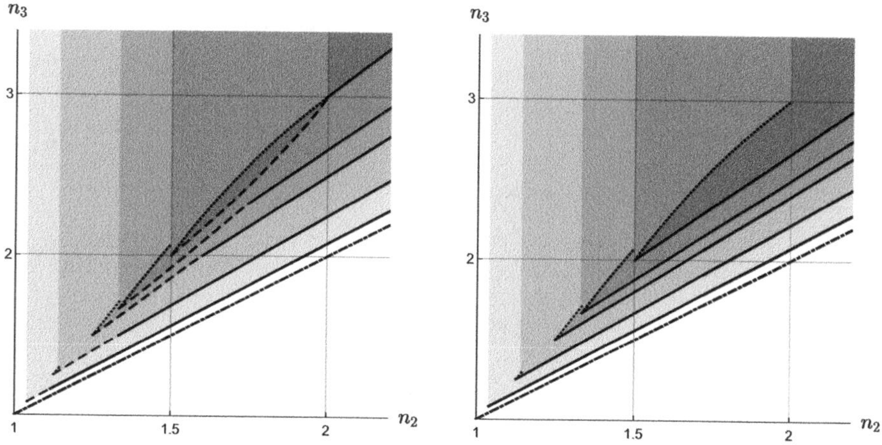


**Figure 6.19.** *Bounds of normalized moments of the APH($n$) class (left) and the APHM($n$) class (right) for* $n = 2, 3, 4, 8, 25$

The proof of theorem 6.28 is provided in section 6.6.3.

The bounds of the APH($n$) class are illustrated in Figure 6.19. The darkest shaded area contains those normalized moment pairs that can be realized with two phases. Points in the second darkest shaded area can be realized with three phases. The dashed segment of the lower bound corresponds to case (a) of [6.41] and it is not linear. The linear segment drawn with solid line corresponds to case (b). The upper bound (case (c)) is represented with a dotted line. The dashed dotted line corresponds to $n_3 = n_2$, which is the lower limit of non-negative random variables.

The structure of the bounds are presented in Figure 6.20. The figure indicates the $n_3 = 2n_2 - 1$ line as well because it plays a special role. The second and third normalized moments of Erlang($n$) distributions lie on this line. We refer to this line as the Erlang-line. The set of normalized moment pairs $\{n_2, n_3\}$, which satisfy the inequalities of theorem 6.28 for a given $n$, will be denoted by $\mathcal{APH}_3(n)$.

We will use the following disjoint subsets of $\mathcal{APH}_3(n) \setminus \mathcal{APH}_3(n-1)$ (see Figure 6.21):

– Subset 1: $\frac{n+1}{n} \leq n_2 < \frac{n}{n-1}$ and $l_n \leq n_3 \leq u_n$,

– Subset 2: $\frac{n}{n-1} \leq n_2 < \frac{n+4}{n+1}$ and $l_n \leq n_3 \leq l_{n-1}$,

– Subset 3: $\frac{n}{n-1} \leq n_2 < \frac{n-1}{n-2}$ and $u_{n-1} \leq n_3$,

– Subset 4: $\frac{n+4}{n+1} \leq n_2 < \frac{n+3}{n}$ and $\frac{n+1}{n} n_2 < n_3 \leq l_{n-1}$,

– Subset 5: $\frac{n+3}{n} \leq n_2$ and $\frac{n+1}{n} n_2 < n_3 \leq \frac{n}{n-1} n_2$.

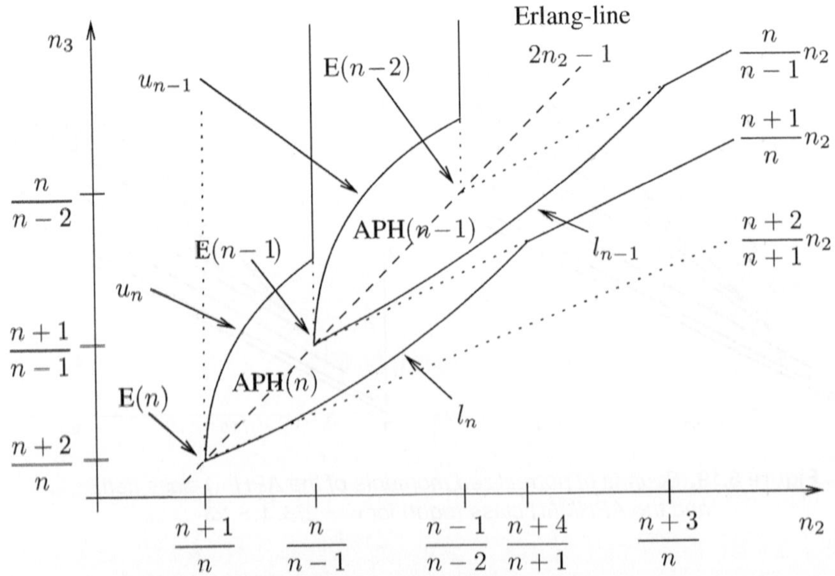


**Figure 6.20.** *Structure of the APH($n$) moment bounds for $n \geq 3$, where E($n$) stands for Erlang($n$)*

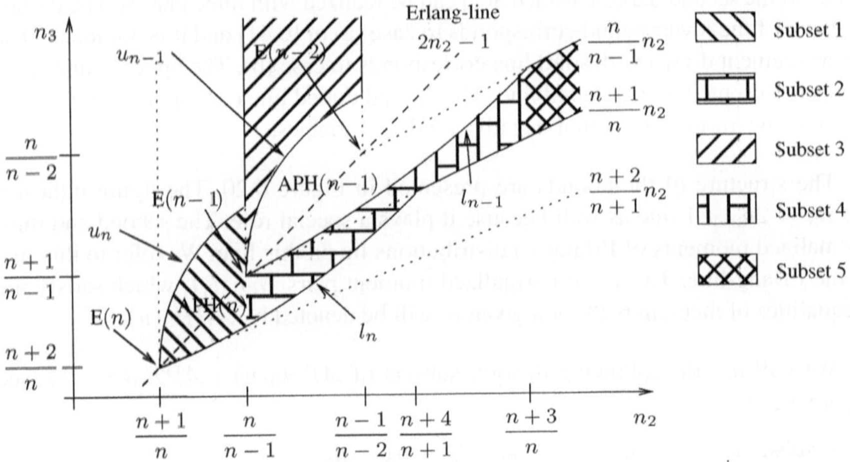


**Figure 6.21.** *The disjoint subsets of $\mathcal{APH}_3(n) \setminus \mathcal{APH}_3(n-1)$ used in the proof of theorem 6.28, where E($n$) stands for Erlang($n$)*

THEOREM 6.29.– *$n_2$ and $n_3$ can be the second and third normalized moments of an APHM($n$) distribution if they satisfy the following constraints: $n_2 \geq \frac{n+1}{n}$ and:*

$$
\begin{aligned}
&n_3 \text{ lower bound: } && \frac{n+2}{n+1}\, n_2 \leq n_3 \\
&n_3 \text{ upper bound: } && \begin{cases} n_3 \leq u_n & \text{if } \dfrac{n+1}{n} \leq n_2 \leq \dfrac{n}{n-1}, \\ n_3 < \infty & \text{if } \dfrac{n}{n-1} < n_2 \end{cases}
\end{aligned}
\qquad [6.44]
$$

*where $u_n$ is the same as in theorem 6.28.*

The proof of theorem 6.29 is provided in section 6.6.4.

As for the case of the APH($n$) class, we illustrate both numerical values (Figure 6.19) and the structure (Figure 6.22) of the moment bounds of the APHM($n$) class. The set of normalized moment pairs $\{n_2, n_3\}$, which satisfy the inequalities of theorem 6.29, is denoted by $\mathcal{APHM}_3(n)$.

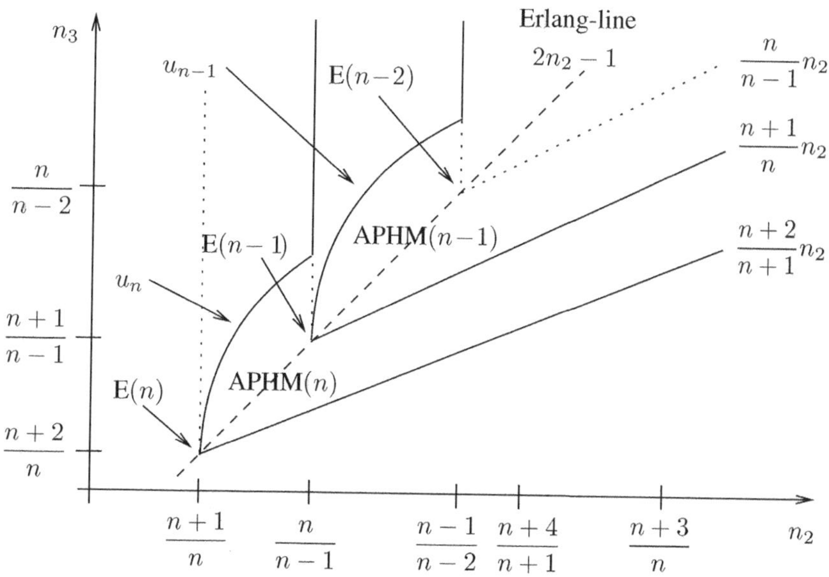


**Figure 6.22.** *Structure of the APHM($n$) moment bounds for $n \geq 3$, where E($n$) stands for Erlang($n$)*

### 6.6.2. *Explicit moment matching with minimal number of parameters*

In this section, we present moment matching procedures that result in minimal size APH($n$) and APHM($n$) distributions with minimal number of parameters (two and three, respectively). That is, the proposed APH($n$) representation is characterized by two parameters for all $n > 1$. The provided explicit methods are applicable over the entire $\mathcal{APH}_3(n) \setminus \mathcal{APH}_3(n-1)$ and the $\mathcal{APHM}_3(n) \setminus \mathcal{APHM}_3(n-1)$ sets.

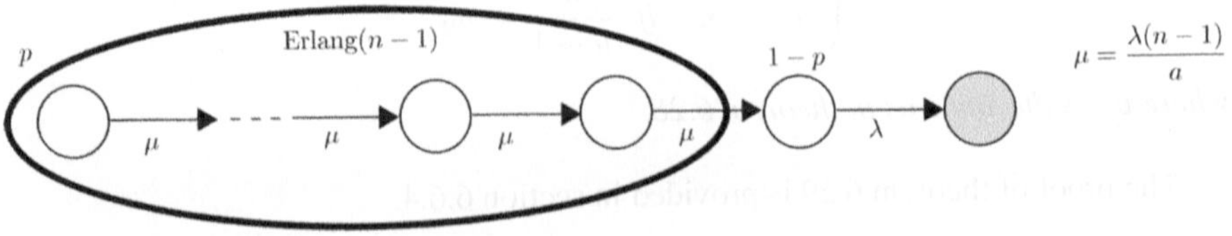


**Figure 6.23.** *Erlang-Exp structure with parameters* $(a, p, \lambda, n)$

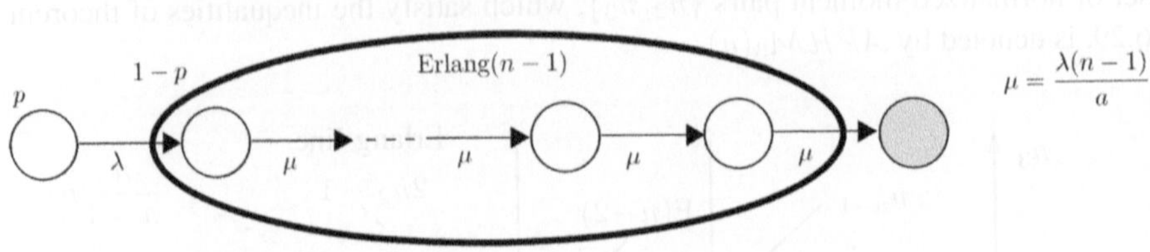


**Figure 6.24.** *Exp-Erlang structure with parameters* $(a, p, \lambda, n)$

THEOREM 6.30.– *Given the normalized moments $n_2$ and $n_3$, such that $\{n_2, n_3\} \in \mathcal{APH}_3(n) \setminus \mathcal{APH}_3(n-1)$, we have the following two cases for an APH($n$) distribution which matches the second and third normalized moments:*

*i) if $n_2 \le \frac{n}{n-1}$ or $n_3 \le 2n_2 - 1$ (i.e. the point is in Subset 1, 2, 4 or 5), then the Erlang-Exp distribution, presented in Figure 6.23, with parameters:*

$$a = \frac{(b\,n_2 - 2)(n-1)\,b}{(b-1)\,n}, \quad p = \frac{b-1}{a}, \qquad [6.45]$$

*and*

$$b = \frac{2(4-n(3n_2-4))}{n_2(4+n-nn_3)+\sqrt{nn_2}\sqrt{12n_2{}^2(n+1)+16n_3(n+1)+n_2(n(n_3-15)(n_3+1)-8(n_3+3))}}$$

*matches the $n_2$ and $n_3$ normalized moments;*

*ii) if $n_2 > \frac{n}{n-1}$ and $n_3 > u_{n-1}$ (i.e. the point is in Subset 3), then the Exp-Erlang distribution, presented in Figure 6.24, with parameters*

$$a = \frac{2(f-1)(n-1)}{(n-1)(n_2 f^2 - 2f + 2) - n}, \quad p = (f-1)a$$

*matches the $n_2$ and $n_3$ normalized moments, where parameter $f$ is defined in the following steps.*

$$K_1 = n-1,\ K_2 = n-2,\ K_3 = 3\,n_2 - 2\,n_3,\ K_4 = n_3 - 3,\ K_5 = n - n_2,$$

$$K_6 = 1 + n_2 - n_3,\ K_7 = n + n_2 - n\,n_2,\ K_8 = 3 + 3\,n_2^2 + n_3 - 3\,n_2\,n_3,$$

$$K_9 = 108\,K_1^2\Big(4\,K_2^2\ K_3\,n^2\,n_2 + K_1^2\,K_2\,K_4^2\,n\,n_2^2 + 4\,K_1\,K_5\,(K_5^2 - 3\,K_2\,K_6\,n\,n_2)$$

$$+\sqrt{-16\,K_1^2\,K_7^6 + (4\,K_1\,K_5^3 + K_1^2\,K_2\,K_4^2\,n\,n_2^2 + 4\,K_2\,n\,n_2\,(K_4\,n^2 - 3\,K_6\,n_2 + K_8\,n))^2}\Big),$$

$$K_{10} = \frac{K_4^2}{4\,K_3^2} - \frac{K_5}{K_1\,K_3\,n_2},\ K_{11} = \frac{2^{1/3}(3\,K_5^2 + K_2\,(K_3 + 2\,K_4)\,n\,n_2)}{K_3\,K_9^{1/3}\,n_2},$$

$$K_{12} = \frac{K_9^{1/3}}{3\,2^{7/3}\,K_1^2\,K_3\,n_2},\ K_{13} = \sqrt{K_{10} + K_{11} + K_{12}},$$

$$K_{14} = \frac{6\,K_1\,K_3\,K_4\,K_5 + 4\,K_2\,K_3^2\,n - K_1^2\,K_4^3\,n_2}{4\,K_1^2\,K_3^3\,K_{13}\,n_2},\ K_{15} = -\frac{K_4}{2\,K_3},$$

$$K_{16} = \sqrt{2\,K_{10} - K_{11} - K_{12} - K_{14}}\,,\ K_{17} = \sqrt{2\,K_{10} - K_{11} - K_{12} + K_{14}}\,,$$

$$K_{18} = 36\,K_5^3 + 36\,K_2\,K_4\,K_5\,n\,n_2 + 9\,K_1\,K_2\,K_4^2\,n\,n_2^2$$

$$-\sqrt{81\,(4\,K_5^3 + 4\,K_2\,K_4\,K_5\,n\,n_2 + K_1\,K_2\,K_4^2\,n\,n_2^2)^2 - 48\,(3\,K_5^2 + 2\,K_2\,K_4\,n\,n_2)^3}\,,$$

$$K_{19} = -\frac{K_5}{K_1\,K_4\,n_2} - \frac{2^{2/3}\,(3\,K_5^2 + 2\,K_2\,K_4\,n\,n_2)}{3^{1/3}\,K_1\,K_4\,n_2\,K_{18}^{1/3}} - \frac{K_{18}^{1/3}}{6^{2/3}\,K_1\,K_4\,n_2},$$

$$K_{20} = 6\,K_1\,K_3\,K_4\,K_5 + 4\,K_2\,K_3^2\,n - K_1^2\,K_4^3\,n_2\,,\ K_{21} = K_{11} + K_{12} + \frac{K_5}{2\,n\,K_1\,K_3},$$

$$K_{22} = \sqrt{\frac{3\,K_4^2}{4\,K_3^2} - \frac{3\,K_5}{K_1\,K_3\,n_2} + \sqrt{4\,K_{21}^2 - \frac{nK_2}{n_2\,K_1^2\,K_3}}},$$

$$f = \begin{cases} K_{13} + K_{15} - K_{17} & \textit{if}\ u_{n-1} < n_3 < 3n_2/2, \\ K_{19} & \textit{if}\ n_3 = 3n_2/2, \\ -K_{13} + K_{15} + K_{16} & \textit{if}\ n_3 > 3n_2/2\ \textit{and}\ K_{20} > 0, \\ K_{15} + K_{22} & \textit{if}\ K_{20} = 0, \\ K_{13} + K_{15} + K_{17} & \textit{if}\ K_{20} < 0. \end{cases}$$

Comments on the theorem:

– Similar to corollary 6.27, only $a$ and $p$ affect the second and third normalized moments while $\lambda$ does not. The required first moment can be set with $\lambda$.

– For a given $n$ and $n_2$, $f$ is a piecewise function of $n_3$ and it is continuous as illustrated in Figure 6.25. In the figure $n = 3, n_2 = 1.7, n_3 > 2.5$ and the limits of the "pieces" are $n_3 = 3n_2/2 = 2.55$ and $n_3 = (45 + 3\sqrt{55})/10 \approx 6.7248595$ (where $K_{20} = 0$), which are indicated by a change in the line style.

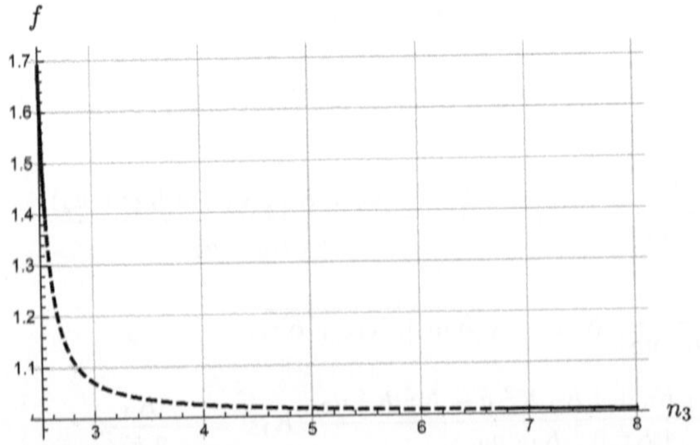


**Figure 6.25.** *$f$ is a continuous function of $n_3$ ($n = 4, n_2 = 1.7$)*

In order to prove theorem 6.30, we need the following corollary.

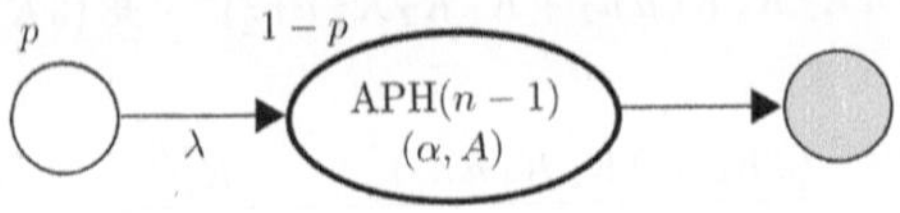


**Figure 6.26.** *Extension of an APH($n-1$) with an additional phase in the front*

COROLLARY 6.31.– *Let $n_2$ and $n_3$ be the second and third normalized moments of an APH($n-1$) distribution. The extension of this APH($n-1$) distribution with an additional phase in the front according to Figure 6.26 results in an APH($n$) distribution with normalized moments:*

$$\hat{n}_2 = \frac{a^2 n_2 + 2p(1+a)}{p^2 + a^2 + 2ap}, \qquad \hat{n}_3 = \frac{a^3 n_3 n_2 + 3p(n_2 a^2 + 2(a+1))}{(p+a)(a^2 n_2 + 2p(a+1))}. \qquad [6.46]$$

*Proof.* Let $m_1$, $m_2$ and $m_3$ denote the first three moments of the APH($n-1$) distribution. The first three moments of the APH($n$) distribution of the given structure are:

$$\hat{m}_1 = p\left(m_1 + \frac{1}{\lambda}\right) + (1-p)m_1, \quad \hat{m}_2 = p\left(m_2 + \frac{2m_1}{\lambda} + \frac{2}{\lambda^2}\right) + (1-p)m_2,$$

$$\hat{m}_3 = p\left(m_3 + \frac{3m_2}{\lambda} + \frac{6m_1}{\lambda^2} + \frac{6}{\lambda^3}\right) + (1-p)m_3.$$

Using the definition of the normalized moments results in the corollary. □

*Proof of theorem 6.30.* Part (i) of the theorem is based on corollary 6.27 where the APH($n-1$) distribution is an Erlang($n-1$) distribution with normalized moments $n_2 = \frac{n}{n-1}$ and $n_3 = \frac{n+1}{n-1}$. Substituting these normalized moments into [6.40] results in implicit equations for $p$ and $a$. Solving the equation for $p$ and $a$ results in [6.45] (see Bobbio et al. [BOB 05] for further details).

Part (ii) of the theorem can be proven similarly applying corollary 6.31. Once again the APH($n-1$) distribution is an Erlang($n-1$) distribution. Substituting its normalized moments into [6.46] results in implicit equations for $p$ and $a$. This case leads to the fourth order polynomial equation $c_4 f^4 + c_3 f^3 + c_2 f^2 + c_1 f + c_0 = 0$, where:

$$c_4 = n_2(3n_2 - 2n_3)(n-1)^2\,, \quad c_3 = 2n_2(n_3 - 3)(n-1)^2\,,$$

$$c_2 = 6(n-1)(n-n_2)\,, \quad c_1 = 4n(2-n)\,, \quad c_0 = n(n-2)\,,$$

and parameter $f$ is the solution which results in a proper APH($n$) distribution. □

Theorem 6.30 uses an Erlang-Exp distribution for the normalized moment pairs in Subset 1, but it is also possible to find an Exp-Erlang distribution for those moment pairs according to the method presented in Part (ii) of the theorem. The moment pairs of Subset 3 are not reachable with size $n$ Erlang-Exp distribution and the moment pairs of Subsets 2, 4 and 5 are not reachable with size $n$ Exp-Erlang distribution.

THEOREM 6.32.– *Given normalized moments $n_2$ and $n_3$ such that $\{n_2, n_3\} \in \mathcal{APHM}_3(n) \setminus \mathcal{APHM}_3(n-1)$, we have the following two cases for an APHM($n$) distribution which matches the second and third normalized moments:*

*i) if $\{n_2, n_3\} \in \mathcal{APH}_3(n)$ and $n_3 \neq \frac{3}{2}n_2$, then the APH($n$) distribution provided in theorem 6.30 matches the $n_2$ and $n_3$ normalized moments;*

*ii) if $\{n_2, n_3\} \in \mathcal{APHM}_3(n) \setminus \mathcal{APH}_3(n)$, then the APHM($n$) distribution presented in Figure 6.27 with parameter $q = \frac{1}{2n_2 - n_3}$ and APH($n$) distribution provided by theorem 6.30 for $n_2^* = \frac{n_2}{2n_2 - n_3}$ and $n_3^* = \frac{n_3}{2n_2 - n_3}$ matches the $n_2$ and $n_3$ normalized moments.*

*Proof.* Part (i) of the theorem is a consequence of theorem 6.30, while Part (ii) is a consequence of corollary 6.33. □

COROLLARY 6.33.– *Consider an APH($n$) distribution with normalized moments $n_2$ and $n_3$. The APHM($n$) distribution constructed according to Figure 6.27 with parameter $q$ has the following normalized moments:*

$$\hat{n}_2 = \frac{n_2}{q}, \quad \hat{n}_3 = \frac{n_3}{q}. \qquad [6.47]$$

*Proof.* The non-normalized moments of the APHM($n$) distribution are related to the non-normalized moments of the APH($n$) distribution as $\hat{m}_i = qm_i$ ($i \geq 1$). From this relation, the corollary is obtained simply by the definition of the normalized moments. □

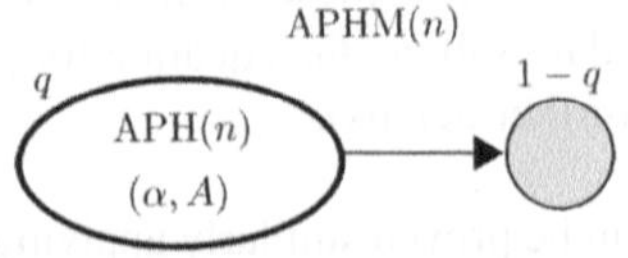


**Figure 6.27.** *Composition of APHM(n) distribution*

Part (ii) of theorem 6.32 generates the APHM($n$) distribution with the required normalized moments $\{n_2, n_3\}$ based on corollary 6.33. The proposed solution is such that the normalized moments of the APH($n$) distribution lie exactly on the Erlang-line ($n_3 = 2n_2 - 1$). This is not the only possibility. One could select other points of $\mathcal{APH}_3(n) \setminus \mathcal{APH}_3(n-1)$, which then can be transformed into an APHM($n$) distribution with normalized moments $\{n_2, n_3\}$.

### 6.6.3. *Proof of theorem 6.28*

We prove theorem 6.28 by induction. For $n = 2$, the theorem is proved in [TEL 02]. We assume that the theorem holds for $n - 1$ ($n \geq 3$), that is:

a1) for any APH($n - 1$) distribution, the second and third normalized moments are in $\mathcal{APH}_3(n - 1)$;

a2) for any $\{n_2, n_3\} \in \mathcal{APH}_3(n - 1)$, there is an APH($n - 1$) distribution whose second and third normalized moments are $\{n_2, n_3\}$,

and based on these assumptions we show that:

s1) for any $\{n_2, n_3\} \in \mathcal{APH}_3(n - 1)$, $0 < a < \infty$, $0 < p < 1$ the transformation of corollary 6.27 results in $\mathcal{T}(\{n_2, n_3\}, a, p) \in \mathcal{APH}_3(n)$;

s2) for any $\{n_2, n_3\} \in \mathcal{APH}_3(n) \setminus \mathcal{APH}_3(n - 1)$, there is an APH($n$) distribution whose second and third normalized moments are $\{n_2, n_3\}$.

We start with showing (s1). Firstly, we prove that $\hat{n}_2 \geq \frac{n+1}{n}$ (this inequality is shown in [ALD 87] as well. Here, we give a different proof).

According to [6.40], $\hat{n}_2$ is a monotone increasing function of $n_2$. Based on assumption (a1), the second normalized moment of an APH($n-1$) distribution fulfills $n_2 \geq \frac{n}{n-1}$. The derivative of $\hat{n}_2$ in [6.40] is:

$$\frac{d}{da}\hat{n}_2 = \frac{2p(an_2 - ap - 1)}{(1+ap)^3}. \qquad [6.48]$$

This derivative becomes 0 at $a = \frac{1}{n_2 - p}$, where the second derivative is non-negative. As a result:

$$\hat{n}_2 \geq \hat{n}_2\big|_{a=\frac{1}{n_2-p}} = 2 - \frac{p}{n_2} \geq 2 - \frac{1}{n_2}, \qquad [6.49]$$

where the right equality is obtained at $p = 1$. Hence, the minimum of $\hat{n}_2$ is obtained at $n_2 = \frac{n}{n-1}$, $p = 1$, $a = n - 1$ and it is $\hat{n}_2 = \frac{n+1}{n}$ (this minimum is provided by the Erlang($n$) distribution).

In the following, we treat the four inequalities of [6.41] separately. The last one, [6.41d], is evident.

*Inequality [6.41a]:* We rewrite [6.40] using $g = 1 + pa$:

$$\hat{n}_2 = \frac{2g + a(g-1)n_2}{g^2}, \qquad \hat{n}_3 = \frac{6g + a(g-1)n_2(n_3 a + 3)}{2g^2 + a(g-1)gn_2}. \qquad [6.50]$$

By definition $1 \leq g < \infty$. From [6.50], we have $a = \frac{(\hat{n}_2 g - 2)g}{n_2(g-1)}$. Substituting it into $\hat{n}_3$ of [6.50] gives:

$$\hat{n}_3 = \frac{3}{g} + \frac{(\hat{n}_2 g - 2)^2}{(g-1)g\hat{n}_2}\frac{n_3}{n_2}. \qquad [6.51]$$

According to [6.51], $\hat{n}_3$ is a monotone increasing function of $n_3/n_2$, hence the minimal $\hat{n}_3$ value of the $\mathcal{APH}_3(n)$ class at $\hat{n}_2$ is obtained when $n_3/n_2$ takes the minimal value of the $\mathcal{APH}_3(n-1)$ class. The minimal $n_3/n_2$ value of the $\mathcal{APH}_3(n-1)$ class is:

$$\min_{\mathcal{APH}_3(n-1)} n_3/n_2 = \frac{n+1}{n}, \qquad [6.52]$$

since at $n_2 = n/(n-1)$, the lower bound of $n_3$ is $l_{n-1} = (n+1)/(n-1)$ and $dl_{n-1}/dn_2 \geq (n+1)/n$. Note that this minimal $n_3/n_2$ ratio is provided by the Erlang($n-1$) distribution. Hence:

$$\hat{n}_3 \geq \hat{n}_3^{min} = \frac{3}{g} + \frac{(\hat{n}_2 g - 2)^2}{(g-1)g\hat{n}_2}\frac{n+1}{n}. \qquad [6.53]$$

Based on [6.53], $\hat{n}_3^{min}$ has the following limits in the $g \in (1, \infty)$ interval:

$$\lim_{g \to 1^+} \hat{n}_3^{min} = \infty, \qquad \lim_{g \to \infty} \hat{n}_3^{min} = \frac{n+1}{n}\hat{n}_2.$$

This way, the minimum of $\hat{n}_3^{min}$ is obtained at $g \to \infty$ when the derivative of $\hat{n}_3^{min}$ with respect to $g$ has no roots in $(1, \infty)$. It is obtained at the root of the derivative of $\hat{n}_3^{min}$ when it has one root in the $(1, \infty)$ interval. The case of having two roots in the $(1, \infty)$ interval cannot occur as is shown below.

The solutions of $\frac{d}{dg}\hat{n}_3^{min} = 0$ are:

$$g_1 = \frac{r + \sqrt{rs}}{r - s}, \qquad g_2 = \frac{r - \sqrt{rs}}{r - s} = \frac{\sqrt{r}(\sqrt{r} - \sqrt{s})}{r - s},$$

where $r = 4(n+1) - 3n\hat{n}_2$ and $s = (n+1)(\hat{n}_2 - 2)^2$. By definition $s \geq 0$ and $r > 0$ if $\hat{n}_2 < 4(n+1)/3n$, and $r - s > 0$ if $\hat{n}_2 < (n+4)/(n+1)$. The following cases need to be considered:

– if $\frac{n+1}{n} < \hat{n}_2 < \frac{n+4}{n+1}$, then $g_1 \in (1, \infty)$, $g_2 \in (0, 1)$;

– if $\frac{n+4}{n+1} < \hat{n}_2 < \frac{4n+4}{3n}$, then $g_1 \in (-\infty, 0)$, $g_2 \in (0, 1)$;

– if $\frac{4n+4}{3n} < \hat{n}_2$, then $r < 0$ and there is no real root.

Root $g_2 \in (0, 1)$ because $\sqrt{r} - \sqrt{s}$ and $r - s$ has the same sign when $\hat{n}_2 < 4(n+1)/3n$ and $1 > \dfrac{r - \sqrt{sr}}{r - s}$. The last inequality holds, because if $s < r$, then we have:

$$s < \sqrt{sr} \quad \Rightarrow \quad r - s > r - \sqrt{sr} \quad \Rightarrow 1 > \frac{r - \sqrt{sr}}{r - s},$$

and if $s > r$, then we have:

$$s > \sqrt{sr} \quad \Rightarrow \quad r - s < r - \sqrt{sr} \quad \Rightarrow 1 > \frac{r - \sqrt{sr}}{r - s}.$$

Figure 6.28 illustrates the behavior of the roots when $n = 4$. The curve corresponding to $g_1$ has a discontinuity at $\frac{n+4}{n+1}$ where it tends to $\infty$ from the left and to $-\infty$ from the right. The other root $g_2 \in (0, 1)$ for any $\hat{n}_2$. The fact that the optimal $g$ value tends to infinity as $\hat{n}_2$ tends to $\frac{n+4}{n+1}$ indicates that the lower bound is continuous at $\hat{n}_2 = \frac{n+4}{n+1}$.

Considering inequality [6.41a], we restrict our attention to the $\hat{n}_2 \in \left(\frac{n+1}{n}, \frac{n+4}{n+1}\right)$ interval. In this interval, $g_1$ falls into $(1, \infty)$. Substituting $g_1$ into [6.53] results in inequality [6.41a].

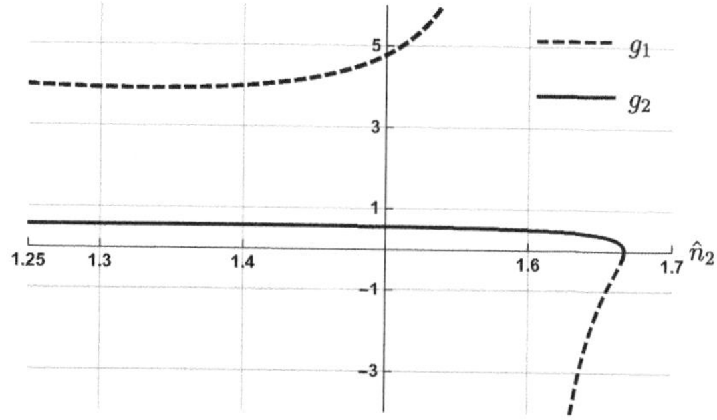


**Figure 6.28.** *The roots of the derivative of* [6.53] *when* $n = 4$

*Inequality [6.41b]:* The derivative of $\hat{n}_3^{min}$ with respect to $g$ has no root in $(1, \infty)$ when $\hat{n}_2 \in \left(\frac{n+4}{n+1}, \infty\right)$. Hence, $\hat{n}_3^{min}$ is a monotone decreasing function of $g$ with:

$$\lim_{g \to \infty} \hat{n}_3^{min} = \frac{n+1}{n} \hat{n}_2.$$

If $\frac{n+1}{n} \le \hat{n}_2 < \frac{n}{n-1}$, then $\frac{n}{n-1} \le n_2 < \frac{n-1}{n-2}$ according to [6.49].

From [6.40]:

$$a = \frac{\hat{n}_2 - 1 \pm \dfrac{\sqrt{n_2(\hat{n}_2 - 2) + p}}{\sqrt{p}}}{n_2 - \hat{n}_2 p}. \qquad [6.54]$$

For a given $(n_2, \hat{n}_2)$ pair, $p \in \{n_2(2 - \hat{n}_2), 1\}$, because $a$ is complex when $p < n_2(2 - \hat{n}_2)$.

Substituting [6.54] into $\hat{n}_3$ of [6.40] results in:

$$\begin{aligned}\hat{n}_3 &= \frac{3n_2^2\hat{n}_2 + 2n_2(\hat{n}_2 - 2)n_3}{n_2^2\hat{n}_2} + \frac{n_2(\hat{n}_2 - 2)n_3 + (4n_3 - 3n_2\hat{n}_2)p}{n_2^2\hat{n}_2}\left(1 \pm \sqrt{\frac{n_2(\hat{n}_2 - 2)}{p} + 1}\right) \\ &\le \frac{3n_2^2\hat{n}_2 + 2n_2(\hat{n}_2 - 2)n_3}{n_2^2\hat{n}_2} + \frac{n_2(\hat{n}_2 - 2)n_3 + (4n_3 - 3n_2\hat{n}_2)p}{n_2^2\hat{n}_2}\left(1 + \sqrt{\frac{n_2(\hat{n}_2 - 2)}{p} + 1}\right).\end{aligned} \qquad [6.55]$$

For $p \in \{n_2(2 - \hat{n}_2), 1\}$, the maximum of $\hat{n}_3$ is reached when $p = 1$, since $\hat{n}_2 - 2 < 0$ and $4n_3 - 3n_2\hat{n}_2 > 0$ (according to [6.52]) when $\frac{n}{n-1} \le n_2 < \frac{n-1}{n-2}$ and $\frac{n+1}{n} \le \hat{n}_2 < \frac{n}{n-1}$. Furthermore, according to [6.55], $\hat{n}_3$ is a monotone function of $n_3$, hence the upper bound of $\hat{n}_3$ is obtained at the upper bound of $n_3$. Substituting the upper bound of $\hat{n}_3$ into [6.55] results in [6.41c].

The proof of (s2) is divided into parts according to the five disjoint subsets of $\mathcal{APH}_3(n) \setminus \mathcal{APH}_3(n-1)$ (see Figure 6.21).

Subset 1: The following property holds for Subset 1. Given an integer $n \geq 3$ and a pair $(n_2, n_3)$ such that:

$$\frac{n+1}{n} \leq n_2 < \frac{n}{n-1} \text{ and } l_n \leq n_3 \leq u_n,$$

there is an APH($n$) distribution of Erlang-Exp structure whose second and third normalized moments are $n_2$ and $n_3$, respectively.

This property of Subset 1 can be proven by the following steps.

1) Given an integer $n$ and a real $n_2$ such that $\frac{n+1}{n} \leq n_2 < \frac{n}{n-1}$, the APH($n$) distribution of Erlang-Exp structure with parameters:

$$a = \frac{(b\,n_2 - 2)(n-1)\,b}{(b-1)\,n} \text{ and } p = \frac{b-1}{a},$$

where:

$$b = \frac{4 + n - n\,l_n}{2\,(n_2 + n\,n_2 - n\,l_n)},$$

has second and third normalized moments $n_2$ and $l_n$, respectively. This step of the proof is a direct consequence of corollary 6.27.

2) Given an integer $n$ and a real $n_2$ such that $\frac{n+1}{n} \leq n_2 < \frac{n}{n-1}$, the APH($n$) distribution of Erlang-Exp structure with parameters:

$$a = b - 1 \text{ and } p = 1,$$

where:

$$b = \frac{n}{\sqrt{(n-1)(n(n_2-1)-1)} + 1},$$

has second and third normalized moments $n_2$ and $u_n$, respectively. This step of the proof follows simply from corollary 6.27.

3) Given an integer $n$ and a real $n_2$ such that $\frac{n+1}{n} \leq n_2 < \frac{n}{n-1}$, the APH($n$) distribution of Erlang-Exp structure with parameters:

$$a = \frac{(b\,n_2 - 2)(n-1)\,b}{(b-1)\,n} \text{ and } p = \frac{b-1}{a},$$

where:

$$b \in \left[ \frac{n}{\sqrt{(n-1)(n(n_2-1)-1)}+1}, \frac{4+n-n\,l_n}{2\,(n_2+n\,n_2-n\,l_n)} \right],$$

has second normalized moment $n_2$ and its third normalized moment is a continuous, monotone decreasing function of $b$, which can be seen from corollary 6.27 and elementary properties of the emerging expressions.

Subset 2: Exactly the same proof applies for Subset 2 as for Subset 1, because $u_n > l_{n-1}$ in the $n_2 \in \left(\frac{n}{n-1}, \frac{n+4}{n+1}\right)$ range (note that $u_n$, defined in [6.43], is not the upper bound in the $n_2 \in \left(\frac{n}{n-1}, \frac{n+4}{n+1}\right)$ range).

Subset 3: An APH($n$) distribution with normalized moments $\{n_2, n_3\}$, where:

$$\frac{n}{n-1} \le n_2 < \frac{n-1}{n-2} \quad \text{and} \quad u_{n-1} \le n_3,$$

can be composed of an APH($n-1$) distribution and an exponential phase according to corollary 6.27 where the normalized moments pair of the APH($n-1$) distribution and the parameters of the transformation are:

$$\left\{ \frac{1}{2-n_2}, \frac{n_2(3-3n_2+n_3)}{(n_2-2)^2} \right\}, \quad p=1, \quad a=\frac{2-n_2}{n_2-1}.$$

The normalized moments pair $\left\{ \frac{1}{2-n_2}, \frac{n_2(3-3n_2+n_3)}{(n_2-2)^2} \right\}$ is in Subset 3 of the APH($n-1$) class.

Subset 5: The proof is similar to the one for Subset 1. Given an integer $n \ge 3$ and the pair $\{n_2, n_3\}$ such that:

$$\frac{n+3}{n} \le n_2 \quad \text{and} \quad \frac{n+1}{n}\, n_2 < n_3 \le \frac{n}{n-1}\, n_2,$$

there is an APH($n$) distribution of Erlang-Exp structure whose second and third normalized moments are $n_2$ and $n_3$, respectively.

We show this property of Subset 5 in the following steps.

1) Given an integer $n$ and a real $n_2$ such that $\frac{n+3}{n} \le n_2$, we have the following limits at $b \to \infty$:

$$a = \frac{(b\,n_2-2)(n-1)\,b}{(b-1)\,n} \to \infty \quad \text{and} \quad p = \frac{b-1}{a} \to \frac{n}{(n-1)n_2},$$

and consequently, the normalized moments of the APH($n$) distribution of Erlang-Exp structure is $n_3 \to \frac{n+1}{n}\, n_2$.

2) Given an integer $n$ and a real $n_2$ such that $\frac{n+3}{n} \leq n_2$, the APH($n$) distribution of Erlang-Exp structure with parameters:

$$a = \frac{(b\,n_2 - 2)(n-1)\,b}{(b-1)\,n} \quad \text{and} \quad p = \frac{b-1}{a},$$

where:

$$b = \frac{4 - 3n + n^2(n_2-1) + \sqrt{17n^2 + n^4(n_2-1)^2 - 6n^3(n_2-1) - 4n^2 n_2 + 12n(n_2-2)}}{2n_2},$$

has second and third normalized moments $n_2$ and $\frac{n}{n-1}\,n_2$, respectively.

3) Given an integer $n$ and a real $n_2$ such that $\frac{n+3}{n} \leq n_2$, the APH($n$) distribution of Erlang-Exp structure with parameters:

$$a = \frac{(b\,n_2 - 2)(n-1)\,b}{(b-1)\,n} \quad \text{and} \quad p = \frac{b-1}{a},$$

where:

$$b \in \left[\frac{4 - 3n + n^2(n_2-1) + \sqrt{17n^2 + n^4(n_2-1)^2 - 6n^3(n_2-1) - 4n^2 n_2 + 12n(n_2-2)}}{2n_2}, \infty\right),$$

has second normalized moment $n_2$ and its third normalized moment is a continuous, monotone decreasing function of $b$.

Subset 4: The same proof applies for Subset 4 as for Subset 5, because $\frac{n}{n-1}\,n_2 > l_{n-1}$ in the $n_2 \in \left(\frac{n+4}{n+1}, \frac{n+3}{n}\right)$ range.

### 6.6.4. *Proof of theorem 6.29*

The proof is composed of two parts. Based on the fact that all APHM($n$) distributions are obtained from an APH($n$) distribution according to transformation of corollary 6.33, we show that:

s1) the second and third normalized moments of an APHM($n$) distribution fulfills [6.44];

s2) for any $\{n_2, n_3\} \in \mathcal{APHM}_3(n)$, there is an APHM($n$) distribution whose second and third normalized moments are $n_2$ and $n_3$.

s1) The moment bounds of the APH($n$) distributions are known from theorem 6.28. The transformation of corollary 6.33 allows to project an $\{n_2, n_3\} \in \mathcal{APH}_3(n)$ point along the $n_3^* = \frac{n_3}{n_2} n_2^*$ line to any $n_2^* \geq n_2$ point. Since the derivative of the upper bound of $\mathcal{APH}_3(n)$ is greater than the associated $n_3/n_2$ ratio (i.e. $du_n/dn_2 > u_n/n_2$) all over the $n_2 \in \left(\frac{n+1}{n}, \frac{n}{n-1}\right)$ interval, the upper limit of the APHM($n$) class is identical with the upper limit of the APH($n$) class.

The derivative of the lower bound of $\mathcal{APH}_3(n)$ is also greater than the associated $n_3/n_2$ ratio (i.e. $dl_n/dn_2 > l_n/n_2$) all over the $n_2 \in \left(\frac{n+1}{n}, \frac{n+4}{n+1}\right)$ interval, but in this case it means that the transformation of corollary 6.33 can transform the points of the lower bound out of $\mathcal{APH}_3(n)$. The lower bound of $\mathcal{APHM}_3(n)$ is provided by the projection of the $\{n_2, n_3\} \in \mathcal{APH}_3(n)$ point with minimal $n_3/n_2$ ratio. This point is $\left\{\frac{n+1}{n}, \frac{n+2}{n}\right\}$, and the lower bound is the $n_3 = \frac{n+2}{n+1}n_2$ line.

s2) For any point $\{n_2, n_3\} \in \mathcal{APH}_3(n)$, there is an APH($n$) distribution with the given second and third normalized moments according to theorem 6.28. For any point $\{n_2, n_3\} \in \mathcal{APHM}_3(n) \setminus \mathcal{APH}_3(n)$, the procedure provided in theorem 6.32 results in an APHM($n$) distribution with the given normalized moments.

## 6.7. Matching any valid odd number of moments with generalized hyper-Erlang distributions

This section describes a procedure to match an odd number of valid moments with PH distributions (the validity of a given set of moments can be checked based on [1.6] and [1.7]). To this end, a special class of PH distributions is introduced, called generalized hyper-Erlang distributions (GHErDs). This class is an extension of the class of hyper-Erlang distributions (see Figure 5.4) in the sense that the involved weights and intensities can be complex or negative real numbers. It is known that even if a GHErD has some complex or negative parameters and if its PDF is valid (real and non-negative), then it corresponds to a PH distribution with Markovian representation with possibly larger size. Such Markovian representation can be obtained using algorithm 5.3.

We present a matching procedure where the user has to provide an odd number of valid moments and an upper bound for the sum of the degrees, which is the rank of the resulting PH distribution. Given these parameters, the method obtains all GHErDs that match the target moments and have a Markovian representation up to the given size.

### 6.7.1. *Moment matching procedure for generalized hyper-Erlang distributions*

The hyper-Erlang distributions can be used to match moments with SCV smaller and larger than one. To further increase the flexibility of moment matching, we generalize the set of hyper-Erlang distributions and discuss the solution of the moment matching problem for that general class.

DEFINITION 6.34.– *A random variable* $X$ *has* generalized hyper-Erlang distribution *of order-*$N$ *if its density function is:*

$$f(x) = \sum_{i=1}^{N} \alpha_i \frac{(\lambda_i x)^{r_i - 1}}{(r_i - 1)!} \lambda_i e^{-\lambda_i x}, \qquad [6.56]$$

*with* $f(x) \geq 0$ *and* $\sum_{i=1}^{N} \alpha_i = 1$ *and the parameters satisfies* $\lambda_i \in \mathbb{C}, \mathcal{R}(\lambda_j) \geq 0, \alpha_j \in \mathbb{C}$ *and* $r_i \in \mathbb{N}$ *for* $i = 1, \ldots, N$.

The main difference between the hyper-Erlang distribution (definition 5.10) and the GHErD is that the $\alpha_i$ and $\lambda_i$ parameters of GHErDs can be complex. Apart from that, the expressions characterizing their distribution and moments are identical.

Having fixed the rank parameters $r_i, i = 1, 2, ..., N$, the number of free parameters is $2N-1$ and they are $\lambda_1, \lambda_2, ..., \lambda_N$ and $\alpha_1, \alpha_2, ..., \alpha_{N-1}$ (since $\alpha_N = 1 - \sum_{i=1}^{N-1} \alpha_i$). Accordingly, an order-$N$ GHErD can be used to match $2N - 1$ moments.

To find an order-$N$ GHErD with moments $m_1, \ldots, m_{2N-1}$, we have to solve the system of polynomial equations given in [5.14] up to $k = 1, \ldots, 2N - 1$.

Due to the structure of the system of polynomial equations, no explicit solution is available for arbitrary $N$. However, there are excellent tools available that are able to solve polynomial systems numerically. For this purpose, we are using PHCpack (see [VER 99]), which is a multi-platform open-source tool and is under continuous development and refinement. We emphasize that we are not applying nonlinear programming or other optimization methods to find the solution of the moment matching problem (as was done in [BUC 10]). Instead, we find *all* the solutions of the system of polynomial equations.

Depending on the moments to match, this polynomial system typically has several solutions, and it is also possible that it has no solutions at all. If it does have solutions, each solution either defines a valid PH distribution, or it does not. If an obtained solution is a hyper-Erlang distribution, that is the solution is real and $\lambda_i, \alpha_i > 0$ for $i = 1, \ldots, N$, then it is a valid solution of the moments matching problem. If an obtained solution is not a hyper-Erlang distribution because some of the $\lambda_i$ and/or $\alpha_i$ parameters are complex, then we need to check if the solution is valid. To this end, we first try to obtain a Markovian representation from the non-Markovian $(\boldsymbol{\alpha}, \boldsymbol{A})$ representation in the form of [5.12] using algorithm 5.3. If the algorithm finds a Markovian representation, then it is a valid solution of the moments matching problem.

If none of the solution of the polynomial system result in a valid solution, then the entries of vector $\boldsymbol{r} = (r_1 \ldots r_N)$ need to be increased to obtain a valid solution. To prove that increasing the entries of vector $\boldsymbol{r}$ will result in a valid solution eventually, we introduce a sub-class of GHErD and apply a related result.

DEFINITION 6.35.– *A random variable $X$ is an order $N$* mixture of Erlang distributions of common order (MECO) *if it is an order $N$ hyper-Erlang distribution with identical $r_i$ parameters.*

That is, MECOs is a subclass of GHErD where all the $r_i$ parameters are identical. In [JOH 89], it is proven that order-$N$ MECO can match any valid $m_1, m_2, ..., m_{2N-1}$ set of moments. A set of moments is valid if it satisfies [1.6] and [1.7]. Since MECO is a sub-class of GHErD, it is ensured that by increasing the entries of the vector $\boldsymbol{r}$, a PH distribution with moments equal to $m_1, m_2, ..., m_{2N-1}$ will be found eventually.

#### 6.7.1.1. *Optimizing the $r_i$ parameters*

Finding the appropriate $\boldsymbol{r}$ can be made automatic as well. In this case, the user just has to enter a single parameter, $R_{\max}$, and the algorithm repeats the moment matching with all vectors $\boldsymbol{r}$ satisfying $\sum_{i=1}^{N} r_i \leq R_{\max}$.

---

**Algorithm 6.1** GHErD fitting algorithm based on moment matching

---

**procedure** FITGHERD($m_1, \ldots, m_{2N-1}, R_{\max}, D(\cdot)$)<br>
    `res` $\leftarrow \{\}$<br>
    **for all** $\boldsymbol{r} = \begin{bmatrix} r_1 & \ldots & r_N \end{bmatrix}$ with $\sum_{i=1}^{N} r_i \leq R_{\max}$ **do**<br>
        $\{\lambda_i, \alpha_i\} \leftarrow$ Solve polynomial equations of [5.14] for $k = 0, \ldots, 2N-1$<br>
        **for** each solution $(\lambda, \alpha)$ **do**<br>
            $(\boldsymbol{\alpha}, \boldsymbol{A}) \leftarrow$ Create non-Markovian representation<br>
            $(\boldsymbol{\beta}, \boldsymbol{B}) \leftarrow$ Transform $(\boldsymbol{\alpha}, \boldsymbol{A})$ to a Markovian representation with algorithm 5.3<br>
            **if** $(\boldsymbol{\beta}, \boldsymbol{B})$ is Markovian **then**<br>
                Add $(\boldsymbol{\beta}, \boldsymbol{B})$ to set `res`<br>
            **end if**<br>
        **end for**<br>
    **end for**<br>
    **if** `res` is $\{\}$ **then**<br>
        **error** "No solutions found, $R_{\max}$ has to be increased."<br>
    **end if**<br>
    $(\boldsymbol{\gamma}, \boldsymbol{G}) \leftarrow \arg\min_{(\boldsymbol{\beta}, \boldsymbol{B}) \in res} D(\boldsymbol{\beta}, \boldsymbol{B})$<br>
    **return** $(\boldsymbol{\gamma}, \boldsymbol{G})$<br>
**end procedure**

---

The steps of the procedure are presented in algorithm 6.1. First, the algorithm solves the moment matching problem with different $\boldsymbol{r}$ vectors up to $\sum_{i=1}^{N} r_i \leq R_{\max}$, and collects all solutions that have a Markovian representation in set `res`. Note that all solutions in `res` match the first $2N-1$ moments of the target distribution. In the second step (last line of the algorithm), the best solution is selected by a distance function $D(\cdot)$. Any distance function can be used that quantifies the distance between two distributions. Two possible distance functions are:

– Moment distance (MD): the sum of the squared relative difference of the moments up to the $K$th moment. Denoting the $k$th moment of the PH distribution resulting from the matching algorithm by $m_k(\boldsymbol{\beta}, \boldsymbol{B})$:

$$D(\boldsymbol{\beta}, \boldsymbol{B}) = \sum_{k=1}^{K} \left( \frac{m_k(\boldsymbol{\beta}, \boldsymbol{B}) - m_k}{m_k} \right)^2. \qquad [6.57]$$

(Note that $m_k(\boldsymbol{\beta}, \boldsymbol{B}) = m_k$ for $k \leq 2N - 1$).

– Relative entrophy (RE): it has been introduced as a measure of goodness of PH approximation in [BOB 94]. By denoting the PDF of the target distribution by $\hat{f}(x)$ and that of the PH distribution by $f(x)$, the RE is:

$$D(\boldsymbol{\beta}, \boldsymbol{B}) = \int_0^{\infty} f(x) \log \frac{f(x)}{\hat{f}(x)} dx. \qquad [6.58]$$

Both distance functions have advantages. The moment distance relies only on the moments, thus the exact shape of the target distribution (i.e. the PDF) is not required. RE, however, may quantify the similarity of the shape of the density functions better. Both distance functions will be evaluated in subsequent numerical examples.

in the case of using the moment distance, the method will be referred to as FitGHErD_MD, otherwise FitGHErD_RE.

### 6.7.2. *Numerical examples*

In this section, we apply the presented fitting method to two well-known traffic measurement traces, the BC-pAug89 and the LBL-TCP-3 traces[1]. The BC-pAug89 trace records 1 million packet arrivals on an Ethernet network, while the LBL-TCP-3 trace captures two hours of wide-area TCP traffic. These traces are not representative of today's network traffic, we nevertheless use them because both are frequently used for demonstration of various traffic fitting algorithms, as a kind of a benchmark.

The moment matching method has been implemented in MATLAB. We used PHCpack v2.3.76[2] to solve the system of polynomial equations which has a MATLAB interface as well.

1 Downloaded from http://ita.ee.lbl.gov/html/contrib/BC.html and from http://ita.ee.lbl.gov/html/contrib/LBL-TCP-3.html, respectively

2 A newer version can be obtained from http://homepages.math.uic.edu/~jan/download.html. In general, our experience with PHCpack is positive. When we got warnings that the polynomial system is ill-conditioned we normalized the trace (such that the mean value is one), which solved this issue.

The solution time of the polynomial system depends on $N$ heavily. We got prompt results in the case of $N = 2$, it took about 0.5s in the case of $N = 3$ and 25s in the case of $N = 4$. As the FitGHErD method relies on the iterative solution of a large number of polynomial equations, we did not increase parameter $N$ beyond 4, which means that we are going to match at most 7 moments in the numerical examples. When $N$ is large, the number of solutions is large as well, however trying to create the corresponding Markovian representations is so fast that it has negligible effect on the execution time. The total execution time of the algorithm depends on parameter $R_{\max}$ as well. If $R_{\max}$ is large, a large number of $\boldsymbol{r}$ vectors are considered, and a large number of polynomial systems are solved.

In the subsequent case studies, we compare the following four moment matching-based PH fitting methods:

1) the method presented in this section, where the PH distribution with the smallest moment distance is selected from the set of all PH distributions resulting from the algorithm;

2) the method of this section, where the PH distribution with the smallest relative entropy is selected;

3) moment matching with a MECO [JOH 89];

4) the method of Bobbio et al. [BOB 05], which is able to match the first three moments only.

#### 6.7.2.1. *The LBL trace*

In this example, $R_{\max}$ is set to 20. According to algorithm 6.1, this means that the moment matching is performed for all vectors $\boldsymbol{r}$ satisfying $\sum_{i=1}^{N} r_i \leq R_{\max}$. For a given vector $\boldsymbol{r}$, the moment matching problem can result in several different valid PH distributions. At the end, we have a large number of PH distributions from which we can select the best according to some distance function. Table 6.7 shows the number of $\boldsymbol{r}$ vectors, the number of valid solutions and the running time.

| No. of moments to match | No. of different $\boldsymbol{r}$ vectors | No. of valid solutions | Running time |
|---|---|---|---|
| 3 | 100 | 88 | 28 sec |
| 5 | 237 | 688 | 337 sec |
| 7 | 408 | 3920 | 810 min |

**Table 6.7.** *The number of different $\boldsymbol{r}$ vectors, the number of valid solutions and the total running time in the LBL example*

The numerical results are shown in Table 6.8. When the first three moments are matched, all methods found the same solution. Even if a large number of $\boldsymbol{r}$ vectors have been checked by FitGHErD, the best solution has been found to be the same according to both distance functions.

| | Method | MD | RE | No. of phases |
|---|---|---|---|---|
| 3 | FitGHErD_MD | 1.786 | 0.3024 | $2\ (\boldsymbol{r} = [1\ 1])$ |
| | FitGHErD_RE | 1.786 | 0.3024 | $2\ (\boldsymbol{r} = [1\ 1])$ |
| | MECO [JOH 89] | 1.786 | 0.3024 | $2\ (\boldsymbol{r} = [1\ 1])$ |
| | ErlExp [BOB 05] | 1.786 | 0.3024 | 2 |
| 5 | FitGHErD_MD | 0.0072 | 0.0984 | $3\ (\boldsymbol{r} = [1\ 1\ 1])$ |
| | FitGHErD_RE | 0.0386 | 0.0953 | $5\ (\boldsymbol{r} = [1\ 1\ 2])$ |
| | MECO [JOH 89] | 0.0072 | 0.0984 | $3\ (\boldsymbol{r} = [1\ 1\ 1])$ |
| 7 | FitGHErD_MD | $8.26 \times 10^{-6}$ | 3.9727 | $20\ (\boldsymbol{r} = [2\ 2\ 3\ 13])$ |
| | FitGHErD_RE | 0.00499 | 0.0727 | $8\ (\boldsymbol{r} = [1\ 1\ 3\ 3])$ |
| | MECO [JOH 89] | 0.00475 | 0.1339 | $8\ (\boldsymbol{r} = [2\ 2\ 2\ 2])$ |

**Table 6.8.** *Results of the fitting of the LBL trace. The first column gives the number of matched moments*

When five moments are matched, the MECO matching method returns a hyper-exponential distribution that was found to be optimal by FitGHErD as well according to the distance of moments. However, FitGHErD was able to find a PH distribution with lower RE. This PH distribution has five phases and the corresponding $\boldsymbol{r}$ vector is $\boldsymbol{r} = [1\ 1\ 2]$. The sum of the elements of vector $\boldsymbol{r}$ is only 4, which means that a new eigenvalue has been introduced and the size of the PH distribution has been increased by 1 to obtain a Markovian representation (this new eigenvalue cancels out, it has no effect on the PDF).

The advantage and the flexibility of FitGHErD is best seen when seven moments are matched. FitGHErD was able to find a PH distribution with significantly lower moment distance, and another one with significantly lower RE as well.

Figures 6.29–6.31 plot the density functions belonging to the methods discussed, both on linear and logarithmic scale. While the tail of the PDF is fitted well by all methods, the plots differ significantly in the body of the PDF. Based on a visual comparison, the solution found by FitGHErD by matching seven moments and selecting the best according to the RE distance seems to capture the shape of the PDF best.

### 6.7.2.2. *The BC trace*

The numerical results corresponding to the fitting of the BC trace are summarized in Table 6.9. When fitting the first three moments, the same hyper-exponential distribution turned out to be optimal by all the methods involved in the comparison. When fitting five moments, FitGHErD has found a PH distribution with slightly lower relative entropy. In the case when seven moments are matched, FitGHErD has found a PH distribution with better moment distance, and another one with significantly better RE than the MECO based method.

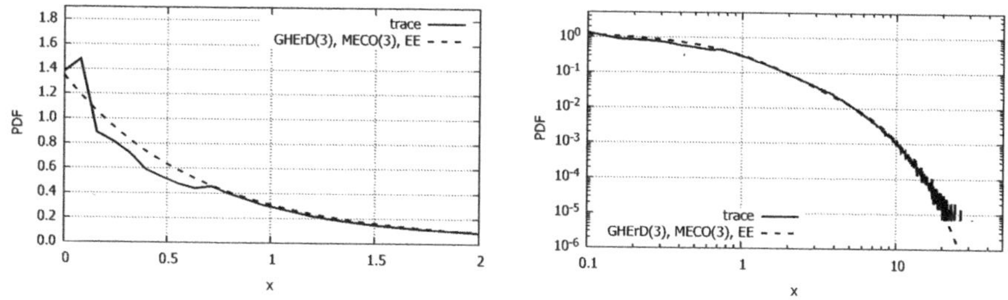


**Figure 6.29.** *Comparison of the density functions matching three moments of the LBL trace*

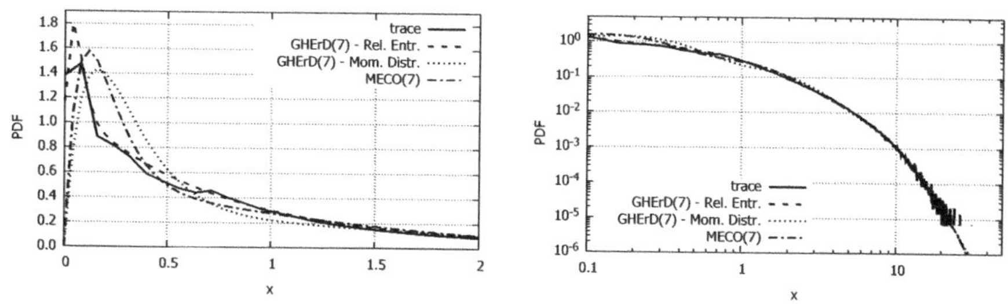


**Figure 6.30.** *Comparison of the density functions matching seven moments of the LBL trace*

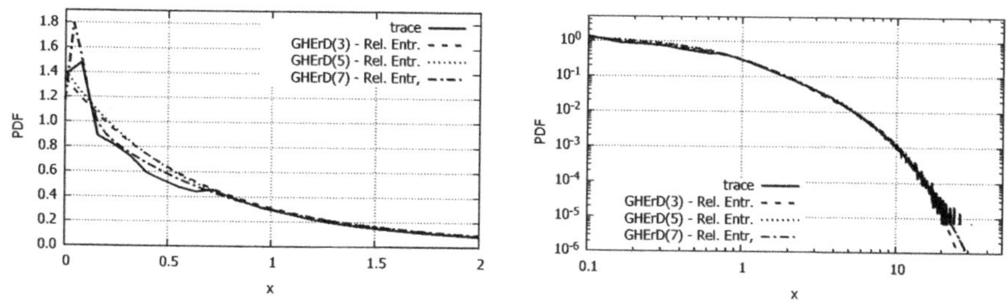


**Figure 6.31.** *Comparison of the results of FitGHErD, matching three, five and seven moments (LBL trace)*

| | Method | MD | RE | No. of phases |
|---|---|---|---|---|
| 3 | FitGHErD_MD | 3.0509 | 0.30244 | 2 ($r = [1\ 1]$) |
| | FitGHErD_RE | 3.0509 | 0.30244 | 2 ($r = [1\ 1]$) |
| | MECO [JOH 89] | 3.0509 | 0.30244 | 2 ($r = [1\ 1]$) |
| | ErlExp [BOB 05] | 3.0509 | 0.30244 | 2 |
| 5 | FitGHErD_MD | 0.00198 | 0.30521 | 14 ($r = [1\ 4\ 8]$) |
| | FitGHErD_RE | 55.4699 | 0.30212 | 5 ($r = [1\ 1\ 2]$) |
| | MECO [JOH 89] | 55.4699 | 0.30212 | 3 ($r = [1\ 1\ 1]$) |
| 7 | FitGHErD_MD | 0.0056 | 0.48178 | 20 ($r = [1\ 2\ 2\ 15]$) |
| | FitGHErD_RE | 0.0072 | 0.185 | 19 ($r = [1\ 1\ 2\ 15]$) |
| | MECO [JOH 89] | 0.0391 | 0.3536 | 16 ($r = [4\ 4\ 4\ 4]$) |

**Table 6.9.** *Results of the fitting of the BC trace. The first column gives the number of matched moments*

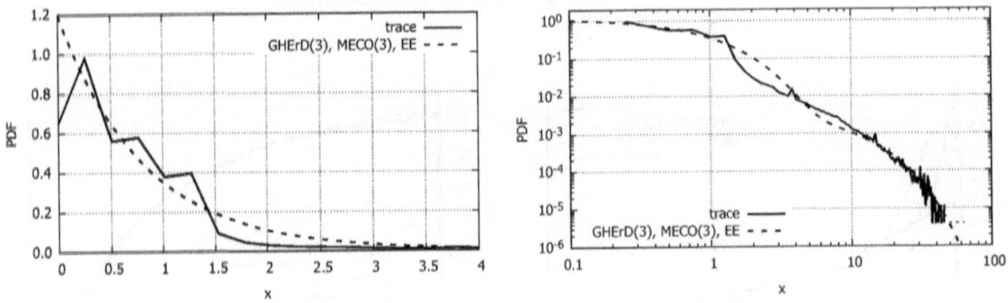


**Figure 6.32.** *Comparison of the density functions matching three moments of the BC trace*

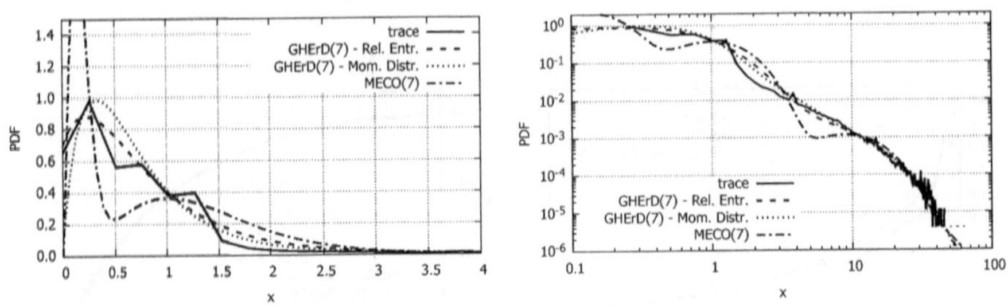


**Figure 6.33.** *Comparison of the density functions matching seven moments of the BC trace*

The density functions corresponding to the investigated cases are depicted in Figures 6.32–6.34. Figure 6.33 demonstrates how different the shapes of the density functions can be even if the first seven moments are the same. The MECO-based method looks to be the least successful in this example, while FitGHErD_RE managed to capture the characteristics of the density function reasonably well.

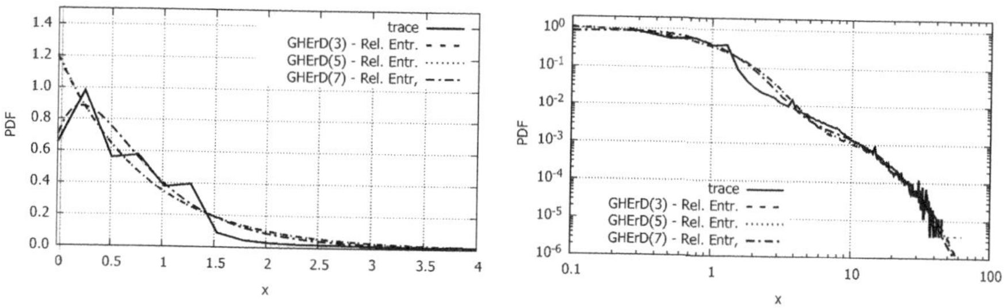


**Figure 6.34.** *Comparison of the results of FitGHErD, matching three, five and seven moments (BC trace)*

## 6.8. Matching probability density function at zero

In this chapter, we discuss matching the behavior of the PDF at zero, that is, setting the exact value of the PDF at zero, or even some derivatives of the PDF at zero. Matching (or approximating) the behavior of the PDF at zero can in some cases help to obtain proper distributions, or distributions that approximate the shape of a given PDF better. This will be illustrated in section 6.8.1.

One approach to match PDF behavior at zero was already outlined in example 6.21, where such moments are computed that guarantee the desired behavior at zero. Here, we describe another approach. We extend the procedure provided in section 6.5 to capture the behavior of the original PDF around zero. This can be done by modifying the set of equations that we use to determine the parameters of the Laplace transform given in [6.31]. We add equations related to the PDF behavior at zero and remove those referring to the highest order moments.

The relation between the parameters of the Laplace transform of the PDF and the PDF behavior at zero can be obtained based on two properties of the Laplace transform. First, the Laplace transform of the $k$th derivative of $f(t)$ is given by:

$$f^{(k)*}(s) = \int_{t=0}^{\infty} e^{-st} f^{(k)}(t)dt = s^k f^*(s) - \sum_{\ell=1}^{k} s^{k-\ell} f^{(\ell-1)}(0). \qquad [6.59]$$

Second, the $k$th derivative of the PDF at zero can be obtained as:

$$f^{(k)}(0) = \lim_{s\to\infty} s f^{(k)*}(s). \qquad [6.60]$$

Equations [6.59] and [6.60] provide a recursive calculation procedure to calculate $f^{(k)}(0)$ as a function of the parameters of the LT function defined in [2.9].

Applying the procedure defined by [6.59] and [6.60] for [2.9], we obtain that the PDF of an ME($n$) at zero assumes the value:

$$f(0) = \lim_{s \to \infty} s f^*(s) = \varphi_{n-1} \qquad [6.61]$$

while the first derivatives at zero can be obtained as:

$$f'(0) = \lim_{s \to \infty} (s(sf^*(s) - f(0))) = \varphi_{n-2} - \varphi_{n-1}\varrho_{n-1}, \qquad [6.62]$$

and:

$$f''(0) = \varphi_{n-3} - \varphi_{n-2}\varrho_{n-1} + \varphi_{n-1}\left(\varrho_{n-1}^2 - \varrho_{n-2}\right).$$

Fixing $f(0)$ (and consequently $\varphi_{n-1}$) means that the number of further moments that can be captured with an ME($n$) distribution is decreased from $2n-1$ to $2n-2$, while matching both $f(0)$ and $f'(0)$ means that we can capture only $2n-3$ moments, and so on.

It can be seen from [6.61] that matching $f(0)$ directly determines $\varphi_{n-1}$. Having determined $\varphi_{n-1}$, [6.62] provides a linear equation that must be satisfied if one aims to capture $f'(0)$. In general, capturing $f^{(k)}(0)$ for $0 \le k \le k_{\max}$ adds simple linear equations to those that are used to match the moments.

### 6.8.1. *Numerical examples*

We provide numerical examples to illustrate the possible advantages of considering the properties of the PDF at zero when constructing ME distributions. In section 6.8.2, we show that there are cases in which considering the moments of the distribution leads to an ME function with improper PDF, while also considering the behavior around zero results in a proper distribution. In section 6.8.3, we provide an example where moment matching gives a proper PDF, but the shape of the PDF can be improved by considering the value of the original PDF at zero instead of considering the highest order moment. In section 6.8.4, we show that even if the original characteristics around zero are not possible to capture, it is possible to set the characteristics of the ME distribution around zero in such a way that the fitting of the shape of the PDF improves considerably.

### 6.8.2. *Moment matching using the behavior around zero*

Consider the lognormal distribution whose PDF is given by:

$$f_{\ln}(t) = \frac{e^{-\frac{(\log(t)-\alpha)^2}{2\beta^2}}}{\sqrt{2\pi}t\beta} \qquad [6.63]$$

with parameters $\alpha = 0$, $\beta = \frac{1}{2}$.

We derive two ME(3) distributions to match the characteristics of the above PDF. The first one, ME($\boldsymbol{\alpha}_1, \boldsymbol{A}_1$), matches five moments and has an improper PDF (it assumes negative values around 0). Its representation is given by the vector–matrix pair:

$$\boldsymbol{\alpha}_1 = \begin{pmatrix} \frac{1}{3} & \frac{1}{3} & \frac{1}{3} \end{pmatrix}, \ \boldsymbol{A}_1 = \begin{pmatrix} -2.67775 & 20.4761 & -14.6936 \\ -0.351307 & -1.18954 & 0.753936 \\ -0.237024 & 0.601207 & -1.34227 \end{pmatrix}.$$

The second one, ME($\boldsymbol{\alpha}_2, \boldsymbol{A}_2$), matches three moments and also $f_{\ln}(0)$ and $f'_{\ln}(0)$, and results in a proper distribution. Its representation is:

$$\boldsymbol{\alpha}_2 = \begin{pmatrix} \frac{1}{3} & \frac{1}{3} & \frac{1}{3} \end{pmatrix}, \ \boldsymbol{A}_2 = \begin{pmatrix} -1.87683 & 21.0592 & -17.4174 \\ -0.395566 & -1.22176 & 0.90445 \\ -0.192765 & 0.633427 & -1.49278 \end{pmatrix}.$$

The non-negativity of ME($\boldsymbol{\alpha}_2, \boldsymbol{A}_2$) can be checked, for example, by BuTools [BUT 15]. In Figure 6.35, we depict the original lognormal PDF and the two matching ME(3) distributions. In Table 6.10, we provide the first five moments of the original and the matching ME distributions. For ME($\boldsymbol{\alpha}_2, \boldsymbol{A}_2$), even if the fourth and fifth moments are not matched, they are close to those of the lognormal distribution and also the shape of the PDF is similar to that of the lognormal distribution.

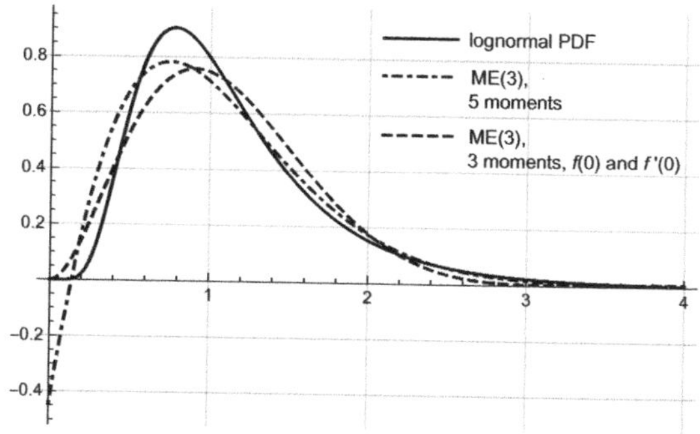


**Figure 6.35.** *Matching parameters of the lognormal distribution with $\alpha = 0, \beta = 1/2$: matching five moments and matching three moments, the value of the PDF at zero and the derivative of the PDF at zero*

| | $m_1$ | $m_2$ | $m_3$ | $m_4$ | $m_5$ |
|---|---|---|---|---|---|
| lognormal | 1.13315 | 1.64872 | 3.08022 | 7.38906 | 22.7599 |
| ME($\boldsymbol{\alpha}_1, \boldsymbol{A}_1$) | 1.13315 | 1.64872 | 3.08022 | 7.38906 | 22.7599 |
| ME($\boldsymbol{\alpha}_2, \boldsymbol{A}_2$) | 1.13315 | 1.64872 | 3.08022 | 7.58974 | 24.5879 |

**Table 6.10.** *Moments of the lognormal distribution with $\alpha = 0$, $\beta = \frac{1}{2}$ and the matching ME distributions*

### 6.8.3. *Improving matrix exponential approximation by matching the behavior around zero*

We consider now the three-parameter Weibull distribution commonly used to model the time to failure. The PDF is of the form:

$$f_{W3}(t) = e^{-\left(\frac{t+\alpha}{\beta}\right)^\gamma}(t+\alpha)^{\gamma-1}\beta^{-\gamma}\gamma,$$

where $\gamma$ is the so-called shape parameter, $\beta$ is the scale parameter and $\alpha$ determines the shift with respect to the two-parameter Weibull distribution.

We construct two ME(2) distributions to match three-parameters of the Weibull distribution, when $\gamma = \frac{1}{2}$, $\beta = 2$ and $\alpha = \frac{1}{2}$. The first one matches the first three moments of the $f_{W3}(t)$ distribution and has a proper PDF, but the shape of the PDF is rather different from $f_{W3}(t)$. Instead the second ME distribution matches two moments and the value of the original PDF at zero. The validity of the matching ME distributions is guaranteed by the fact that the moments are in the area defined by [6.65]. In Figure 6.36, we show the original PDF and the PDF of the two matching distributions. The representations of the two ME(2) distributions are:

$$\boldsymbol{\alpha}_3 = \begin{pmatrix} \frac{1}{2} & \frac{1}{2} \end{pmatrix}, \ \boldsymbol{A}_3 = \begin{pmatrix} -0.228593 & -0.0909786 \\ -0.0833333 & -0.0833333 \end{pmatrix},$$

and:

$$\boldsymbol{\alpha}_4 = \begin{pmatrix} \frac{1}{2} & \frac{1}{2} \end{pmatrix}, \ \boldsymbol{A}_4 = \begin{pmatrix} -0.716667 & -0.116667 \\ -0.0833333 & -0.0833333 \end{pmatrix}.$$

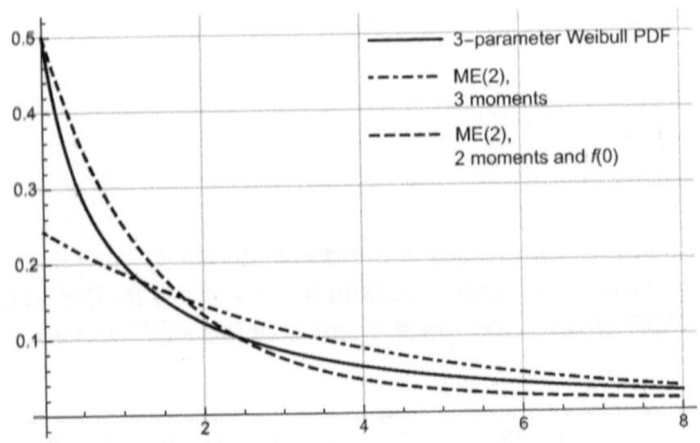


**Figure 6.36.** *Approximating the three-parameter Weibull distribution: matching three moments and matching two moments and the value of the PDF at zero*

In Figure 6.37, we show instead two ME(3) distributions. The first one captures the first five moments of the three-parameter Weibull distribution while the second one captures the first three moments, PDF at zero and the first derivative of the PDF at zero. It can be seen that not even five moments are enough to have an ME distribution whose PDF is similar to that of the original distribution. Instead, taking into account the behavior around zero improves the visual fitting of the PDF. The representations of the two ME(3) distributions are:

$$\boldsymbol{\alpha}_5 = \begin{pmatrix} \frac{1}{3} & \frac{1}{3} & \frac{1}{3} \end{pmatrix}, \quad \boldsymbol{A}_5 = \begin{pmatrix} -0.289888 & -0.0963572 & -0.15042 \\ 0.0163624 & -0.0565627 & -0.0129118 \\ -0.127473 & -0.0545485 & -0.0981994 \end{pmatrix},$$

and:

$$\boldsymbol{\alpha}_6 = \begin{pmatrix} \frac{1}{3} & \frac{1}{3} & \frac{1}{3} \end{pmatrix}, \quad \boldsymbol{A}_6 = \begin{pmatrix} -0.688889 & 0.725397 & -1.20317 \\ 0.338519 & -0.720053 & 0.83709 \\ -0.44963 & 0.608942 & -0.948201 \end{pmatrix}.$$

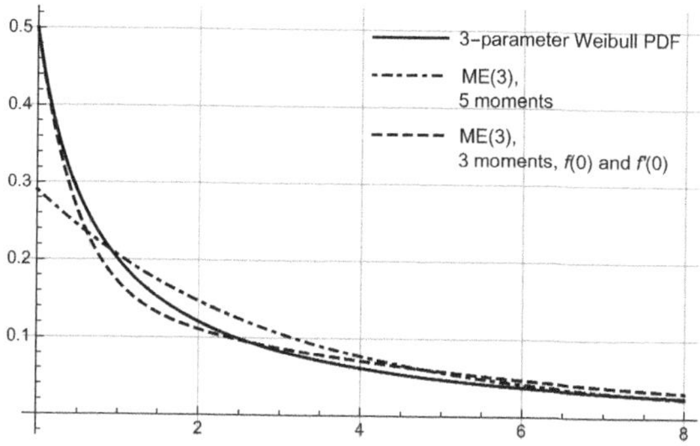


**Figure 6.37.** *Approximating the three-parameter Weibull distribution: matching five moments and matching three moments, the value and the first derivative of the PDF at zero*

In Table 6.11, we provide the first five moments of the original three-parameter Weibull distribution and those of the ME distributions depicted in Figures 6.36 and 6.37. The ME($\boldsymbol{\alpha}_4, \boldsymbol{A}_4$) distribution gives a bad fit already of the third moment. It is interesting to note that the distributions which fit the density at zero, ME($\boldsymbol{\alpha}_4, \boldsymbol{A}_4$) and ME($\boldsymbol{\alpha}_6, \boldsymbol{A}_6$), apparently better fits the shape of the PDF than the only moments matching distributions, ME($\boldsymbol{\alpha}_3, \boldsymbol{A}_3$) and ME($\boldsymbol{\alpha}_5, \boldsymbol{A}_5$), but at the same time they perform worse moments fitting.

| | $m_1$ | $m_2$ | $m_3$ | $m_4$ | $m_5$ |
|---|---|---|---|---|---|
| 3-parameter Weibull | 6 | 152 | 9264 | $1.044 \times 10^6$ | $1.888 \times 10^8$ |
| ME($\boldsymbol{\alpha}_3, \boldsymbol{A}_3$) (Figure 6.36) | 6 | 152 | 9264 | 848870 | $9.929 \times 10^7$ |
| ME($\boldsymbol{\alpha}_4, \boldsymbol{A}_4$) (Figure 6.36) | 6 | 152 | 6576 | 384384 | $2.812 \times 10^7$ |
| ME($\boldsymbol{\alpha}_5, \boldsymbol{A}_5$) (Figure 6.37) | 6 | 152 | 9264 | $1.044 \times 10^6$ | $1.888 \times 10^8$ |
| ME($\boldsymbol{\alpha}_6, \boldsymbol{A}_6$) (Figure 6.37) | 6 | 152 | 9264 | 878976 | $1.079 \times 10^8$ |

**Table 6.11.** *Moments of the three-parameter Weibull distribution and moments of the matching ME distributions*

### 6.8.4. *Improving matrix exponential approximation by approximating the behavior around zero*

We consider again a lognormal distribution, provided in [6.63], now with parameters $\alpha = 1, \beta = 1.8$. In this case, matching five moments results in an ME(3) with PDF which is considerably further from the original one. Its representation is:

$$\boldsymbol{\alpha}_7 = \begin{pmatrix} \frac{1}{3} & \frac{1}{3} & \frac{1}{3} \end{pmatrix}, \ \boldsymbol{A}_7 = \begin{pmatrix} -0.0286601 & -0.028144 & -0.0281441 \\ -0.00912991 & -0.0109089 & -0.010908 \\ -0.0394053 & -0.0376263 & -0.0376272 \end{pmatrix}.$$

Matching $f(0)$ and four moments, or $f'(0)$ and $f(0)$ and three moments results in ME(3)s that are not proper distributions. By observing the original PDF (Figure 6.38), one can see that it starts from 0, but then it immediately has a steep peak with a gradual monotonic decrease. By setting the characteristics of the ME(3) distribution around 0 as:

$$f(0) = 0.45, f'(0) = -0.32 \qquad [6.64]$$

and matching three moments results in a much more satisfactory fit of the original one (see Figure 6.38). The representation of this ME(3) is:

$$\boldsymbol{\alpha}_8 = \begin{pmatrix} \frac{1}{3} & \frac{1}{3} & \frac{1}{3} \end{pmatrix}, \ \boldsymbol{A}_8 = \begin{pmatrix} -0.433521 & -0.383529 & -0.387345 \\ 29.9426 & 26.2806 & 26.5629 \\ -29.9911 & -26.3291 & -26.6114 \end{pmatrix}.$$

The choice given in [6.64] is motivated by the fact that the line $-0.32x + 0.45$ approximates well the initial decay of the lognormal PDF right after the peak. In general, numerical values for $f(0)$ and $f'(0)$ can be obtained by applying a linear fit to the original PDF.

In Table 6.12, we report the first five moments of the lognormal distribution and those of the ME(3) distributions. The trade off for fitting the shape of the PDF better is higher the distance of the fourth and fifth moments of the ME(3) from the original moments.

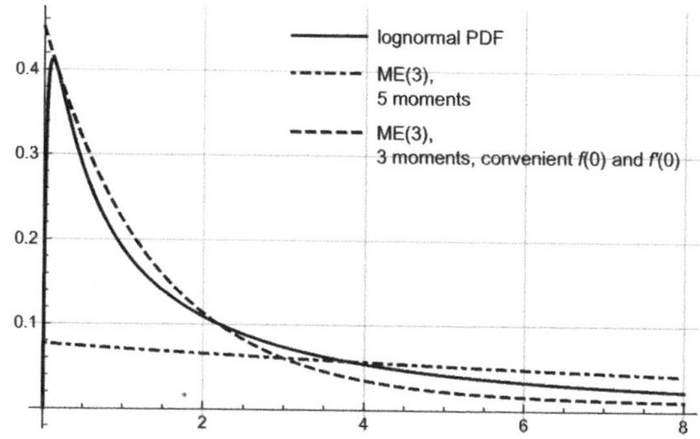


**Figure 6.38.** *Approximating the lognormal distribution with $\alpha = 1, \beta = 1.8$: matching five moments and matching three moments and approximately the shape of the PDF around zero by convenient values of $f(0)$ and $f'(0)$*

| | $m_1$ | $m_2$ | $m_3$ | $m_4$ | $m_5$ |
|---|---|---|---|---|---|
| lognormal | 13.73 | 4817 | $4.314 \times 10^7$ | $9.882 \times 10^{12}$ | $5.745 \times 10^{19}$ |
| ME($\boldsymbol{\alpha}_7, \boldsymbol{A}_7$) | 13.73 | 4817 | $4.314 \times 10^7$ | $9.882 \times 10^{12}$ | $5.745 \times 10^{19}$ |
| ME($\boldsymbol{\alpha}_8, \boldsymbol{A}_8$) | 13.73 | 4817 | $4.314 \times 10^7$ | $6.095 \times 10^{11}$ | $1.078 \times 10^{16}$ |

**Table 6.12.** *Moments of the lognormal distribution with $\alpha = 1, \beta = 1.8$ and moments of the matching ME(3) distributions*

## 6.9. Moment bounds of PH(2) distributions

To compute the moment bounds of PH distributions, it is convenient to use the normalized moments defined in [1.5], because they are invariant to the mean of the distribution. Besides, given a PH distribution represented by $(\boldsymbol{\alpha}, \boldsymbol{A})$ with mean equal to $m_1$, the first moment of the PH distribution represented by $(\boldsymbol{\alpha}, c\boldsymbol{A})$ is $= m_1/c$, while its normalized moments are equal to those calculated from $(\boldsymbol{\alpha}, \boldsymbol{A})$.

Note also that the SCV is equal to $n_2 - 1$. Furthermore, for any distribution one must have $n_3 \geq n_2$.

For the continuous time case, it is known that APH(2) distributions, PH(2) distributions and ME(2) distributions form the same class, that is:

$$\text{APH}(2) \equiv \text{PH}(2) \equiv \text{ME}(2).$$

Accordingly, they share the following moment bounds:

$$n_2: \quad \frac{3}{2} \leq n_2$$

$$n_3: \quad \frac{3\left(\sqrt{2}(2-n_2)^{3/2}+3n_2-4\right)}{n_2} \leq n_3 \leq 6 - \frac{6}{n_2} \quad \text{if } n_2 \leq 2 \qquad [6.65]$$

$$\frac{3n_2}{2} < n_3 \qquad \text{if } n_2 > 2$$

The above moments can be obtained by considering the canonical form of PH(2) distributions (section 5.6) with parameters that lead to extremal (minimal or maximal) normalized moments. The set of moments that can be realized with a PH(2) distribution is depicted in Figure 6.39. Solid lines indicate borders that are part of the set. A border without solid line can be approached arbitrarily, but values close to the boundary lead to very large intensities in the infinitesimal generator.

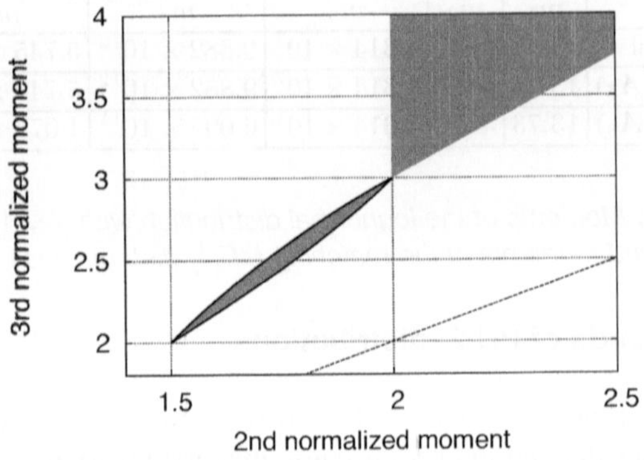


**Figure 6.39.** *Region of normalized moments that can be realized with PH(2) distributions. The dashed line indicates the lowest possible $n_3$ for any distribution*

In the discrete case, it is not possible to transform the moments in a way that the moment bounds are independent of $m_1$. Next we present the bounds for discrete distributions in terms of the first moment $m_1$, the SCV and the third factorial moment $f_3$.

Another difference is that in the discrete case the size 2 acyclic DPH distributions (ADPH(2)) form a proper subset of the DPH(2) distributions. We have:

$$\text{ADPH}(2) \subset \text{DPH}(2) \equiv \text{MG}(2).$$

The DPH(2)\ADPH(2) distributions (those DPH(2) distributions which are not ADPH(2) distributions) have one positive and one negative eigenvalues. The moment bounds for ADPH(2) distributions are the following [TEL 02]:

$$
\begin{aligned}
m_1: \quad & 1 \le m_1, \\
SCV: \quad & \frac{\{m_1\}(1-\{m_1\})}{m_1^2} \le SCV && \text{if } 1 \le m_1 < 2, \\
& \frac{1}{2} - \frac{1}{m_1} \le SCV && \text{if } 2 \le m_1, \\
f_3: \quad & g \le f_3 \le \frac{3f_2(f_2 - 2m_1 + 2)}{2(m_1 - 1)} && \text{if } 1 \le m_1 < 2 \text{ and} \\
& && \frac{\{m_1\}(1-\{m_1\})}{m_1^2} \le SCV < 1 - \frac{1}{m_1}, \\
& g \le f_3 \le 6m_1^2(m_1 - 1)SCV && \text{if } 2 \le m_1 \text{ and} \\
& && \frac{1}{2} - \frac{1}{m_1} \le SCV < \frac{\{m_1\}(\{m_1\}-1)}{m_1^2}, \\
& g \le f_3 \le \frac{3f_2(f_2 - 2m_1 + 2)}{2(m_1 - 1)} && \text{if } 2 \le m_1 \text{ and} \\
& && \frac{\{m_1\}(\{m_1\}-1)}{m_1^2} \le SCV < 1 - \frac{1}{m_1}, \\
& \frac{3f_2(f_2 - 2m_1 + 2)}{2(m_1 - 1)} \le f_3 && \text{if } 1 \le m_1 \text{ and } 1 - \frac{1}{m_1} \le SCV,
\end{aligned}
\qquad [6.66]
$$

where:

$$
\begin{aligned}
g &= \frac{6}{e^3}\left(m_1 e(3f_2 + 2m_1)(f_2 - 2m_1 + 2) - 2f_2^2(f_2 - \sqrt{2d})\right), \\
d &= 2m_1^2 - 2m_1 - f_2, \\
e &= 2m_1 + \sqrt{2d}, \\
f_2 &= SCVm_1^2 - m_1 + m_1^2.
\end{aligned}
$$

In order to illustrate the process of determining the moment bounds, we devise a semi-numerical procedure to determine the bounds for those DPH(2) distributions that are not covered by the bounds given in [6.66], that is, those with a positive and a negative eigenvalue. The canonical form of these distributions is given as [PAP 13] (see section 5.6):

$$
\boldsymbol{\beta} = \begin{pmatrix} p_1 & 1 - p_1 \end{pmatrix}, \quad \boldsymbol{B} = \begin{pmatrix} 1 - b_1 & b_1 \\ b_2 & 0 \end{pmatrix}.
$$

Their first moment, the SCV and the third factorial moment are:

$$m_1 = \frac{b_1 - b_2 p_1 + b_2 + p_1}{b_1 - b_1 b_2}, \quad [6.67]$$

$$SCV = \frac{b_2\left(b_1^2 + (b_1-1)b_2 + b_1 + 2\right) - (b_1-2)(b_2-1)^2 p_1 + (b_2-1)^2\left(-p_1^2\right)}{(b_1 - b_2 p_1 + b_2 + p_1)^2}, \quad [6.68]$$

$$f_3 = \frac{6(b_2-1)p_1(b_1(b_2(b_1b_2+3)-1)+1) - 6b_2(b_1b_2(b_1(b_1+b_2+2)+3)+1)}{b_1^3(b_2-1)^3}. \quad [6.69]$$

Based on [6.67], a given mean $m_1$ is realized if:

$$p_1 = \frac{b_2 + b_1((b_2-1)m_1+1)}{b_2-1},$$

from which the SCV is:

$$SCV = \frac{(b_1 - b_1 b_2)m_1^2 - (b_2 b_1 + b_1 + 2)m_1 + 2}{b_1(b_2-1)m_1^2}. \quad [6.70]$$

The $SCV$ in [6.70] is a decreasing function of $b_1$ and an increasing function of $b_2$. Accordingly, the minimal $SCV$ is realized with $b_1 = 1$ and $b_2 = 0$, but this results in a value for $p_1$ with $0 \le p_1 \le 1$ if and only if $1 \le m_1 \le 2$. In this case, the minimal $SCV$ is:

$$\frac{\{m_1\}\left(1 - \{m_1\}\right)}{m_1^2}$$

which is in accordance with theorem 6.4. If $m_1 > 2$, then the smallest SCV is achieved for $p = 1, b_1 = 1/(m_1 - 1), b_2 = 0$ and it is:

$$\frac{(m_1-1)(m_2-1)}{m_1^2}.$$

There is no upper bound on $SCV$, since for $b_1 = b_2$, $p_1 \to 0$ and $b_2 \to 0$, the $SCV$ tends to infinity.

Given the mean and the SCV ($m_1$,$SCV$), we can obtain $p_1$ and $b_1$ from [6.67] and [6.68], which leads to:

$$p_1 = \frac{(b_2-1)m_1(m_1(b_2 SCV + b_2 - 2) + b_2 + 4) + 2}{m_1\left(b_2^2 + (b_2-1)^2(SCV+1)m_1 - 1\right)}, \quad [6.71]$$

$$b_1 = \frac{2 - 2m_1}{m_1((b_2-1)(SCV+1)m_1 + b_2 + 1)}. \quad [6.72]$$

From [6.71], [6.72] and [6.69], we obtain the third factorial moment $f_3$ as a function of $m_1, SCV$ and $b_2$:

$$f_3 = \left(3m_1 \left(m_1 \left((b_2-1)(SCV+1)^2 m_1^2 - 4(b_2-1)(SCV+1)m_1 + 4b_2 SCV + 3b_2 - 2SCV - 5\right) + 2\right)\right) / \left(2(b_2-1)(m_1-1)\right). \quad [6.73]$$

Equation [6.73] is an increasing function of $b_2$ if $SCV \le 1-1/m_1$ and a decreasing function of $b_2$ otherwise. This implies that, given $m_1$ and $SCV$, the expression in [6.73] applied with the smallest possible and the largest possible values of $b_2$ provides the extremal values for $f_3$. By a possible value we mean a value of $b_2$ which, together with the values of $m_1$ and $SCV$, leads to $0 \le p_1 \le 1$ and $0 \le b_1 \le 1$ when applied in [6.71] and [6.72]. $b_2 = 0$ is a possible value and hence it gives the minimal $f_3$ if $SCV \le 1 - 1/m_1$ and the maximal $f_3$ otherwise. Furthermore, we must have:

$$b_2 \le \frac{m_1(SCVm_1 + m_1 - 3) + 2}{m_1(SCVm_1 + m_1 + 1)}$$

in order to obtain $0 < b_1 \le 1$. Based on these, one can develop a numerical procedure to determine the largest value of $b_2$ for which $0 \le p_1 \le 1$. Calculating $f_3$ for this value of $b_2$ gives the maximal $f_3$ if $SCV \le 1 - 1/m_1$ and the minimal $f_3$ otherwise. The result of the procedure is illustrated in Figure 6.40 where the set of realizable moments is depicted for two values of the mean (since the results differ significantly for $m_1 < 2$ and $m_1 > 2$), and the moment region of the acyclic class is also provided based on [6.66].

## 6.10. Moment bounds of PH(3) distributions

Explicit symbolic moment bounds are not known for PH(3) and ME(3) distributions, but it is possible to investigate the moment bounds numerically. The set of PH(3) distributions is a proper subset of the set of ME(3) distributions. In this chapter, we numerically illustrate the moment bounds of these two sets.

ME(3) and PH(3) distributions are uniquely defined by their first five moments. For a given set of first five moments, the corresponding Laplace transform can be obtained by Padé approximation as described in section 6.5 and from the Laplace transform a non-Markovian vector–matrix representation $(\boldsymbol{\gamma}, \boldsymbol{G})$ can be obtained according to section 5.3.4.

If the $-\boldsymbol{\gamma} e^{\boldsymbol{G}t}\boldsymbol{G}\mathbb{1}$ matrix exponential function is negative for some $t \ge 0$, then the given set of five moments does not belong to an ME(3) distribution. If the matrix exponential function is non-negative for $t \ge 0$, then five moments represent an ME(3) distribution, which might or might not be a PH(3) distribution.

If the $(\boldsymbol{\gamma}, \boldsymbol{G})$ non-Markovian representation can be transformed to a canonical PH(3) representation using the method given in algorithm 5.2, then the five moments represent a PH(3) distribution. If a given set of five moments represent an ME(3) (PH(3)) distribution, then we say that the set of moments are inside the moment bounds of the ME(3) (PH(3)) class.

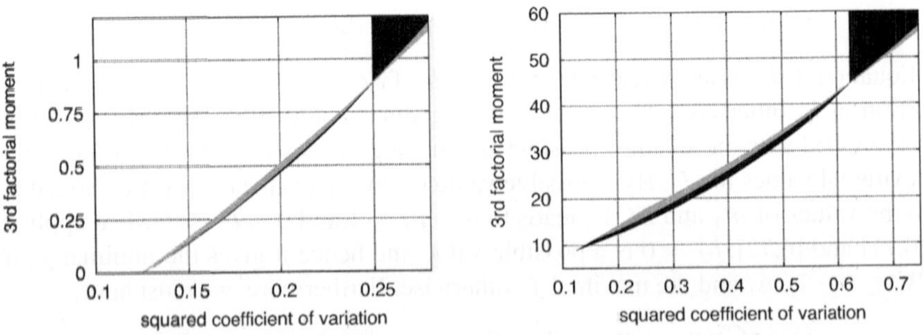


**Figure 6.40.** *Set of realizable moments for DPH(2) distributions with $m_1 = 4/3$ on the left and with $m_1 = 8/3$ on the right. The darker area shows the set of moments corresponding to the ADPH(2) class while the lighter area shows the set of moments corresponding to DPH(2)\ADPH(2) distributions*

As previously, to reduce the number of parameters, we work with normalized moments [1.5]. The subsequent numerical results are divided into investigations of the $n_2, n_3$ domain with arbitrary $n_4, n_5$, and investigations of the $n_4, n_5$ domain with given $n_2, n_3$.

### 6.10.1. *The second and third normalized moments*

The $n_2, n_3$ normalized moment bounds of the PH(3) class are not completely known yet. There is a result for the valid range of the APH(3) class [BOB 05], and there is a numerically checked conjecture that the related borders of the PH(3) class coincide with the ones of the APH(3) class [HOR 09]. Here, we compare the borders of the ME(3) class with these borders of the PH(3) class.

To check if an $(n_2, n_3)$ pair is inside the range of the ME(3) class is rather difficult. The procedure summarized in the introduction of section 6.10 checks if $\{n_2, n_3, n_4, n_5\}$ defines an ME(3) distribution. Using that, for a given $(n_2, n_3)$ pair a natural procedure would be to check the ME(3) membership of $\{n_2, n_3, x, y\}$, where $x$ and $y$ run through the positive quarter plain. Unfortunately, this procedure is infeasible, because it is practically impossible to find valid $(n_4, n_5)$ pairs with such a search.

To find $(n_2, n_3)$ pairs inside the ME(3)\PH(3) range, we use the following function with complex roots:

$$f(t) = a\, e^{-\lambda t}(1 + \cos(\omega t + \phi)), \tag{6.74}$$

where $a$ is a normalizing constant ($\int_t e^{-\lambda t}(1 + \cos(\omega t + \phi))dt = 1/a$), $\lambda$ is a scaling factor and $\omega$ and $\phi$ are shape parameters. The above function corresponds to a subset of the ME(3) class. For $\lambda = 1$:

$$n_2 = \frac{2\left(\sqrt{1+\omega^2} + \cos(\phi + \arctan(\omega))\right)\left(\left(1+\omega^2\right)^{\frac{3}{2}} + \cos(\phi + 3\arctan(\omega))\right)}{\left(1 + \omega^2 + \cos(\phi + 2\arctan(\omega))\right)^2},$$

$$n_3 = \frac{3\left(\sqrt{1+\omega^2} + \cos(\phi + \arctan(\omega))\right)\left(\left(1+\omega^2\right)^2 + \cos(\phi + 4\arctan(\omega))\right)}{\left(1 + \omega^2 + \cos(\phi + 2\arctan(\omega))\right)\left((1+\omega^2)^{\frac{3}{2}} + \cos(\phi + 3\arctan(\omega))\right)}.$$

If the above equations are satisfied by a pair of real $\phi$ and $\omega$ values for a given $(n_2, n_3)$ pair, then the given $(n_2, n_3)$ pair can be realized by an ME(3) distribution.

Figure 6.41 depicts the moment region of the ME(3) subclass (gray plus black area) characterized by [6.74] and the region of the PH(3) class (black) on the $(n_2, n_3)$ plane. Our numerical investigations suggest that the moment regions of whole ME(3) class are the same as the regions obtained by considering only the form given in [6.74]. The left most point of this region where $n_2 = 1.200902$ gives the ME(3) distribution with minimal $n_2$ or, equivalently, with minimal coefficient of variation, and this point corresponds to the minimal coefficient of variation of the ME(3) class reported in [ÉLT 06]. The $n_3$ normalized moment of the PH(3) class, and consequently the ME(3) class, is only lower bounded when $n_2 > 1.5$.

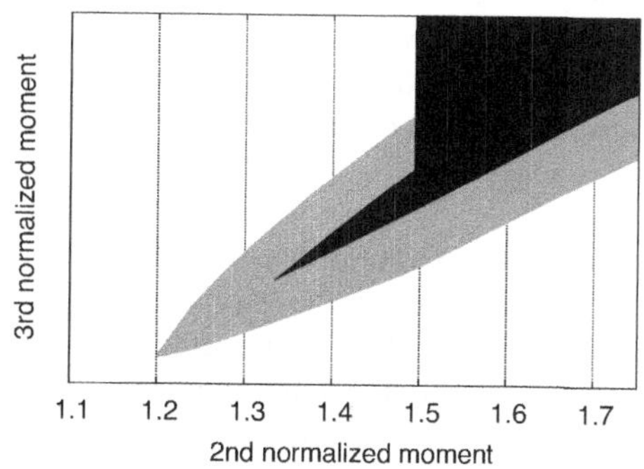


**Figure 6.41.** *The region of second and third normalized moments of the PH(3) and the ME(3) classes*

### 6.10.2. *The fourth and fifth normalized moments*

In this section, we study the bounds of the $n_4$ and $n_5$ normalized moments for a given $(n_2, n_3)$ pair of second and third normalized moments.

As mentioned in the previous section, it is practically impossible to find the $(n_4, n_5)$ region of ME(3) or PH(3) distributions with a given $(n_2, n_3)$ pair using general search algorithms for the $\{n_2, n_3, x, y\}$ normalized moments set. To overcome this difficulty, we generate a single $(\hat{n}_4, \hat{n}_5)$ pair that is inside the region of ME(3) or PH(3) distributions with normalized moments $n_2$ and $n_3$, and explore the $(n_4, n_5)$ moment bounds starting from $(\hat{n}_4, \hat{n}_5)$.

To find a $(\hat{n}_4, \hat{n}_5)$ pair, we apply the following procedure. If the $(n_2, n_3)$ pair is inside the moment bounds of PH(3), we generate a size 3 Exp-Erlang or Erlang-Exp distribution (introduced in section 6.6) with normalized moments $(n_2, n_3)$. If the $(n_2, n_3)$ pair is inside the moment bounds of ME(3)\PH(3), then we construct an ME(3) distribution in the form of [6.74] with normalized moments $(n_2, n_3)$. In both cases, $\hat{n}_4$ and $\hat{n}_5$ are the fourth and fifth normalized moments of the obtained distribution.

We start by considering cases for which $n_2 = 1.45$. Based on the results presented in section 6.10.1, for this value of $n_2$ the interval of realizable third normalized moments is $(1.6517, 2.1498)$ with ME(3) while it is $(1.8457, 1.9573)$ with PH(3). First, we look at the middle point of the $n_3$ interval that can be realized with a PH(3), that is, $n_3 = 1.9015$. Figure 6.42 depicts the $(n_4, n_5)$ moment bounds for both PH(3) and ME(3) distributions. In the figures, we have $n_4$ on the x-axis and $n_5$ on the y-axis. Further, the lighter gray region contains the points that are realized with ME(3) or PH(3) distributions with one real and a pair of complex eigenvalues, while the darker gray area contains points where the ME(3) or PH(3) distributions are realized with three real eigenvalues. It is clear from Figure 6.42 that the moment bounds of ME(3) distributions are much wider than the moment bounds of PH(3) distributions. In Figure 6.43, we concentrate on the lower peak of the regions depicted in Figure 6.42. The ME(3) class is somewhat more flexible in this subregion as well, and the flexibility is increased both for distributions with one real and two complex eigenvalues and for distributions with three real eigenvalues.

Now we turn our attention to $n_3$ values that cannot be realized by a PH(3) with $n_2 = 1.45$. In particular, Figure 6.44 depicts the $(n_4, n_5)$ moment bounds for $n_2 = 1.45$, $n_3 = 1.725$ and $n_2 = 1.45$, $n_3 = 2.1$ which lie, respectively, beneath and above the $n_3$ interval that can be realized with PH(3). By comparison with Figure 6.42, it is clear that approaching the possible minimum and maximum values of $n_3$, the $(n_4, n_5)$ moment bounds not only changes its shape but it is shrinking as well. To illustrate this shrinking further, Figure 6.45 depicts the $(n_4, n_5)$ moment bounds for $n_2 = 1.45$, $n_3 = 2.1249$ and $n_2 = 1.45$, $n_3 = 2.1373$ for which the moment bounds get narrower and shorter.

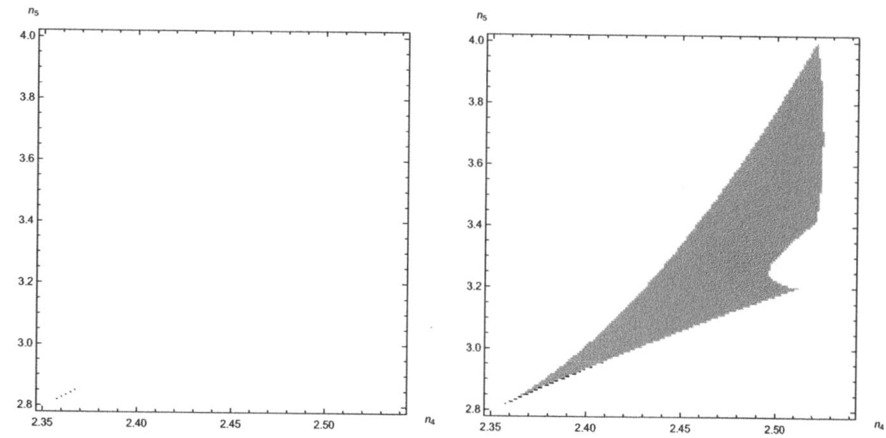

**Figure 6.42.** $(n_4, n_5)$ *moment bounds of PH(3) distributions (on the left, rather small region around* $(2.36, 2.82)$*) and ME(3) distributions (on the right) in the case of* $n_2 = 1.45$ *and* $n_3 = 1.9015$

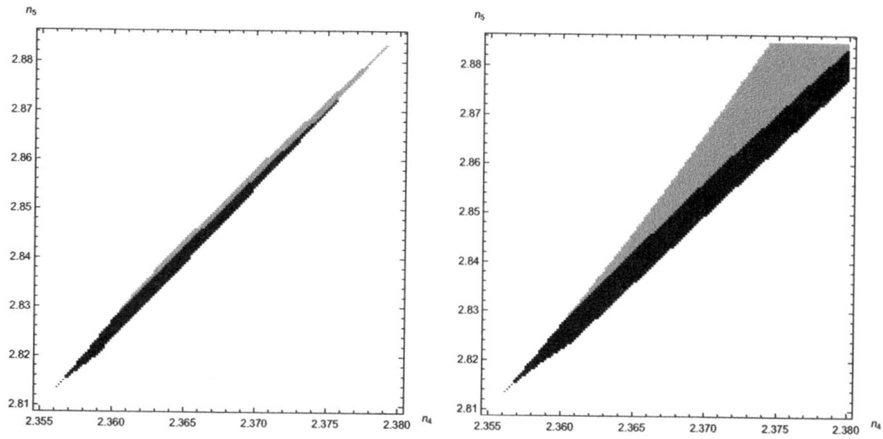

**Figure 6.43.** *Enlarged plot of* $(n_4, n_5)$ *moment bounds of PH(3) distributions (on the left) and ME(3) distributions (on the right) in the case of* $n_2 = 1.45$ *and* $n_3 = 1.9015$*. The gray area indicates PH(3) distributions with real eigenvalues and the black area indicates PH(3) distributions with complex eigenvalues*

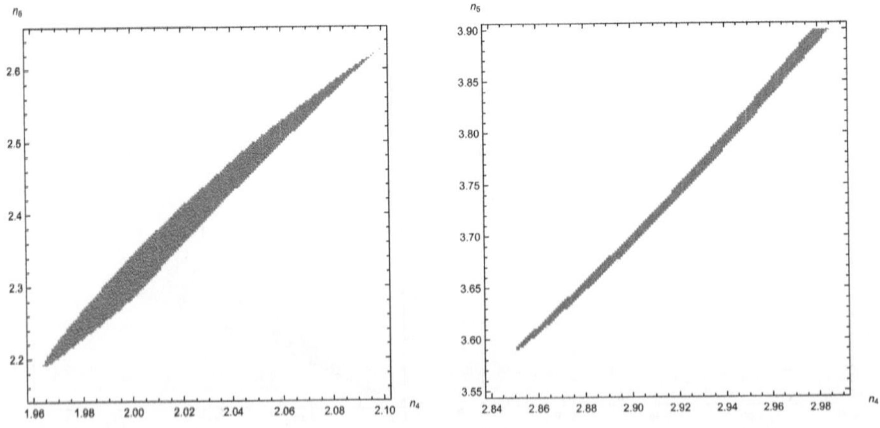


**Figure 6.44.** $(n_4, n_5)$ *normalized moment bounds of ME(3) distributions for* $n_2 = 1.45$ *and* $n_3 = 1.725$ *(on the left) and* $n_2 = 1.45$ *and* $n_3 = 2.1$ *(on the right)*

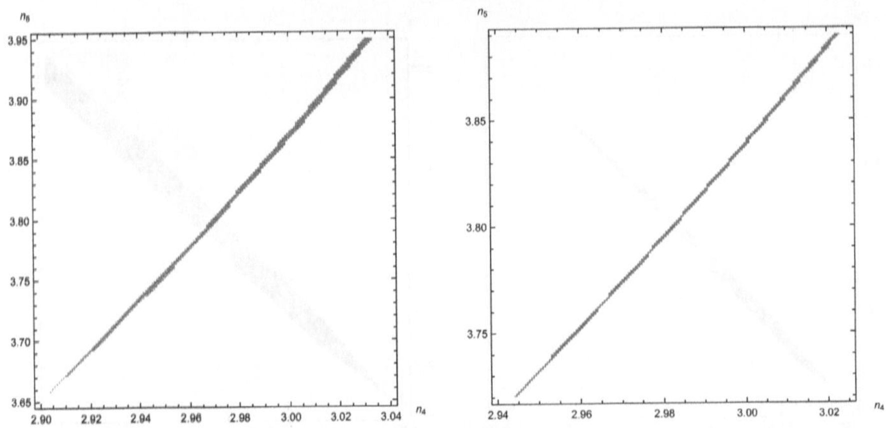


**Figure 6.45.** $(n_4, n_5)$ *normalized moment bounds of ME(3) distributions for* $n_2 = 1.45$ *and* $n_3 = 2.1249$ *(on the left) and* $n_2 = 1.45$ *and* $n_3 = 2.1373$ *(on the right)*

Next, we investigate a few cases with $n_2 = 1.6$. We start with two values of $n_3$, 1.9 and 2.0, that cannot be realized with a PH(3). The $n_4, n_5$ moment bounds are depicted in Figure 6.46. Diverging from the minimal $n_3$ value, that is, by increasing the actual value of $n_3$, the moment bounds become larger. Diverging further from the minimal $n_3$ value, we choose $n_3 = 2.2$ which can be realized by PH(3).

Figure 6.47 depicts the lower peak of the $(n_4, n_5)$ moment bounds for PH(3) and ME(3). This figure reports new qualitative properties. It indicates that the $(n_4, n_5)$ moment bounds can be composed of more than two areas and the regions are not convex.

The $n_2 = 1.6$, $n_3 = 2.2$ case is further illustrated by Figure 6.48, and there is no upper bound for $n_4$ and $n_5$. Instead, Figure 6.49 illustrates how the $(n_4,n_5)$ moment bounds change and move by increasing $n_3$ to 2.3.

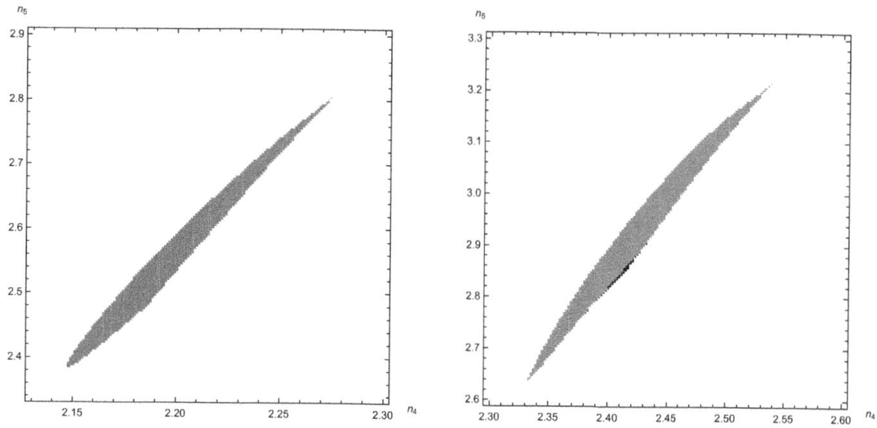


**Figure 6.46.** *$(n_4, n_5)$ moment bounds of ME(3) distributions for $n_2 = 1.6$ and $n_3 = 1.9$ (on the left) and for $n_2 = 1.6$ and $n_3 = 2.0$ (on the right)*

Next we investigate cases with $n_2 = 2.2$. Figure 6.50 depicts the moment bounds for $n_3 = 2.7333$, which cannot be realized by PH(3), and $n_3 = 2.9333$, which is the lower limit for PH(3), that is, a single $(n_4, n_5)$ point can be realized with PH(3).

For $n_3 = 3.1333$, the regions are shown in Figure 6.51, and for $n_3 = 3.3333$, the regions are shown in Figure 6.52. For $n_3 = 3.1333$, there are upper bounds for $n_4$ and $n_5$, which are not present for $n_3 = 3.3333$.

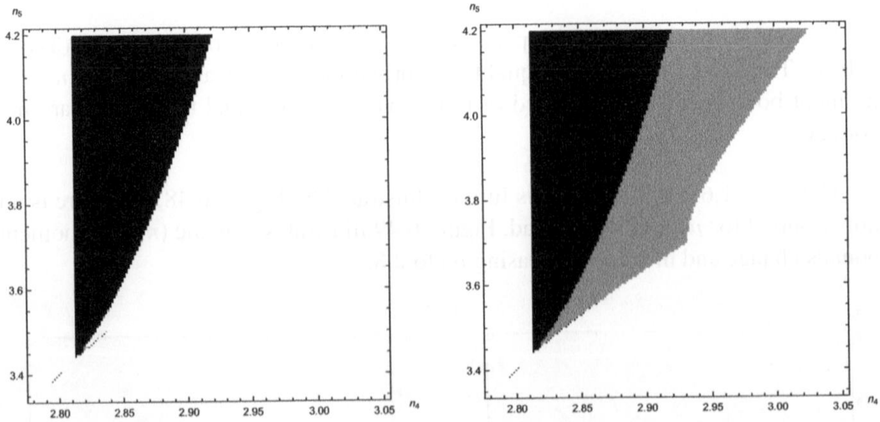


**Figure 6.47.** *Lower peak of the $n_4, n_5$ moment bounds of PH(3) distributions (on the left) and ME(3) distributions (on the right) in the case of $n_2 = 1.6$ and $n_3 = 2.2$*

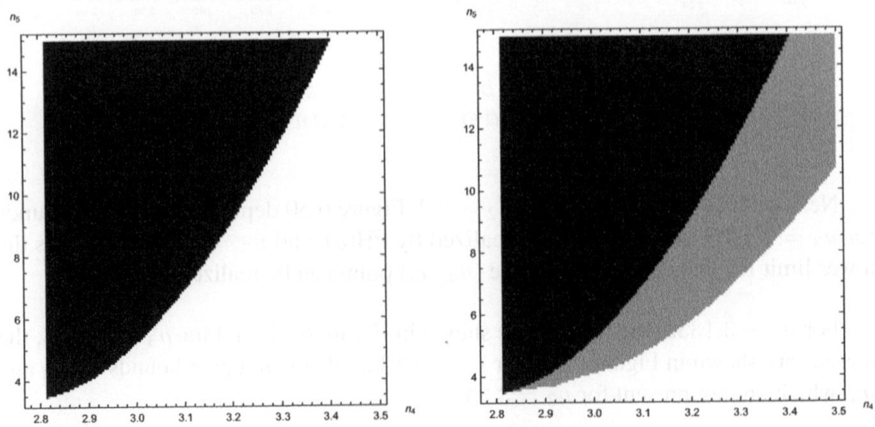


**Figure 6.48.** *$(n_4, n_5)$ moment bounds of PH(3) distributions (on the left) and ME(3) distributions (on the right) in the case of $n_2 = 1.6$ and $n_3 = 2.2$*

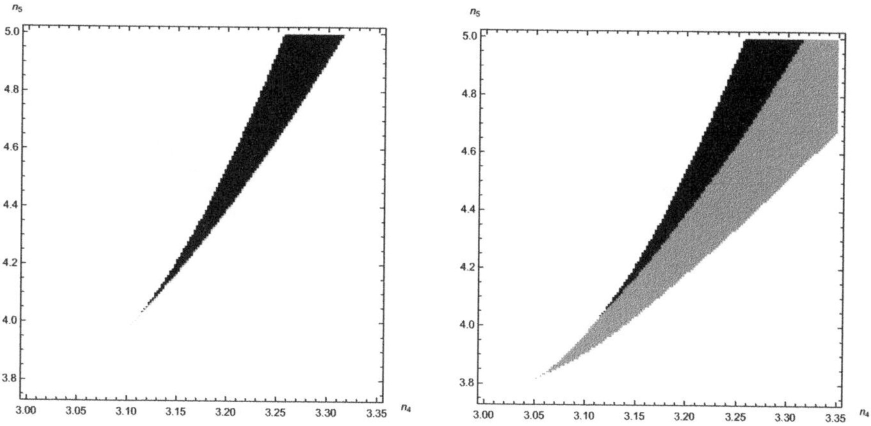


**Figure 6.49.** *$(n_4, n_5)$ moment bounds of PH(3) distributions (on the left) and ME(3) distributions (on the right) in the case of $n_2 = 1.6$ and $n_3 = 2.3$*

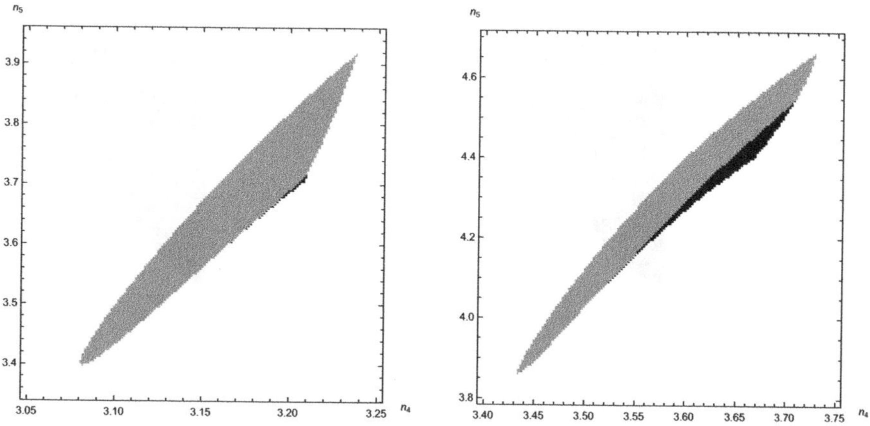


**Figure 6.50.** *$(n_4, n_5)$ moment bounds of ME(3) distributions for $n_3 = 2.7333$ (on the left) and $n_3 = 2.9333$ (on the right) in the case of $n_2 = 2.2$*

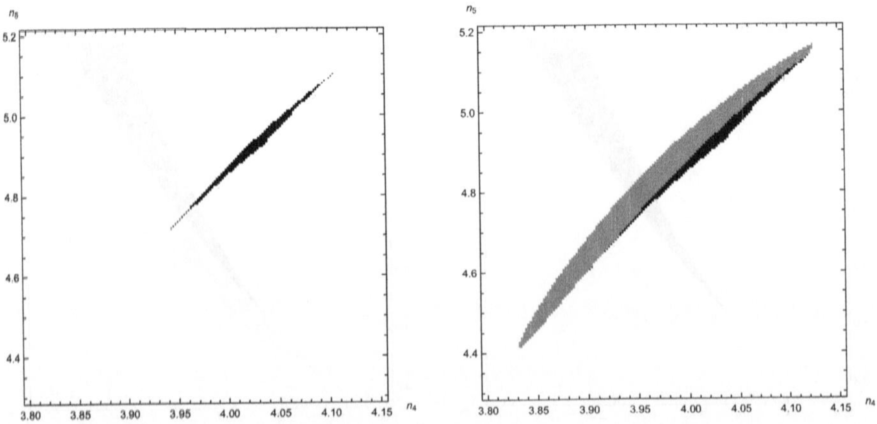


**Figure 6.51.** *($n_4, n_5$) moment bounds of PH(3) distributions (on the left) and ME(3) distributions (on the right) in the case of $n_2 = 2.2$ and $n_3 = 3.1333$*

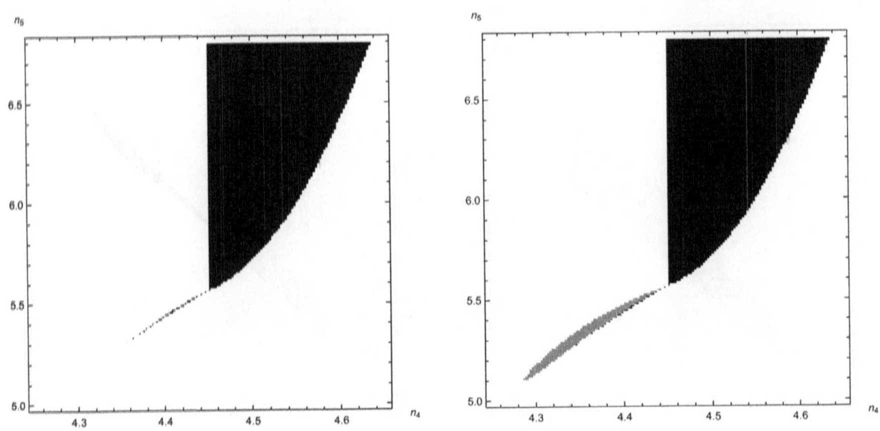


**Figure 6.52.** *($n_4, n_5$) moment bounds of PH(3) distributions (on the left) and ME(3) distributions (on the right) in the case of $n_2 = 2.2$ and $n_3 = 3.3333$*

# 7

# Distribution Fitting

The problem in which a distribution is given by its PDF (or CDF) and an approximating phase type distribution must be found is commonly referred to as phase type distribution fitting or parameter estimation. One approach to address this problem is to define a *distance measure*, which describes how similar the distributions are, and solving the numerical optimization problem in which the goal is to find an initial vector – infinitesimal generator pair which minimizes the distance measure.

Throughout this section, we will refer to a benchmark composed of nine distributions introduced in [BOB 94] with relevance in reliability analysis and quantitative evaluation problems. The distributions are the following (in parenthesis we give the abbreviation by which we refer to them from now on):

– Weibull: $\hat{f}(x) = \frac{\beta}{\eta}\left(\frac{x}{\eta}\right)^{\beta-1} e^{-\left(\frac{t}{\eta}\right)^{\beta}} \begin{cases} \eta = 1, \beta = 1.5, & \text{(W1)} \\ \eta = 1, \beta = 0.5, & \text{(W2)} \end{cases}$

– lognormal: $\hat{f}(x) = \frac{1}{x\sigma\sqrt{2\pi}} e^{-\frac{(\ln x - \mu)^2}{2\sigma^2}} \begin{cases} \mu = 1, \sigma = 1.8, & \text{(L1)} \\ \mu = 1, \sigma = 0.8, & \text{(L2)} \\ \mu = 1, \sigma = 0.2, & \text{(L3)} \end{cases}$

– uniform on $[0, 1]$: $\hat{f}(x) = \mathcal{I}_{\{0 \leq x \leq 1\}}$, (U1)

– uniform on $[1, 2]$: $\hat{f}(x) = \mathcal{I}_{\{1 \leq x \leq 2\}}$, (U2)

– shifted exponential: $\hat{f}(x) = \frac{1}{2}e^{-x} + \frac{1}{2}e^{-(x-1)}\mathcal{I}_{\{x \geq 1\}}$, (SE)

– matrix exponential: $\hat{f}(x) = \left(1 + \frac{1}{(2\pi)^2}\right)(1 - \cos(2\pi x))e^{-x}$. (ME*)

Table 7.1 provides the mean, the variance and the third central moment (i.e. $\hat{m}_1, \hat{c}_2$ and $\hat{c}_3$) for all distributions of the benchmark.

| | W1 | W2 | L1 | L2 | L3 | U1 | U2 | SE | ME* |
|---|---|---|---|---|---|---|---|---|---|
| $\hat{m}_1$ | 0.903 | 2 | 13.7 | 3.74 | 2.77 | 0.5 | 1.5 | 1.5 | 1.045 |
| $\hat{c}_2$ | 0.376 | 20 | 4629 | 12.6 | 0.313 | $0.08\dot{3}$ | $0.08\dot{3}$ | 1.25 | 0.953 |
| $\hat{c}_3$ | 0.246 | 592 | $4.29 \times 10^7$ | 164 | 0.108 | 0 | 0 | 2 | 1.99 |

**Table 7.1.** *Mean, variance and third central moment of the distributions of the benchmark*

## 7.1. Distance measures

Denoting by $\hat{f}(t)$ the PDF to fit and by $f(t)$ the PDF of the approximation, the most used distance measures are [BOB 94]:

– relative entrophy:

$$D_{\text{rel}} = \int_0^\infty \hat{f}(t) \log\left(\frac{\hat{f}(t)}{f(t)}\right) dt, \qquad [7.1]$$

– absolute and squared PDF difference:

$$D_{\text{pdf}} = \int_0^\infty \left|f(t) - \hat{f}(t)\right| dt, \; D_{\text{spdf}} = \int_0^\infty \left(f(t) - \hat{f}(t)\right)^2 dt, \qquad [7.2]$$

– absolute and squared CDF difference:

$$D_{\text{cdf}} = \int_0^\infty \left|F(t) - \hat{F}(t)\right| dt, \; D_{\text{scdf}} = \int_0^\infty \left(F(t) - \hat{F}(t)\right)^2 dt. \qquad [7.3]$$

The above distance measures are included in PhFit [HOR 02], which is a tool providing acyclic phase type approximations given the PDF or samples of a distribution. A detailed description of the numerical optimization procedure implemented in PhFit is given in [HOR 00b].

In Figure 7.1, we illustrate the application of $D_{\text{rel}}, D_{\text{pdf}}$ and $D_{\text{cdf}}$ distance measures to the fitting of W1 distribution. The fitting was carried out by PhFit with size $n = 4$. The results using $D_{\text{spdf}}$ and $D_{\text{scdf}}$ are omitted because are close to those using $D_{\text{pdf}}$ and $D_{\text{cdf}}$.

For the initial interval of the PDF up to $t = 3$, the approximations based on $D_{\text{rel}}$ and $D_{\text{pdf}}$ have similar performance, while $D_{\text{cdf}}$ is able to capture the PDF less. For the tail of the distribution ($t > 3$), $D_{\text{rel}}$ gets closest to the behavior of the W1 distribution and $D_{\text{pdf}}$ performs worst.

According to our experience, the three measures, $D_{\text{rel}}, D_{\text{pdf}}$ and $D_{\text{cdf}}$, usually perform similar at the initial part of the PDF up to the point where the asymptotic

tail behavior starts to dominate the PDF. After this point, there can be significant differences between the distance measures and, in general, it is difficult to tell in advance which one provides better approximation of the tail. Overall, in most cases $D_{\mathrm{rel}}$ leads to the most balanced performance considering both the initial part and the tail of the approximating PDF.

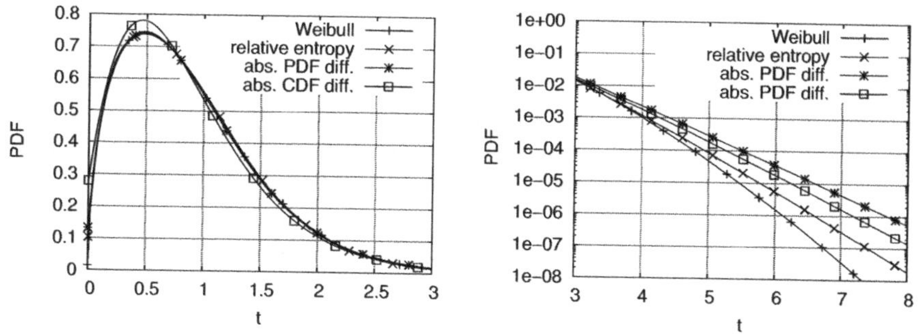


**Figure 7.1.** *Comparison of applying different distance measure to fit distribution W1 with size 4 APH distribution*

All measures tend to ignore special tail behaviors and give more consideration to the initial part of the PDF. Applying phase type fitting to heavy-tailed distributions gained interest in the 1990s. In [HOR 00b], other distance measures were studied in order to accentuate the tail behavior. One such measure is the relative PDF absolute area difference given by:

$$D_{\mathrm{rpdf}} = \int_0^{\infty} \frac{|f(t) - \hat{f}(t)|}{\hat{f}(t)}\, dt.$$

It turns out, however, that in most cases measures that "concentrate" also on the tail have bad performance at the initial part of the PDF. For this reason, in [HOR 00b], a method was introduced that separates the fitting of the initial part and that of the tail of the PDF. This method is also implemented in PhFit [HOR 02], and is presented in section 7.4.

Having performed the fitting according to a given distance measure, the result can be evaluated according to other indicators as well. Bobbio and Telek [BOB 94] suggest:

– the relative error of the first moment:

$$e_1 = \frac{|m_1 - \hat{m}_1|}{\hat{m}_1},$$

where $\hat{m}_1$ is the first moment of the PDF to fit;

– the relative error of the higher central moments given by:

$$e_i = \frac{|c_i - \hat{c}_i|}{\hat{c}_i}, \quad \text{for } i \geq 2,$$

where $\hat{c}_i$ and $c_i$ denote the $i$th central moment of the distribution to fit and that of the approximating phase type distribution, respectively; accordingly, $e_2$ is the relative error of the variance.

An important difference between general indicators, such as $e_i$, and distance measures is that a distance measure evaluates to zero if and only if the two involved PDFs are identical. On the other hand, even when an indicator is zero, for example, $e_i = 0$ for a given $i$, the two distributions can have significantly different PDFs.

At last, we point out a relation between the distance measure RE, $D_{\text{rel}}$, and maximum likelihood estimation. $D_{\text{rel}}$ can be written as:

$$D_{\text{rel}} = \int_0^\infty \hat{f}(t) \log\left(\frac{\hat{f}(t)}{f(t)}\right) dt = \int_0^\infty \hat{f}(t) \log(\hat{f}(t)) dt - \int_0^\infty \hat{f}(t) \log(f(t))\, dt,$$

where the first term depends only on the PDF to fit, and hence constant. Accordingly, optimizing the parameters by minimizing $D_{\text{rel}}$ corresponds to maximizing:

$$\int_0^\infty \hat{f}(t) \log(f(t))\, dt. \qquad [7.4]$$

In this integral, every possible outcome $X$ is considered by its density $\hat{f}(X)$ multiplied by the logarithm of the density (i.e. $\log(f(X))$) of reproducing it by the approximation. Accordingly, [7.4] is equal to the limit of:

$$\frac{\text{log-likelihood}}{\text{number of samples}}$$

as the number of samples tends to infinity. In other words, fitting according to $D_{\text{rel}}$ and maximum likelihood estimation are based on the same principle.

## 7.2. Fitting based on maximum likelihood

In this section, we assume that we are given $K$ samples $\mathcal{T} = \{t_1, ..., t_K\}$ from a distribution and we aim to find the best fitting PH distribution based on the maximum likelihood principle. That is, we search the PH distribution with representation $(\boldsymbol{\alpha}, \boldsymbol{A})$ that reproduces the samples with the highest possible likelihood.

The likelihood of the data is defined as:

$$\mathcal{L}_{\boldsymbol{\alpha},\boldsymbol{A}}(\mathcal{T}) = \prod_{k=1}^{K} \boldsymbol{\alpha} e^{\boldsymbol{A} t_k} \boldsymbol{a} \qquad [7.5]$$

where $\boldsymbol{a} = -\boldsymbol{A}\mathbb{1}$. Our goal is to find the $(\boldsymbol{\alpha}, \boldsymbol{A})$ pair of size $n$ that maximizes the likelihood of the set $\mathcal{T}$.

This is a nonlinear optimization problem, where the goal function is the likelihood function. This optimization problem can be handled with several different approaches. Some approaches are based on direct optimization of the goal function, while some methods are based on the stochastic interpretation of the underlying stochastic model.

### 7.2.1. *The expectation maximization algorithm*

The expectation maximization (EM) method optimizes the likelihood function based on the stochastic interpretation of the underlying Markov chain model whose time to absorption defines the PH distribution. The method is based on the observation that the likelihood would be easier to maximize if the stochastic evolution of the Markov chain was known. Specifically, we utilize the following hidden parameters associated with sample $t_k$ (which is also referred to as $k$th measurement interval):

– $J_{k,n}$ is the $n$th state visited by the Markov chain during the $k$th measurement interval;

– $\theta_{k,n}$ is the sojourn time of the $n$th sojourn of the Markov chain (in state $J_{k,n}$) during the $k$th measurement interval;

– $n_k$ is the number of sojourns during the $k$th measurement interval;

where the sojourn times determines the $k$th sample by $t_k = \sum_{n=1}^{n_k} \theta_{k,n}$.

Based on these hidden variables, the likelihood can be expressed as:

$$\begin{aligned}\mathcal{L} = \prod_{k=1}^{K} & \underbrace{\alpha_{J_{k,1}}}_{\text{initial state}} \underbrace{e^{-\theta_{k,1} A_{J_{k,1}}}}_{\text{sojourn in the first state}} \underbrace{A_{J_{k,1} J_{k,2}}}_{\text{first state transition}} \\ & \cdot e^{-\theta_{k,2} A_{J_{k,2}}} A_{J_{k,2} J_{k,3}} \cdot \dots \cdot e^{-\theta_{k,n_k} A_{J_{k,n_k}}} a_{J_{k,n_k}} \\ = \prod_{k=1}^{K} & \alpha_{J_{k,1}} \prod_{n=1}^{n_k - 1} e^{-\theta_{k,n} A_{J_{k,n}}} A_{J_{k,n} J_{k,n+1}} \cdot e^{-\theta_{k,n_k} A_{J_{k,n_k}}} a_{J_{k,n_k}},\end{aligned}$$

where $A_i = -A_{ii}$ is the departure rate of state $i$ of the CTMC, and the log-likelihood is:

$$\begin{aligned}\log \mathcal{L} = \sum_{k=1}^{K} & \log \alpha_{J_{k,1}} + \sum_{n=1}^{n_k - 1} -\theta_{k,n} A_{J_{k,n}} \\ & + \log A_{J_{k,n} J_{k,n+1}} - \theta_{k,n_k} A_{J_{k,n_k}} + \log a_{J_{k,n_k}}.\end{aligned}$$

It is not necessary to know each individual hidden variable to express the log-likelihood. It is sufficient to know the following aggregate parameters for $i \in \{1, 2, \ldots, n\}$:

– $\Theta_{k,i} = \sum_{n=1}^{n_k} \theta_{k,n} \mathcal{I}_{\{J_{k,n}=i\}}$: the total time spent in state $i$ from the initial state until reaching the absorbing state during the $k$th process;

– $\Theta_i = \sum_{k=1}^{K} \Theta_{k,i}$: the total time spent in state $i$ during the $K$ processes;

– $M_{k,i,j} = \sum_{n=1}^{n_k-1} \mathcal{I}_{\{J_{k,n}=i, J_{k,n+1}=j\}}$: the number of state transitions from state $i$ to state $j$ for $j \in \{1, 2, \ldots, n\} \setminus i$ during the $k$th process;

– $M_{i,j} = \sum_{k=1}^{K} M_{k,i,j}$: the aggregate number of state transitions from state $i$ to state $j$ for $j \in \{1, 2, \ldots, n\}$ during the $K$ processes;

– $K_i = \sum_{k=1}^{K} \mathcal{I}_{\{J_{k,1}=i\}}$: the number of processes which start from state $i$;

– $L_i = \sum_{k=1}^{K} \mathcal{I}_{\{J_{k,n_k}=i\}}$: the number of processes which move to the absorbing state from state $i$.

These aggregate measures satisfy:

$$
t_k = \sum_{i=1}^{n} \Theta_{k,i}, \qquad \sum_{k=1}^{K} t_k = \sum_{i=1}^{n} \Theta_i,
$$

$$
n_k = \sum_i \sum_{j, j \neq i} M_{k,i,j}, \qquad \sum_{k=1}^{K} n_k = \sum_i \sum_{j, j \neq i} M_{i,j},
$$

$$
K = \sum_i K_i, \qquad K = \sum_i L_i.
$$

With these aggregate parameters, the log-likelihood simplifies to:

$$
\log \mathcal{L} = \sum_{i=1}^{n} K_i \log \alpha_i - \sum_{i=1}^{n} \Theta_i A_i + \sum_{i=1}^{n} \sum_{j=1, j \neq i}^{n} M_{i,j} \log A_{i,j} + \sum_{i=1}^{n} L_i \log a_i. \qquad [7.6]
$$

#### 7.2.1.1. *The maximization step of the expectation maximization method*

The EM method is an iterative procedure whose iterations are composed of maximization and expectation steps.

The maximization step of the EM method aims to find the optimal value of the model parameters based on the hidden variables. More precisely, the maximization step computes the maximum of [7.6] based on the aggregate parameters. Due to the fact that the first term of [7.6] depends only on $\alpha_i$ and the rest of the terms depend on $A_{i,j}$ and $a_i$, we can separate the optimization problem into parts.

THEOREM 7.1.– *Assuming* $\alpha_i > 0$ *for* $i \in \{1, \ldots, n\}$ *and* $\sum_{i=1}^{n} \alpha_i = 1$, *the maximum of:*

$$\sum_{i=1}^{n} K_i \log \alpha_i$$

*as a function of* $\alpha_i$, *is obtained at* $\alpha_i = \frac{K_i}{K}$.

*Proof.* Let $k_i = \frac{K_i}{K}$. Both $(k_1, \ldots, k_n)$ and $(\alpha_1, \ldots, \alpha_n)$ define probability measures, whose RE, also referred to as Kullback–Leibler divergence, is defined as $\sum_{i=1}^{n} k_i \log \frac{k_i}{\alpha_i}$. The RE is non-negative and its minimum is zero, which is obtained only when the two measures are identical, that is, $k_i = \alpha_i$ for $i \in \{1, \ldots, n\}$. That is

$$\begin{aligned}
\min_{\alpha_1,\ldots,\alpha_n} \sum_{i=1}^{n} k_i \log \frac{k_i}{\alpha_i} &= \sum_{i=1}^{n} k_i \log k_i - \max_{\alpha_1,\ldots,\alpha_n} \sum_{i=1}^{n} k_i \log \alpha_i \\
&= \sum_{i=1}^{n} k_i \log k_i - \frac{1}{K} \max_{\alpha_1,\ldots,\alpha_n} \sum_{i=1}^{n} K_i \log \alpha_i
\end{aligned}$$

obtained at $\alpha_i = k_i = \frac{K_i}{K}$. □

THEOREM 7.2.– *The maximum of:*

$$-\sum_{i=1}^{n} \Theta_i A_i + \sum_{i=1}^{n} \sum_{j,j \neq i} M_{i,j} \log A_{i,j}$$

*as a function of* $A_{i,j}$ *is obtained at:*

$$A_{i,j} = \frac{\sum_{k=1}^{K} M_{k,i,j}}{\Theta_i}.$$

*Proof.* Using $A_i = \sum_{j=1,j\neq i}^{n} A_{ij} + a_i$, for $i \neq j$ we have:

$$\begin{aligned}
\frac{\partial}{\partial A_{i,j}} \log \mathcal{L} &= \frac{\partial}{\partial A_{i,j}} - \sum_{i=1}^{n} \Theta_i A_i + \sum_{i=1}^{n} \sum_{j,j\neq i} M_{i,j} \log A_{i,j} \\
&= \frac{\partial}{\partial A_{i,j}} - \sum_{i=1}^{n} \Theta_i A_{i,j} + \sum_{i=1}^{n} \sum_{j,j\neq i} M_{i,j} \log A_{i,j} \\
&= -\Theta_i + M_{i,j} \frac{1}{A_{i,j}},
\end{aligned}$$

The optimal $A_{i,j}$ parameter is obtained when:

$$0 = \sum_{k=1}^{K} \Theta_{k,i} - \frac{M_{k,i,j}}{A_{i,j}}. \qquad [7.7]$$

□

THEOREM 7.3.– *The maximum of:*

$$-\sum_{i=1}^{n} \Theta_i A_i + \sum_{i=1}^{n} L_i \log a_i$$

*as a function of $a_i$ is obtained at $a_i = \frac{L_i}{\Theta_i}$.*

*Proof.* Similar to the proof of theorem 7.2:

$$\begin{aligned}\frac{\partial}{\partial a_i} \log \mathcal{L} &= \frac{\partial}{\partial a_i} - \sum_{i=1}^{n} \Theta_i A_i + \sum_{i=1}^{n} L_i \log a_i = \frac{\partial}{\partial a_i} - \sum_{i=1}^{n} \Theta_i a_i + \sum_{i=1}^{n} L_i \log a_i \\ &= -\Theta_i + L_i \frac{1}{a_i},\end{aligned}$$

and the optimal $a_i$ parameter is obtained where the derivative is zero. □

#### 7.2.1.2. *The expectation step of the expectation maximization method*

In the E-step, we compute the expected value of the hidden parameters based on $\boldsymbol{\alpha}$, $\boldsymbol{A}$ and the observed data $\mathcal{T} = (t_1, \ldots, t_K)$.

For $\mathbb{E}(K_i|\mathcal{T})$, we have:

$$\begin{aligned}\mathbb{E}(K_i|\mathcal{T}) &= \mathbb{E}\left(\sum_{k=1}^{K} \mathcal{I}_{\{J_{k,1}=i\}}|\mathcal{T}\right) = \sum_{k=1}^{K} \mathbb{E}(\mathcal{I}_{\{J_{k,1}=i\}}|t_k) \\ &= \sum_{k=1}^{K} \mathrm{P}(\mathcal{J}(0) = i|t_k) \qquad [7.8]\end{aligned}$$

where

$$\begin{aligned}\mathrm{P}(\mathcal{J}(0) = i|t_k) &= \lim_{\Delta\to 0} \mathrm{P}(\mathcal{J}(0) = i|\mathcal{J}(t_k) < n+1, \mathcal{J}(t_k+\Delta) = n+1) \\ &= \lim_{\Delta\to 0} \sum_{\ell=1}^{n} \mathrm{P}(\mathcal{J}(0) = i|\mathcal{J}(t_k) = \ell, \mathcal{J}(t_k+\Delta) = n+1) \\ &= \lim_{\Delta\to 0} \frac{\sum_{\ell=1}^{n} \mathrm{P}(\mathcal{J}(0) = i, \mathcal{J}(t_k) = \ell, \mathcal{J}(t_k+\Delta) = n+1)}{\sum_{\ell=1}^{n} \mathrm{P}(\mathcal{J}(t_k) = \ell, \mathcal{J}(t_k+\Delta) = n+1)} \\ &= \lim_{\Delta\to 0} \frac{\sum_{\ell=1}^{n} \mathrm{P}(\mathcal{J}(0) = i, \mathcal{J}(t_k) = \ell, \mathcal{J}(t_k+\Delta) = n+1)}{\sum_{m=1}^{n}\sum_{\ell=1}^{n} \mathrm{P}(\mathcal{J}(0) = m, \mathcal{J}(t_k) = \ell, \mathcal{J}(t_k+\Delta) = n+1)} \\ &= \lim_{\Delta\to 0} \frac{\alpha_i \mathbf{e}_i e^{\boldsymbol{A}t_k} \boldsymbol{a}\Delta + \sigma(\Delta)}{\boldsymbol{\alpha} e^{\boldsymbol{A}t_k} \boldsymbol{a}\Delta + \sigma(\Delta)} \\ &= \frac{\alpha_i \mathbf{e}_i e^{\boldsymbol{A}t_k} \boldsymbol{a}}{\boldsymbol{\alpha} e^{\boldsymbol{A}t_k} \boldsymbol{a}}, \qquad [7.9]\end{aligned}$$

where the $\sigma(\Delta)$ error term is such that $\lim_{\Delta\to 0}\sigma(\Delta)/\Delta = 0$.

Similarly, for $\mathbb{E}(L_i|\mathcal{T})$, we write:

$$\mathbb{E}(L_i|\mathcal{T}) = \mathbb{E}\left(\sum_{k=1}^{K}\mathcal{I}_{\{J_{k,n_k}=i\}}|\mathcal{T}\right) = \sum_{k=1}^{K}\mathbb{E}\left(\mathcal{I}_{\{J_{k,n_k}=i\}}|t_k\right)$$

$$= \sum_{k=1}^{K} \mathrm{P}\left(\mathcal{J}(t_k^-) = i|t_k\right) \quad [7.10]$$

where

$$\mathrm{P}\left(\mathcal{J}(t_k^-) = i|t_k\right) = \lim_{\Delta\to 0} \mathrm{P}(\mathcal{J}(t_k) = i|\mathcal{J}(t_k+\Delta) = n+1)$$

$$= \lim_{\Delta\to 0}\frac{\mathrm{P}(\mathcal{J}(t_k) = i, \mathcal{J}(t_k+\Delta) = n+1)}{\mathrm{P}(\mathcal{J}(t_k+\Delta) = n+1)}$$

$$= \lim_{\Delta\to 0}\frac{\sum_{m=1}^{n}\mathrm{P}(\mathcal{J}(0) = m, \mathcal{J}(t_k) = i, \mathcal{J}(t_k+\Delta) = n+1)}{\sum_{m=1}^{n}\sum_{\ell=1}^{n}\mathrm{P}(\mathcal{J}(0) = m, \mathcal{J}(t_k) = \ell, \mathcal{J}(t_k+\Delta) = n+1)}$$

$$= \lim_{\Delta\to 0}\frac{\boldsymbol{\alpha}\mathbf{e}_i e^{\mathbf{A}t_k}\mathbf{e}_i{}^T a_i\Delta + \sigma(\Delta)}{\boldsymbol{\alpha} e^{\mathbf{A}t_k}\boldsymbol{a}\Delta + \sigma(\Delta)}$$

$$= \frac{\boldsymbol{\alpha} e^{\mathbf{A}t_k}\mathbf{e}_i{}^T a_i}{\boldsymbol{\alpha} e^{\mathbf{A}t_k}\boldsymbol{a}}. \quad [7.11]$$

For the sojourn time in state $i$, we have $\mathbb{E}(\Theta_i|\mathcal{T}) = \sum_{k=1}^{K}\mathbb{E}(\Theta_{k,i}|t_k)$, where:

$$\mathbb{E}(\Theta_{k,i}|t_k) = \mathbb{E}\left(\sum_{n=1}^{n_k}\theta_{k,n}\mathcal{I}_{\{J_{k,n}=i\}}|t_k\right) = \mathbb{E}\left(\int_{t=0}^{t_k}\mathcal{I}_{\{\mathcal{J}(t)=i|t_k\}}dt\right)$$

$$= \int_{t=0}^{t_k}\mathbb{E}\left(\mathcal{I}_{\{\mathcal{J}(t)=i|t_k\}}\right)dt = \int_{t=0}^{t_k}\mathrm{P}(\mathcal{J}(t) = i|t_k)dt$$

$$= \lim_{\Delta\to 0}\int_{t=0}^{t_k}\mathrm{P}(\mathcal{J}(t) = i|\mathcal{J}(t_k) < n+1, \mathcal{J}(t_k+\Delta) = n+1)dt$$

$$= \lim_{\Delta\to 0}\int_{t=0}^{t_k}\frac{\sum_{\ell=1}^{n}\mathrm{P}(\mathcal{J}(t) = i, \mathcal{J}(t_k) = \ell, \mathcal{J}(t_k+\Delta) = n+1)}{\sum_{\ell=1}^{n}\mathrm{P}(\mathcal{J}(t_k) = \ell, \mathcal{J}(t_k+\Delta) = n+1)}dt$$

$$= \lim_{\Delta\to 0}\int_{t=0}^{t_k}\frac{\sum_{k=1}^{n}\sum_{\ell=1}^{n}\mathrm{P}(\mathcal{J}(0) = k, \mathcal{J}(t) = i, \mathcal{J}(t_k) = \ell, \mathcal{J}(t_k+\Delta) = n+1)}{\sum_{k=1}^{n}\sum_{\ell=1}^{n}\mathrm{P}(\mathcal{J}(0) = k, \mathcal{J}(t_k) = \ell, \mathcal{J}(t_k+\Delta) = n+1)}dt$$

$$= \lim_{\Delta\to 0}\frac{\int_{t=0}^{t_k}\boldsymbol{\alpha} e^{\mathbf{A}t}\mathbf{e}_i{}^T\mathbf{e}_i e^{\mathbf{A}(t_k-t)}\boldsymbol{a}dt\Delta + \sigma(\Delta)}{\boldsymbol{\alpha} e^{\mathbf{A}t_k}\boldsymbol{a}\Delta + \sigma(\Delta)}$$

$$= \frac{\int_{t=0}^{t_k}\boldsymbol{\alpha} e^{\mathbf{A}t}\mathbf{e}_i{}^T\mathbf{e}_i e^{\mathbf{A}(t_k-t)}\boldsymbol{a}dt}{\boldsymbol{\alpha} e^{\mathbf{A}t_k}\boldsymbol{a}}. \quad [7.12]$$

In a similar manner, $\mathbb{E}(M_{i,j}|\mathcal{T}) = \sum_{k=1}^{K} \mathbb{E}(M_{k,i,j}|t_k)$ and:

$$\mathbb{E}(M_{k,i,j}) = \frac{\int_{t=0}^{t_k} \boldsymbol{\alpha} e^{\boldsymbol{A}t} \mathbf{e}_i{}^T A_{i,j} \mathbf{e}_j e^{\boldsymbol{A}(t_k-t)} \boldsymbol{a} dt}{\boldsymbol{\alpha} e^{\boldsymbol{A}t_k} \boldsymbol{a}}. \qquad [7.13]$$

### 7.2.1.3. *Numerical computation of $\mathbb{E}(\Theta_{k,i})$ and $\mathbb{E}(M_{k,i,j})$*

The efficiency of the numerical evaluation of the EM method can be improved based on the following transformation. To compute the numerator of $\mathbb{E}(\Theta_{k,i})$, a convolution integral has to be evaluated, which can be simplified as:

$$\int_{t=0}^{t_k} \boldsymbol{\alpha} e^{\boldsymbol{A}t} \mathbf{e}_i{}^T \mathbf{e}_i e^{\boldsymbol{A}(t_k-t)} \boldsymbol{a}\, dt = \int_{t=0}^{t_k} \mathbf{e}_i e^{\boldsymbol{A}^T t} \boldsymbol{\alpha}^T \boldsymbol{a}^T e^{\boldsymbol{A}^T (t_k-t)} \mathbf{e}_i{}^T\, dt$$

$$= \mathbf{e}_i \begin{pmatrix} \mathbf{0} & \boldsymbol{I} \end{pmatrix} e^{\begin{pmatrix} \boldsymbol{A}^T & \boldsymbol{\alpha}^T \boldsymbol{a}^T \\ \mathbf{0} & \boldsymbol{A}^T \end{pmatrix} t_k} \begin{pmatrix} \boldsymbol{I} \\ \mathbf{0} \end{pmatrix} \mathbf{e}_i{}^T = \mathbf{e}_{n+i}\, e^{\boldsymbol{M}t_k} \mathbf{e}_i{}^T$$

where the size of $\boldsymbol{M} = \begin{pmatrix} \boldsymbol{A}^T & \boldsymbol{\alpha}^T \boldsymbol{a}^T \\ \mathbf{0} & \boldsymbol{A}^T \end{pmatrix}$ is $2n \times 2n$, the size of the matrix blocks is $n \times n$ and the size of the unit vectors in the right most expression is $2n$. That is, the convolution integral can be replaced by the evaluation of a matrix exponential of double size [VAN 78]. In a similar manner, the expected value of $M_{k,i,j}$ is:

$$\mathbb{E}(M_{k,i,j}) = \int_{t=0}^{t_k} \boldsymbol{\alpha} e^{\boldsymbol{A}t} \mathbf{e}_i{}^T A_{i,j} \mathbf{e}_j e^{\boldsymbol{A}(t_k-t)} \boldsymbol{a}\, dt$$

$$= A_{i,j} \int_{t=0}^{t_k} \mathbf{e}_i e^{\boldsymbol{A}^T t} \boldsymbol{\alpha}^T \boldsymbol{a}^T e^{\boldsymbol{A}^T (t_k-t)} \mathbf{e}_j{}^T\, dt$$

$$= A_{i,j} \mathbf{e}_i \begin{pmatrix} \mathbf{0} & \boldsymbol{I} \end{pmatrix} e^{\boldsymbol{M}t_k} \begin{pmatrix} \boldsymbol{I} \\ \mathbf{0} \end{pmatrix} \mathbf{e}_j{}^T = A_{i,j} \mathbf{e}_{n+i}\, e^{\boldsymbol{M}t_k} \mathbf{e}_j{}^T.$$

### 7.2.1.4. *The application of the expectation maximization method*

#### 7.2.1.4.1. An iteration of the expectation maximization method

The EM method aims to update the model parameters, $(\boldsymbol{a}, \boldsymbol{A})$, with the optimal value obtained from the hidden model. If $K_i$, $L_i$, $\Theta_i$, and $M_{i,j}$ were known, then the parameters would be set to $\alpha_i = \frac{K_i}{K}$, $a_i = \frac{L_i}{\Theta_i}$ and $A_{i,j} = \frac{M_{i,j}}{\Theta_i}$. Due to the fact that those hidden parameters are not known, only their expectations, the EM method sets the parameters to $\alpha_i = \frac{\mathbb{E}(K_i)}{K}$, $a_i = \frac{\mathbb{E}(L_i)}{\mathbb{E}(\Theta_i)}$ and $A_{i,j} = \frac{\mathbb{E}(M_{i,j})}{\mathbb{E}(\Theta_i)}$.

That is, in the $\omega$th iteration the EM method updates $\alpha_i$, $a_i$ and $A_{i,j}$ as follows:

$$\alpha_i(\omega+1) = \frac{\mathbb{E}(K_i)}{K} = \alpha_i(\omega)\frac{1}{K}\sum_{k=1}^{K}\frac{\mathbf{e}_i e^{\boldsymbol{A}(\omega)t_k}\boldsymbol{a}(\omega)}{\boldsymbol{\alpha}(\omega)e^{\boldsymbol{A}(\omega)t_k}\boldsymbol{a}(\omega)},$$

$$a_i(\omega+1) = \frac{\mathbb{E}(L_i)}{\sum_{k=1}^{K}\mathbb{E}(\Theta_{k,i})} = a_i(\omega)\frac{\sum_{k=1}^{K}\boldsymbol{\alpha}(\omega)e^{\boldsymbol{A}(\omega)t_k}\mathbf{e}_i{}^T}{\sum_{k=1}^{K}\mathbf{e}_{n+i}\, e^{\boldsymbol{M}(\omega)t_k}\mathbf{e}_i{}^T},$$

$$A_{i,j}(\omega+1) = \frac{\sum_{k=1}^{K}\mathbb{E}(M_{k,i,j})}{\sum_{k=1}^{K}\mathbb{E}(\Theta_{k,i})} = A_{i,j}(\omega)\frac{\sum_{k=1}^{K}\mathbf{e}_{n+i}\, e^{\boldsymbol{M}(\omega)t_k}\mathbf{e}_j{}^T}{\sum_{k=1}^{K}\mathbf{e}_{n+i}\, e^{\boldsymbol{M}(\omega)t_k}\mathbf{e}_i{}^T}.$$

Consequently, the elements of $\alpha_i(\omega+1)$, $a_i(\omega+1)$ and $A_{i,j}(\omega+1)$ are the product of $\alpha_i(\omega)$, $a_i(\omega)$ and $A_{i,j}(\omega)$, respectively, and an actual guess-dependent value. If an element of $\alpha_i(0)$, $a_i(0)$ or $A_{i,j}(0)$ is zero, then the given element remains zero in all iterations. This property also affects the number of parameters of the optimization problem. The EM method optimizes only the non-zero parameters in $(\boldsymbol{\alpha}, \boldsymbol{A})$. If all elements of $(\boldsymbol{\alpha}, \boldsymbol{A})$ are non-zero, then the method optimizes $n^2 + n - 1$ parameters, which is more than the number of free parameters of a PH($n$) distribution, $2n - 1$.

It is a beneficial theoretical property of the EM method that it increases the likelihood in each iteration step [ASM 96], that is:

$$\mathcal{L}_{\boldsymbol{\alpha}(\omega+1),\boldsymbol{A}(\omega+1)}(\mathcal{T}) \geq \mathcal{L}_{\boldsymbol{\alpha}(\omega),\boldsymbol{A}(\omega)}(\mathcal{T}).$$

#### 7.2.1.4.2. Implementation of the EM method

The fact that the likelihood increases in each iteration step does not ensure convergence to the global optimum. Numerical investigations suggest that $\mathcal{L}_{\boldsymbol{\alpha},\boldsymbol{A}}(\mathcal{T})$ might have multiple local optima and the EM method can get captured by any of them depending on the initial guess of the procedure, $(\boldsymbol{\alpha}(0), \boldsymbol{A}(0))$. This way the likelihood value at the end of the iteration is bounded from above by the optimum, where the iteration is converging to.

In general, the number of local minima increases with the number of parameters (the non-zero values in $(\boldsymbol{\alpha}(0), \boldsymbol{A}(0))$), which increases the probability of getting captured in a local optimum with low likelihood value. To cope with this non-desired property, it is recommended to run the procedure $L$ times, starting from $L$ different initial guesses, where $L$ increases with the number of parameters.

The optimal setting of the initial guesses is an open research problem. As a first proposal, uniformly distributed and properly normalized initial guesses can be obtained according to the following procedure [ASM 96]:

1) select the zero elements of $\boldsymbol{\alpha}, \boldsymbol{A}, \boldsymbol{a}$;

2) set all non-zero elements of $\alpha$, $A$, $a$ to be uniformly distributed pseudo random numbers in $(0, 1)$;

3) for $i = 1$ to $n$, $A_{ii} = -a_i - \sum_{j=1, j \neq i}^{n} A_{ij}$;

4) $\mathbb{E}(\mathcal{T}) = \sum_{k=1}^{K} t_k / K$, $m = \alpha(-A)^{-1}\mathbb{1}$;

5) $\alpha = \frac{\alpha}{\alpha \mathbb{1}}$, $A = \frac{m}{\mathbb{E}(\mathcal{T}) A}$, $a = \frac{m}{\mathbb{E}(\mathcal{T}) a}$;

where $\mathbb{E}(\mathcal{T})$ is the sample mean, and the last step ensures that the mean of the generated initial guess is equal to the sample mean.

#### 7.2.1.5. *Special continuous phase type classes*

By properly selecting the zero elements in the initial guess of the EM method, the optimization procedure is restricted to special subclasses of the PH distributions. For example, when $A_{ij} = 0$ for all $i \neq j$, then the EM procedure optimizes the likelihood of hyper-exponential distributions. Similarly, when $A_{ij} = 0$ for all $i > j + 1$ and $i < j$, $\alpha_i = 0$ for all $i > 1$ and $a_i = 0$ for all $i < n$, then the procedure optimizes the likelihood of hypo-exponential distributions.

The zero elements of the CF1, CF2, CF3 canonical forms of APH distributions can also be enforced in the initial guess of the procedure, which is maintained during the EM iterations. Despite of that, the EM method is not able to ensure the order conditions of the canonical forms. E.g., running the EM procedure with an initial guess whose zero elements follow the structure of CF1 does not ensure that the diagonal elements of $A$ are properly ordered according to the CF1 form (see Figure 5.5).

In these special cases, the computation of the matrix exponential in [7.12] and [7.13] might simplify significantly. Utilizing a special structure of the background Markov chain are sometimes indicated in the name of the procedure, for example, EM method for hyper-exponential distribution.

Some structural properties of the background Markov chain can also be involved in the maximization and the expectation steps of the EM method. One of the most widely used example for this situation is the EM method for hyper-Erlang distributions [THÜ 06]. In this case, the procedure utilizes that the PDF of the Erlang distribution is known and easy to compute, which makes the EM iteration for hyper-Erlang distributions much faster than the one for PH distributions with general matrix structure.

### 7.2.2. *Likelihood optimization for acyclic phase type distributions*

According to theorem 5.11, any APH distribution can be transformed to the CF1 form in Figure 5.5 with the associated $(\alpha, A)$ matrix representation and PDF given in [5.15] and [5.16], respectively. In this section, we introduce the direct likelihood optimization method for APH distributions in CF1 form, which is implemented in the PhFit tool [HOR 02].

The setting is similar to the previous section: we assume that we are given $K$ samples $\mathcal{T} = \{t_1, \ldots, t_K\}$, and we search the APH distribution in CF1 form (given in [5.15]) with representation $(\boldsymbol{\alpha}, \boldsymbol{A})$ of size $n$ that reproduces $\mathcal{T}$ with the highest possible likelihood. The parameters of the representation are $\boldsymbol{p} = (p_1 \ldots p_n)$ and $\boldsymbol{\lambda} = (\lambda_1 \ldots \lambda_n)$.

Similar to [7.5], the log-likelihood of the data is defined as:

$$\log \mathcal{L}_{\boldsymbol{p},\boldsymbol{\lambda}}(\mathcal{T}) = \log \prod_{k=1}^{K} \boldsymbol{\alpha} e^{\boldsymbol{A} t_k} \boldsymbol{a} = \sum_{k=1}^{K} \log \sum_{i=1}^{n} p_i f_i(t_k), \qquad [7.14]$$

where $\boldsymbol{a} = -\boldsymbol{A}\mathbb{1}$, $f_i(t) = \lambda_i e^{-\lambda_i t} * \lambda_{i+1} e^{-\lambda_{i+1} t} * \ldots * \lambda_n e^{-\lambda_n t}$ and the structure of $\boldsymbol{A}$ is restricted according to [5.15], which we used in the second step.

The direct optimization procedure uses the derivatives of the log-likelihood function according to the model parameters:

$$\begin{aligned} \frac{\partial}{\partial p_j} \log \mathcal{L}_{\boldsymbol{p},\boldsymbol{\lambda}}(\mathcal{T}) &= \frac{\partial}{\partial p_j} \sum_{k=1}^{K} \log \sum_{i=1}^{n} f(t_k) = \sum_{k=1}^{K} \frac{\partial}{\partial p_j} \log \sum_{i=1}^{n} p_i f_i(t_k) \\ &= \sum_{k=1}^{K} \frac{f_j(t_k)}{\sum_{i=1}^{n} p_i f_i(t_k)} = \sum_{k=1}^{K} \frac{f_j(t_k)}{f(t_k)}. \end{aligned} \qquad [7.15]$$

For the derivative according to $\lambda_j$, we have:

$$\frac{\partial}{\partial \lambda_j} \log f(t) = \frac{\partial}{\partial \lambda_j} \log \sum_{i=1}^{n} p_i f_i(t) = \frac{\sum_{i=1}^{j} p_i \frac{\partial}{\partial \lambda_j} f_i(t)}{f(t)}, \qquad [7.16]$$

because $f_i(t)$ is independent of $\lambda_j$ for $i > j$. Let:

$$f_{ij}(t) = \lambda_i e^{-\lambda_i t} * \ldots * \lambda_{j-1} e^{-\lambda_{j-1} t} * \lambda_{j+1} e^{-\lambda_{j+1} t} * \ldots * \lambda_n e^{-\lambda_n t},$$

that is $f_i(t) = f_{ij}(t) * \lambda_j e^{-\lambda_j \tau}$. For $i \le j$, we write:

$$\begin{aligned} \frac{\partial}{\partial \lambda_j} f_i(t) &= \frac{\partial}{\partial \lambda_j} \int_{\tau=0}^{t} \lambda_j e^{-\lambda_j \tau} f_{ij}(t-\tau) d\tau \\ &= \int_{\tau=0}^{t} e^{-\lambda_j \tau} f_{ij}(t-\tau) d\tau - \int_{\tau=0}^{t} \lambda_j \tau e^{-\lambda_j \tau} f_{ij}(t-\tau) d\tau \\ &= \frac{1}{\lambda_j} \lambda_j e^{-\lambda_j t} * f_{ij}(t) - \frac{1}{\lambda_j} \lambda_j e^{-\lambda_j t} * \lambda_j e^{-\lambda_j t} * f_{ij}(t) \\ &= \frac{1}{\lambda_j} f_i(t) - \frac{1}{\lambda_j} \lambda_j e^{-\lambda_j t} * f_i(t) = \frac{1}{\lambda_j} \left( f_i(t) - g_{ij}(t) \right), \end{aligned}$$

where $f_{ij}(t)$ is independent of $\lambda_j$,

$$g_{ij}(t) = \lambda_j e^{-\lambda_j t} * f_i(t),$$

and in the second step we used $\lambda_j e^{-\lambda_j t} * \lambda_j e^{-\lambda_j t} = \lambda_j^2 t e^{-\lambda_j t}$, which is the density function of the Erlang(2, $\lambda_j$) distribution. Finally,

$$\frac{\partial}{\partial \lambda_j} \log \mathcal{L}_{\boldsymbol{p},\boldsymbol{\lambda}}(\mathcal{T}) = \sum_{k=1}^{K} \frac{\partial}{\partial \lambda_j} \log f(t_k) = \sum_{k=1}^{K} \frac{\sum_{i=1}^{j} p_i \frac{\partial}{\partial \lambda_j} f_i(t_k)}{f(t_k)}$$

$$= \frac{1}{\lambda_j} \sum_{k=1}^{K} \frac{\sum_{i=1}^{j} p_i \left(f_i(t_k) - g_{ij}(t_k)\right)}{f(t_k)}. \qquad [7.17]$$

In these expressions, $f(t)$ and $f_i(t)$ can be computed from $f(t) = -\boldsymbol{\alpha} e^{\boldsymbol{A}t} \boldsymbol{A}\mathbb{1}$ and $f_i(t) = -\mathbf{e}_i e^{\boldsymbol{A}t} \boldsymbol{A}\mathbb{1}$, where $\boldsymbol{\alpha}$ and $\boldsymbol{A}$ are defined in [5.15]. For the computation of $g_{ij}(t)$, we introduce matrix $\boldsymbol{A_j}$ of size $n+1$ as:

$$\boldsymbol{A_j} = \begin{pmatrix} -\lambda_1 & \lambda_1 & & & \\ & -\lambda_2 & \lambda_2 & & \\ & & \ddots & \ddots & \\ & & & -\lambda_n & \lambda_n \\ & & & & -\lambda_j \end{pmatrix}, \qquad [7.18]$$

and compute $g_{ij}(t)$ from $g_{ij}(t) = -\mathbf{e}_i e^{\boldsymbol{A_j}t} \boldsymbol{A_j}\mathbb{1}$.

The likelihood optimization problem can be formulated as:

$$\max_{\boldsymbol{p},\boldsymbol{\lambda}} \log \mathcal{L}_{\boldsymbol{p},\boldsymbol{\lambda}}(\mathcal{T})$$

$$\text{subject to } \boldsymbol{p} \geq \mathbf{0}, \boldsymbol{p}\mathbb{1} = 1 \text{and } 0 < \lambda_i \leq \lambda_{i+1} \text{for } i = 1, \ldots, n-1.$$

In this optimization problem, the constraints are linear functions of the parameters, but the log-likelihood function is not.

The PhFit tool implements a greedy linear search for finding the solution of this nonlinear optimization problem in consecutive iteration steps. In a small neighborhood of $(\boldsymbol{p}, \boldsymbol{\lambda})$, we have:

$$\log \mathcal{L}_{\boldsymbol{p}+\boldsymbol{\Delta},\boldsymbol{\lambda}+\boldsymbol{\delta}}(\mathcal{T}) \approx \mathcal{L}_{\boldsymbol{p},\boldsymbol{\lambda}}(\mathcal{T}) + \frac{\partial \mathcal{L}_{\boldsymbol{p},\boldsymbol{\lambda}}(\mathcal{T})}{\partial \boldsymbol{p}} \boldsymbol{\Delta} + \frac{\partial \mathcal{L}_{\boldsymbol{p},\boldsymbol{\lambda}}(\mathcal{T})}{\partial \boldsymbol{\lambda}} \boldsymbol{\delta},$$

where

$$\frac{\partial \mathcal{L}_{\boldsymbol{p},\boldsymbol{\lambda}}(\mathcal{T})}{\partial \boldsymbol{p}} = \begin{pmatrix} \frac{\partial \mathcal{L}_{\boldsymbol{p},\boldsymbol{\lambda}}(\mathcal{T})}{\partial p_1} & \cdots & \frac{\partial \mathcal{L}_{\boldsymbol{p},\boldsymbol{\lambda}}(\mathcal{T})}{\partial p_n} \end{pmatrix}$$

and $\frac{\partial \mathcal{L}_{\boldsymbol{p},\boldsymbol{\lambda}}(\mathcal{T})}{\partial \boldsymbol{\lambda}}$ is defined similarly.

Based on this local linearization of $\log \mathcal{L}_{\boldsymbol{p}+\boldsymbol{\Delta},\boldsymbol{\lambda}+\boldsymbol{\delta}}(\mathcal{T})$, an iteration step of the greedy linear search is:

$$\max_{\boldsymbol{\Delta},\boldsymbol{\delta}} \frac{\partial \mathcal{L}_{\boldsymbol{p},\boldsymbol{\lambda}}(\mathcal{T})}{\partial \boldsymbol{p}} \boldsymbol{\Delta} + \frac{\partial \mathcal{L}_{\boldsymbol{p},\boldsymbol{\lambda}}(\mathcal{T})}{\partial \boldsymbol{\lambda}} \boldsymbol{\delta}$$

$$\text{subject to } \boldsymbol{\Delta} \leq \varepsilon \mathbb{1},\ \boldsymbol{\delta} \leq \varepsilon \mathbb{1},\ \boldsymbol{p} + \boldsymbol{\Delta} \geq \mathbf{0},\ (\boldsymbol{p} + \boldsymbol{\Delta})\mathbb{1} = 1$$

$$\text{and } 0 < \lambda_i + \delta_i \leq \lambda_{i+1} + \delta_{i+1} \text{for } i = 1, \ldots, n-1.$$

This iteration step modifies the parameter values from $(\boldsymbol{p}, \boldsymbol{\lambda})$ to $(\boldsymbol{p} + \boldsymbol{\Delta}, \boldsymbol{\lambda} + \boldsymbol{\delta})$. The validity of this new set of parameters needs to be checked because the linear approximation of $\log \mathcal{L}_{\boldsymbol{p}+\boldsymbol{\Delta},\boldsymbol{\lambda}+\boldsymbol{\delta}}(\mathcal{T})$ might differ significantly from its exact value. One way to check the validity of the approximation is to evaluate the exact value of $\log \mathcal{L}_{\boldsymbol{p}+c\boldsymbol{\Delta},\boldsymbol{\lambda}+c\boldsymbol{\delta}}(\mathcal{T})$ for $c = 0.1, 0.2, \ldots, 1$ and take parameters for the next iteration at the maximum of those values.

## 7.3. Fitting based on families of polynomials

Families of polynomials can be used to approximate continuous PDFs. By appropriate transformations, these polynomials can be converted into PH or ME distributions.

This approach was used for the first time in [DUF 07], applying Jacobi polynomials; the main idea was to map the interval [-1,1], on which the Jacobi polynomials form an orthogonal class, to the interval $[0, \infty)$ by an appropriate change of variable. The resulting distributions are in the class of ME distributions. However, the procedure requires complicated calculations with numerical problems.

In this section, we present a more straightforward approach based on Bernstein polynomials (BPs).

### 7.3.1. *Bernstein polynomials and Bernstein expolynomials*

The degree-$n$ BP [LOR 53] approximating a function $f$ on $[a, b]$ is defined as:

$$B_n(x) = \sum_{i=0}^{n} f\left(a + \frac{i}{n}(b-a)\right) \binom{n}{i} \frac{(x-a)^i (b-x)^{n-i}}{(b-a)^n}. \quad [7.19]$$

BPs exhibit a series of favorable properties which motivate their use for distribution approximation:

– *Globality:* The approximant is global in the sense that a single polynomial is used to approximate on the entire interval $[a, b]$.

– *Positivity:* Since all the polynomials $(x-a)^i(b-x)^{n-i}, i = 0, 1, \ldots, n$, are positive on the interval $[a, b]$ and we also assume that the function $f$ is positive (in our case $f$ is a PDF), the polynomial in [7.19] is positive on $[a, b]$.

– *Simple derivation:* A BP is determined in a straightforward manner by the samples of the approximated function without any optimization process.

– *Convergence of the approximation:* If the approximated function is continuous, then the degree-$n$ BP converges uniformly to it as $n$ is increased. Moreover, the approximation error can be bounded by a Lipschitz inequality [LOR 53].

Conversely, the following requirements are not guaranteed by BPs but can be accommodated by a suitable adaptation.

– *Divergence at infinity:* A BP is either a constant function or it diverges to plus or minus infinity at infinity. For this reason, it cannot be used to approximate infinite support distributions and it is not feasible for the approximation of transient probabilities of CTMCs if the required interval overlaps with the steady-state regime.

– *Unit-integral:* BP approximation does not guarantee that the approximant preserves the integral of the approximated function. This is clearly a problem when fitting distributions because the approximant is expected to have unit-integral.

The last problem can be managed simply by normalizing the approximant. The problem of divergence can be solved instead by applying the change of variable $x \to e^{-x}$ to [7.19] resulting in the following sum of exponentials with negative exponents:

$$E_n(x) = \sum_{i=0}^{n} f\left(-\log\left(e^{-a} - \frac{i}{n}\left(e^{-a} - e^{-b}\right)\right)\right) \times \binom{n}{i} \frac{(e^{-a} - e^{-x})^i (e^{-x} - e^{-b})^{n-i}}{(e^{-a} - e^{-b})^n}, \qquad [7.20]$$

which converges as $x$ tends to infinity. The functions defined in [7.20], which we call Bernstein expolynomials (BEs), maintain the favorable properties of BPs. It is easy to see that a function in the form of [7.20] belongs to the ME family.

Both BPs and BEs are equal to the approximated function at the endpoints of the interval. The important difference between BPs and BEs is that the latter can be used to approximate a function on the interval $[0, \infty)$. Applying [7.20] with $b = \infty$, the limiting value of the BE will be equal to the limiting value of the approximated function.

Another difference between the two approximations is the way they sample the approximated function. BPs result in an equidistant sampling of the considered interval, that is, the value of the approximated function is taken at the points $a + \frac{i}{n}(b-a)$, $i = 0, 1, \ldots, n$. In order to consider the sampling applied by the expolynomial approximation, let us assume $a = 0, b = \infty$. In this case, the approximated function is sampled at the values $-\log(i/n), i = n, n-1, \ldots, 0$. This implies that more samples are taken from the beginning of the considered interval. This effect can be mitigated by approximating a scaled version of the function, namely $f(x/c)$, and bringing back the approximant to the original scale in the end. Without going into details, we illustrate this effect in two figures. Assuming that $n = 10$ and the interval to be approximated is [0,10], Figure 7.2 depicts the positions where the value of the approximated function is taken for different values of $c$. The larger $c$ is, the closer the sampling points are to $a$ (except the last sampling point which is always at $b$). Instead, as $c$ tends to 0, the sampling tends to be equidistant. In Figure 7.3, the effect of $c$ for $n = 10, a = 0, b = \infty$ is shown.

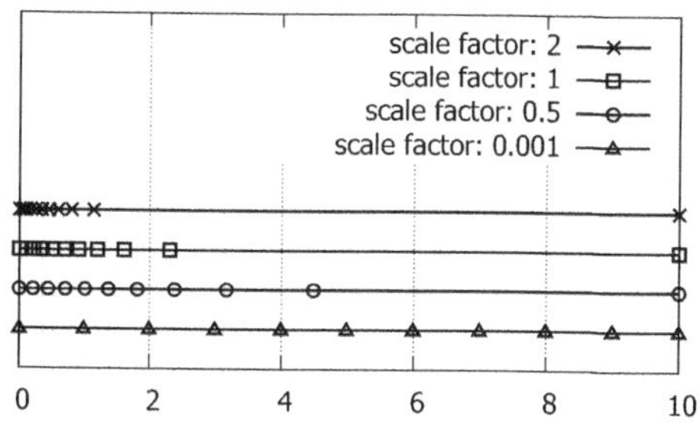


**Figure 7.2.** *Sampling points for different scale factors if $n = 10$ and the interval to fit is [0,10]*

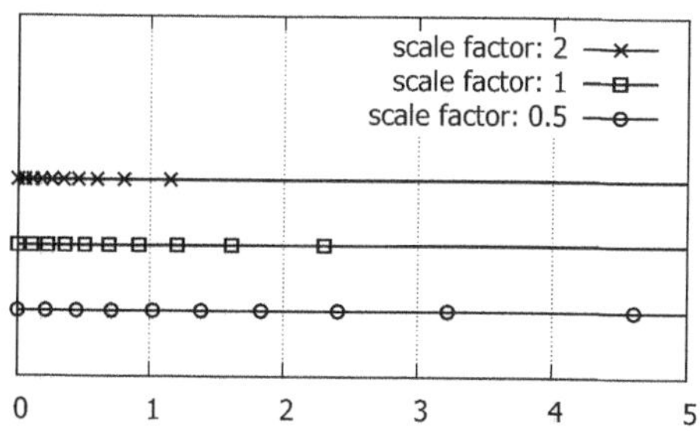


**Figure 7.3.** *Sampling points for different scale factors if $n = 10$ and the interval to fit is [0,∞] (also the value of the function at infinity is used)*

When the functions in [7.20] are applied to fit the PDF of a distribution, the interval of the approximation must be set to [0,∞]. This guarantees that the limiting value of the approximant at infinity is 0, which is a necessary condition for a PDF. It is easy to check that the approximating PDF is in the class of ME distributions.

### 7.3.2. *Application of Bernstein expolynomials to distribution fitting*

In this section, we illustrate the application of BEs to distribution fitting through the benchmark introduced on the first page of Chapter 7. Figures 7.4–7.8 depict the PDF of the distributions and the corresponding approximations for various degrees of BE. For all the experiments, we have chosen a scaling such that the number of samples less than the median of the distribution is the same as the number of samples larger than the median. Moreover, as the mean is of crucial importance in almost all modeling situations, we modified the approximants in such a way that its mean is equal to the mean of the original distribution. This can be done easily by scaling the approximant by an appropriate constant. Consequently, for all approximants, $e_1 = 0$.

By visual inspection, the following considerations can be made. The approximant is not good for distributions whose tail is heavier then exponential (distribution W2 and L1). These two distributions are hard to approximate using general purpose PH fitting methods and can be handled only by methods that pay special attention to the tail (see section 7.4). In the case of distribution L1, we obtain an approximant that is particularly further from the original function. This is due to the adjustment of the mean. Without the adjustment, the mean of the BE approximant is far from the mean of L1 but visually it looks better than the one depicted in Figure 7.5, adjusting the mean makes the visual impression worse.

Distributions with low coefficient of variation (L1, U1 and U2) require a high degree approximation to capture their low variability. Distributions with sharp changes are in general hard to approximate by PH distributions and, consequently, low-degree BE approximations do not provide good approximations. In most of the cases, as foreseen by the theory of BPs, increasing the degree of the approximant results in better fitting.

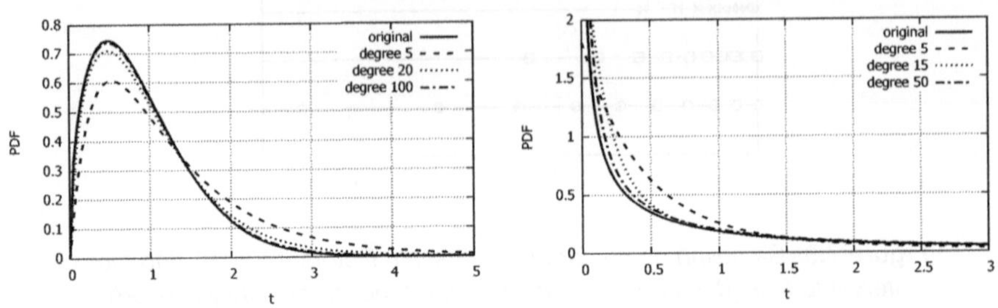


**Figure 7.4.** *BE approximation of distributions W1 (left) and W2 (right)*

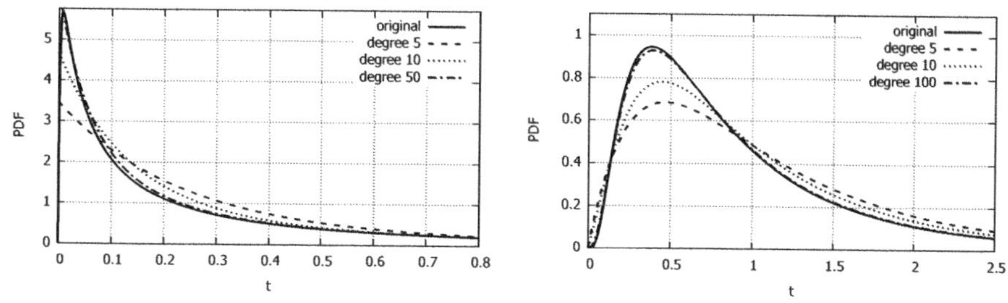


**Figure 7.5.** *BE approximation of distributions L1 (left) and L2 (right)*

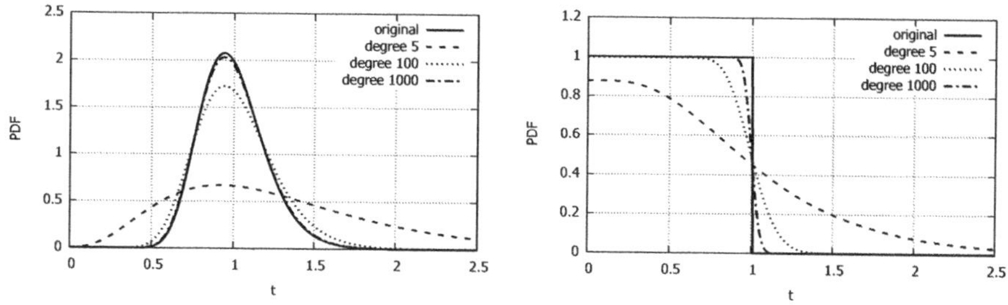


**Figure 7.6.** *BE approximation of distributions L3 (left) and U1 (right)*

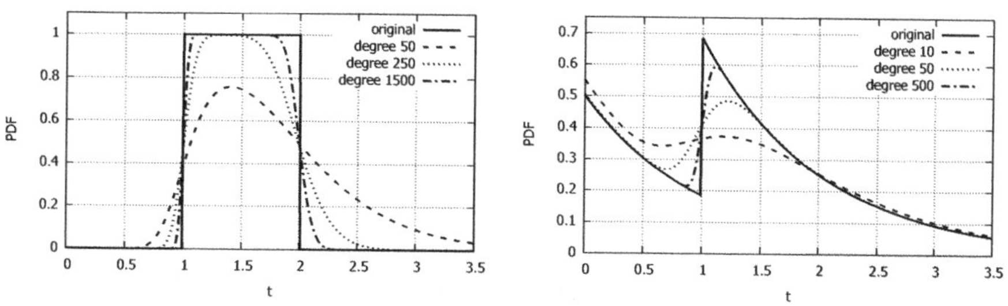


**Figure 7.7.** *BE approximation of distributions U2 (left) and SE (right)*

Table 7.2 reports error measures obtained by applying BE approximations of degree 10, 20 and 100, respectively, to each function of the benchmark. For distributions U1 and U2, the third central moment, $\hat{c}_3$, is 0 due to the symmetry of the uniform distribution and for this reason the tables report the third central moment of the approximants instead of the relative error.

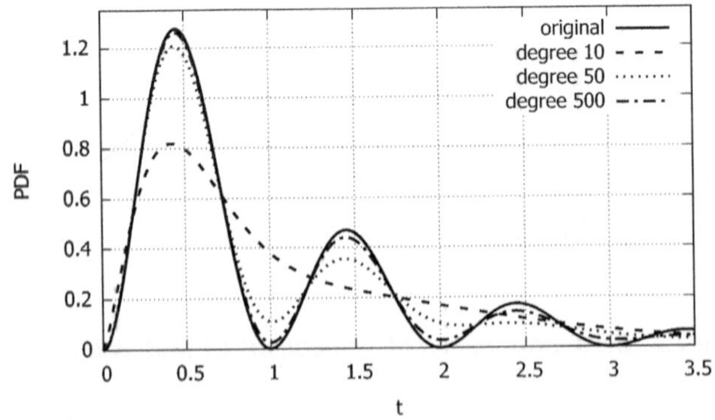


**Figure 7.8.** *BE approximation of distribution ME**

| | $e_2$ | $e_3$ | | $e_2$ | $e_3$ | | $e_2$ | $e_3$ |
|---|---|---|---|---|---|---|---|---|
| $W1_{10}$ | 0.3500 | 1.9312 | $L2_{10}$ | 0.1522 | 0.8746 | $U2_{10}$ | 4.5881 | 0.1510 |
| $W1_{20}$ | 0.2080 | 1.0049 | $L2_{20}$ | 0.1593 | 0.8740 | $U2_{20}$ | 2.8260 | 0.0602 |
| $W1_{100}$ | 0.0458 | 0.0936 | $L2_{100}$ | 0.1423 | 0.8646 | $U2_{100}$ | 0.5303 | 0.0062 |
| $W2_{10}$ | 0.2877 | 0.9346 | $L3_{10}$ | 4.4687 | 4.8666 | $SE_{10}$ | 0.3343 | 0.4270 |
| $W2_{20}$ | 0.2606 | 0.9299 | $L3_{20}$ | 2.3452 | 1.1749 | $SE_{20}$ | 0.2164 | 0.2205 |
| $W2_{100}$ | 0.4333 | 0.9423 | $L3_{100}$ | 0.4878 | 0.5199 | $SE_{100}$ | 0.0703 | 0.0950 |
| $L1_{10}$ | 0.9554 | 0.9999 | $U1_{10}$ | 0.5569 | 0.0576 | $ME^*_{10}$ | 0.0270 | 0.1490 |
| $L1_{20}$ | 0.9476 | 0.9999 | $U1_{20}$ | 0.2906 | 0.0299 | $ME^*_{20}$ | 0.1926 | 0.2016 |
| $L1_{100}$ | 0.9382 | 0.9999 | $U1_{100}$ | 0.0615 | 0.0055 | $ME^*_{100}$ | 0.0357 | 0.1788 |

**Table 7.2.** *Relative errors of the variance and of the third central moment of BE approximations of degree 10, 20 and 100*

As already observed, BE approximation suffers from heavy tails: In the case of distributions W2 and L1 there are large relative errors of the variance and of the third central moment and even increasing the degree of the approximation does not decrease the error.

In the case of low coefficient of variation (distributions L3, U1 and U2), low-degree approximations are not able to capture the variance of the original distribution but increasing the degree improves the fitting.

It is interesting to note that visual inspection and numerical error measures can contradict each other as in the case of distribution ME* for which the degree 100 approximation seems much better than the degree 10 approximation but the relative errors of the moments portray the opposite.

Next we give a brief comparison of the Jacobi polynomial–based method presented in [DUF 07] and the one described in this section. The method of [DUF 07] requires the setting of four parameters for which no guidelines are presented. For the distribution U2, Dufresne [DUF 07] provides a good setting of the four parameters, and for this reason we use U2 in the comparison. In Figure 7.9, we depicted the approximations obtained by the two methods presented in [DUF 07]. Method A does not guarantee positivity and the PDFs given by this method have negative values. Moreover, an oscillating behavior can be observed around 0. Method B provides proper PDFs and gives better approximation than the BE approximant of the same degree. The drawback however of the Jacobi polynomial–based method is that it requires heavy numerical computations and involves numerical problems. Indeed, we were not able to compute the Jacobi polynomial based approximant for degrees higher than 20. Moreover, the computation of the approximant of degree 20 took about 2 min while the computation of the BE approximant is immediate.

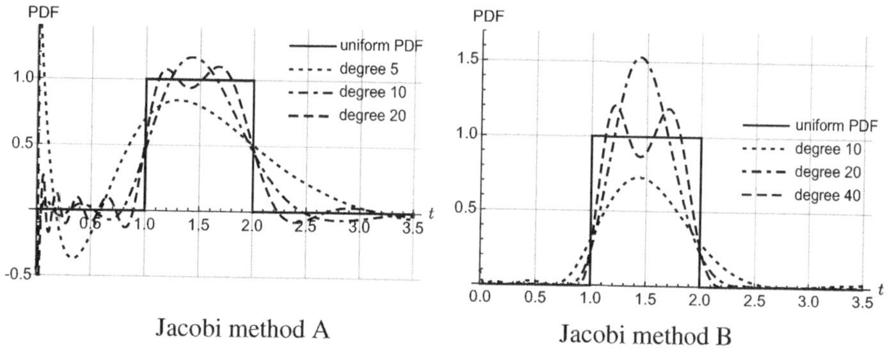


**Figure 7.9.** *Approximation of U2 by the methods proposed in [DUF 07] with different degrees (the squaring applied by method B doubles the degree of the approximation)*

## 7.4. Fitting heavy-tailed distributions

Informally, heavy-tailed distributions are those in which large values have relatively high probability. Such phenomenon was observed in various contexts such as packet size in telecommunication networks and many other situations where Zipf's law applies. One commonly used formal definition is the following. The distribution of a random variable $Y$ is said to be heavy-tailed if its CCDF can be written in the form:

$$1 - F_Y(x) = x^{-\alpha} L(x),$$

where $L(x)$ is slowly varying as $x \to \infty$, that is, $\lim_{x \to \infty} L(ax)/L(x) = 1$ for $a > 0$. In other words, the CCDF follows approximately a power law. Consequently, if $Y$ is heavy tailed with parameter $\alpha$, then only its first $\lceil n < \alpha \rceil$ moments are finite.

Fitting heavy-tailed distributions with PH distributions is not straightforward: every moment of a PH distribution is finite and its CCDF has exponential decay. This observation was confirmed in [HOR 00b] where several distance measures were tested to fit heavy-tailed distributions and none of which lead to satisfactory results.

Nevertheless, it is possible to approximate heavy-tailed distributions by PH distributions in a given finite interval by ad hoc approaches. In section 7.4.1, we show such an ad hoc method to fit heavy-tailed distributions with monotone decreasing PDF. This method can be combined, as illustrated in section 7.4.2, with a general PH fitting approach in order to overcome the constraint of monotonicity.

### 7.4.1. *Fitting heavy-tailed distributions with monotone decreasing density function*

The following method to fit heavy-tailed distributions with monotone decreasing PDF was proposed in [FEL 98]. It results in a hyper-exponential distribution whose CCDF is "very close" to the original CCDF at a given set of points.

The CCDF of a size $n$ hyper-exponential distribution is:

$$\sum_{i=1}^{n} p_i e^{-\lambda_i x},$$

which is determined by $2n-1$ parameters due to $\sum_{i=1}^{n} p_i = 1$. Accordingly, the input of the procedure is a set of $2n-1$ points that we denote by $t_1, t_2, ..., t_{2n-1}$ and assume to be ordered as $t_1 > t_2 > ... > t_{2n-1}$.

The parameters $p_1$ and $\lambda_1$ are determined based on:

$$p_1 e^{-\lambda_1 t_1} = \hat{F}(t_1) \text{ and } p_1 e^{-\lambda_1 t_2} = \hat{F}(t_2),$$

where $\hat{F}(x)$ is the CCDF we aim to fit. The above results in:

$$\lambda_1 = \frac{1}{t_1 - t_2} \ln \frac{\hat{F}(t_2)}{\hat{F}(t_1)} \text{ and } p_1 = \hat{F}(t_1) \left( \frac{\hat{F}(t_2)}{\hat{F}(t_1)} \right)^{\frac{t_1}{t_1 - t_2}}.$$

The assumption behind setting $p_1$ and $\lambda_1$ directly based on $\hat{F}(t_1)$ and $\hat{F}(t_2)$ is that the other terms of the hyper-exponential distribution will have a negligible effect at $t_1$ and $t_2$.

Having determined $p_1, ..., p_{i-1}$ and $\lambda_1, ..., \lambda_{i-1}$ (with $i = 2, 3, ..., n-1$), the next pair of parameters $p_i, \lambda_i$ is determined based on $\hat{F}(t_{2i-1})$ and $\hat{F}(t_{2i})$, considering also the effect of the already known parameters $p_1, \lambda_1, p_2, \lambda_2, ..., p_{i-1}, \lambda_{i-1}$. To this end, let:

$$\hat{F}_i(t) = \hat{F}(t) - \sum_{j=1}^{i-1} p_j e^{-\lambda_j t}.$$

The parameters $p_i$ and $\lambda_i$ (with $i < n$) are determined based on

$$p_i e^{-\lambda_i t_{2i-1}} = \hat{F}_i(t_{2i-1}) \text{ and } p_i e^{-\lambda_i t_{2i}} = \hat{F}_i(t_{2i}),$$

which results in

$$\lambda_i = \frac{1}{t_{2i-1} - t_{2i}} \ln \frac{\hat{F}_i(t_{2i})}{\hat{F}_i(t_{2i-1})} \text{ and } p_i = \hat{F}_i(t_{2i-1}) \left( \frac{\hat{F}_i(t_{2i})}{\hat{F}_i(t_{2i-1})} \right)^{\frac{t_{2i-1}}{t_{2i-1}-t_{2i}}}. \quad [7.21]$$

In order to have a good approximation, we assume that the terms associated with parameters $p_{i+1}, \lambda_{i+1}, ..., p_n, \lambda_n$ make a negligible contribution to the CCDF at $t_{2i-1}$ and $t_{2i}$.

$p_n$ is computed as:

$$p_n = 1 - \sum_{i=1}^{n-1} p_i, \quad [7.22]$$

and the last intensity $\lambda_n$ is determined using $t_{2n-1}$ as:

$$\lambda_n = -\frac{1}{t_{2n-1}} \ln \frac{\hat{F}_n(t_{2n-1})}{p_n}. \quad [7.23]$$

There is no guarantee that the obtained parameters are valid (there can be negative initial probabilities and/or intensities) and that the approximation is satisfactory. In general, if the CCDF $\hat{F}(t)$ is "heavier" than exponential and reasonably regular, then it is possible to choose the points $t_1, t_2, ..., t_{2n-1}$ such that a good approximation is obtained. More precise guidelines are provided in [FEL 98].

We use a distribution with Pareto-tail to illustrate the method. The PDF is given by:

$$\hat{f}(t) = \frac{b^\alpha e^{-b/t}}{\Gamma(\alpha)} x^{-(\alpha+1)} \text{ with } b = 2, \alpha = 1.5 \quad [7.24]$$

is not monotone and hence we concentrate only on the tail behavior. Choosing the points where we aim to match the CCDF as:

$$t = \begin{pmatrix} 10^6 & 10^5 & 10^4 & 10^3 & 10^2 \end{pmatrix},$$

we obtain the parameters:

$$p = \left(9.87574 \cdot 10^{-8} \;\; 9.89693 \cdot 10^{-5} \;\; 0.999901\right)$$

and

$$\lambda = \left(3.83763 \cdot 10^{-6} \;\; 3.88559 \cdot 10^{-4} \;\; 0.0621099\right),$$

and the CCDF as depicted in Figure 7.10 where we indicated the contribution of each phase to the distribution.

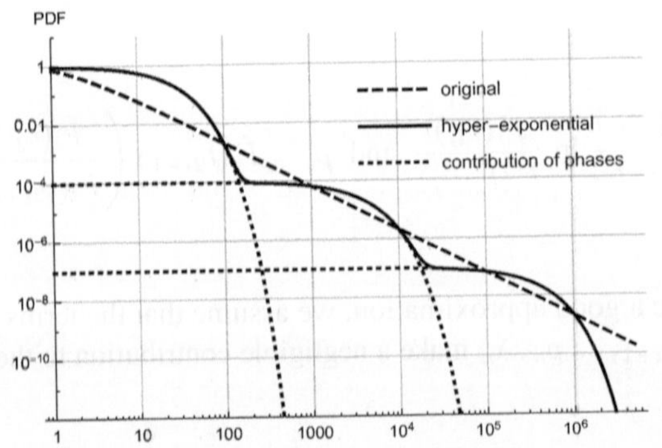


**Figure 7.10.** *CCDF fitted to a Pareto-tail with a hyper-exponential distribution of size three; dotted curves show the separate contribution of each phase*

By increasing the number of phases and selecting the matching points closer to each other, the oscillation of the fitting CCDF around the original one can be decreased significantly. Using the scheme:

$$t_i = Lc_1^{i-1}, i = 1, 3, ..., 2n-1, \quad t_i = c_2 Lc_1^{i-1}, i = 2, 4, ..., 2n-2$$

with parameters

$$n = 10, L = 10^6, \; c_1 = 1/4, \; c_2 = 4/5,$$

we obtain the CCDF given in Figure 7.11. With this setting, the approximating CCDF cannot be distinguished visually from the original Pareto-tail between $t_{2n-1} = Lc_1^{2n-2} = 3.8147$ and $t_1 = L = 10^6$.

### 7.4.2. *Fitting heavy-tailed distributions with possibly non-monotone density function*

In order to overcome the limitation of the procedure described in section 7.4.1, namely, that the PDF must be monotonic, we combine it with a more general approach.

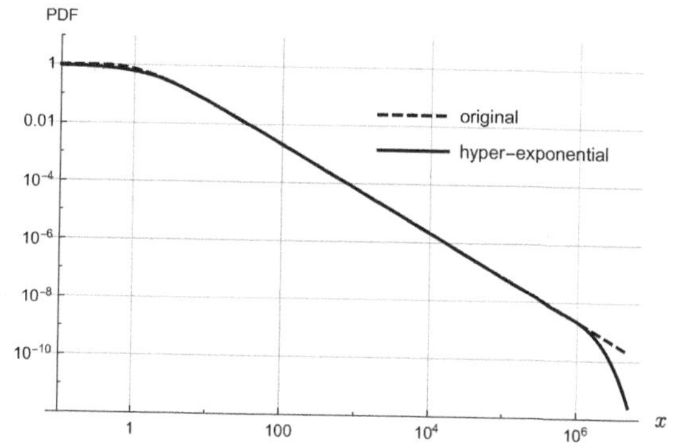


**Figure 7.11.** *CCDF fitted to a Pareto-tail with a hyper-exponential distribution of size 10*

Here, we present one such possibility combining a hyper-exponential and an acyclic PH distribution.

The mixture of a size $n_t$ hyper-exponential distribution, used to approximate the tail behavior, and a size $n_m$ acyclic PH distribution in CF1 form, used to capture the "main" part, results in the following initial probability vector:

$$\boldsymbol{\alpha} = \begin{pmatrix} p_1 & p_2 & \dots & p_{n_t} & q_1 & q_2 & \dots & q_{n_m} \end{pmatrix}$$

and infinitesimal generator

$$\boldsymbol{A} = \begin{pmatrix} -\lambda_1 & 0 & \dots & & & & & & \\ 0 & -\lambda_2 & 0 & \dots & & & & & \\ & & \ddots & & & & & & \\ & \dots & 0 & -\lambda_{n_t} & 0 & \dots & & & \\ & & \dots & 0 & -\gamma_1 & \gamma_1 & 0 & \dots & \\ & & & \dots & 0 & -\gamma_2 & \gamma_2 & 0 & \dots \\ & & & & & & \ddots & & \\ & & & & & & \dots & 0 & -\gamma_{n_m} \end{pmatrix}.$$

The parameters of the hyper-exponential part are determined as described in section 7.4.1 with a small difference. The constraint $\sum_{i=1}^{n} p_i = 1$ becomes $\sum_{i=1}^{n_t} p_i < 1$. Consequently, we use $2n_t$ points where we aim to match the original CCDF and apply [7.21] to determine $p_{n_t}$ and $\lambda_{n_t}$ instead of [7.22] and [7.23]. In order to proceed with the parameters of the CF1 part, we need valid initial probabilities and intensities ($p_i, \lambda_i, 1 \leq i \leq n_t$ must be real and positive with $\sum_{i=1}^{n_t} p_i < 1$) for the hyper-exponential part. If this is not the case, either the value of $n_t$ and/or the $2n_t$ matching points must be changed and the hyper-exponential parameters redetermined.

Once we have suitable values for $p_i, \lambda_i, 1 \leq i \leq n_t$, the remaining CF1 form parameters, $q_i, \gamma_i, 1 \leq i \leq n_m$, must be determined. To this end several approaches are possible, but two points must be considered in any case. First, the initial probabilities of the CF1 part must satisfy:

$$\sum_{i=1}^{n_m} q_i = 1 - \sum_{i=1}^{n_t} p_i$$

in order to have the total initial probability equal to one. Second, the contribution of the hyper-exponential part to the distribution must be taken into account. The parameters of the CF1 part, $q_i, \gamma_i, 1 \leq i \leq n_m$, must be chosen in such a way that their contribution is close in some sense to the non-normalized PDF:

$$\hat{f}_{nn}(t) = \hat{f}(t) - \sum_{i=1}^{n_t} p_i \lambda_i e^{-\lambda_i t}.$$

Having determined $\hat{f}_{nn}(t)$, possible approaches to specify $q_i, \gamma_i, 1 \leq i \leq n_m$ are:

– calculate the first three moments of $\hat{f}_{nn}(t)$ and apply moment matching according to the method in section 6.6,

– create samples according to $\hat{f}_{nn}(t)$ and determine the parameters based on maximum likelihood estimation as in section 7.2,

– choose a distance measure (section 7.1) and find the parameters by direct numerical optimization.

We illustrate the last of the above approaches for the Pareto distribution given in [7.24]. The applied distance measure is relative entropy [7.1], that is, the main part of the distribution is fitted according to the maximum likelihood principle. As for the number of phases, we perform the fitting with $n_m = 4, n_t = 4$ and with $n_m = 8, n_t = 6$. The tail is approximated in the range given by $t_1 = 2 \times 10^6$ and $t_{2n_t} = 10$. The resulting PDFs are shown in Figure 7.12. As expected, increasing the number of phases improves the fitting both in the main part and in the tail.

## 7.5. Continuous versus discrete phase type distributions in practical fitting problems

The support of a DPH distribution is the set of positive integers, while that of a PH is the set of positive real numbers. All properties and algorithms were derived by now accordingly. However, when DPH distributions are used to model durations, an additional parameter needs to be introduced that represents the time step associated with each discrete step. Denoting this time step by $\delta$, the support of a DPH distribution becomes $\{\delta, 2\delta, 3\delta, ...\}$. This parameter $\delta$, which we call the *scale factor* of the DPH distribution, can be seen as a new degree of freedom, since its choice impacts the shape and properties of a DPH distribution over the time axis. In the following, *scaled* DPH *distributions* will refer to those DPH distributions in which one step is associated with a time step of length $\delta$ and, accordingly, their support is $\{\delta, 2\delta, 3\delta, ...\}$.

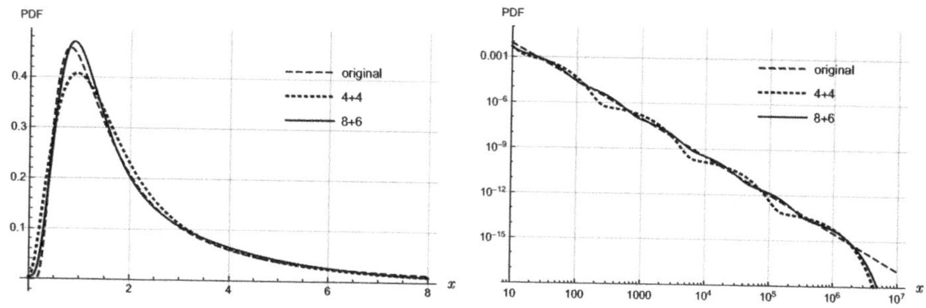


**Figure 7.12.** *Main part (left) and tail (right) of PDFs fitted to a distribution with Pareto-tail with various number of phases dedicated to the main part and to the tail behavior ($n_m = 4, n_t = 4$ and $n_m = 8, n_t = 6$)*

In the continuous case, the minimum SCV is a function of the size only. It is $1/n$ in the case of size equal to $n$ (theorem 6.1). On the other hand, in the discrete case it depends both on the size $n$ and on the mean $m$ (theorem 6.4). Assuming that we aim to fit a distribution with mean equal to 2.5, if the scale factor is 0.1, then the mean corresponds to $m = 25$ steps, while using scale factor 0.5, the mean corresponds to $m = 5$ steps. Hence, it is clear that when DPH distributions are used to approximate a given continuous distribution, the scale factor can largely affect the goodness of the fit. This chapter discusses this issue.

We start out by showing that PH distributions can be obtained as the limit of DPH distributions as $\delta \to 0$.

### 7.5.1. *Limit behavior as the scale factor tends to zero*

Given a CTMC with infinitesimal generator $\boldsymbol{Q}$, the transition probability matrix over an interval of length $\delta$ can be written as:

$$e^{\boldsymbol{Q}\delta} = \sum_{i=0}^{\infty} (\boldsymbol{Q}\delta)^i / i! = \boldsymbol{I} + \boldsymbol{Q}\delta + \sigma(\delta),$$

hence the first-order approximation of $e^{\boldsymbol{Q}\delta}$ is $\boldsymbol{\Pi}(\delta) = \boldsymbol{I} + \boldsymbol{Q}\delta$, which is a proper stochastic matrix if $\delta < 1/\max_{i,j} |\boldsymbol{Q}_{ij}|$. Moreover, $\boldsymbol{\Pi}(\delta)$ is the exact transition probability matrix of the CTMC assuming that at most one transition occurs in the interval of length $\delta$.

We can approximate the behavior of the CTMC at times $(\delta, 2\delta, 3\delta, \ldots, k\delta, \ldots)$ using the DTMC with transition probability matrix $\mathbf{\Pi}(\delta)$. The approximate transition probability matrix at time $t = k\delta$ is:

$$\mathbf{\Pi}(\delta)^k = (\boldsymbol{I} + \boldsymbol{Q}\delta)^{\frac{t}{\delta}}.$$

The following theorem proves that the above first order approximation becomes exact as $\delta \to 0$.

THEOREM 7.4.– *For $\delta \to 0$, we have:*

$$\lim_{\delta \to 0} (\boldsymbol{I} + \boldsymbol{Q}\delta)^{\frac{t}{\delta}} = e^{\boldsymbol{Q}t}.$$

*Proof.* The scalar version of the applied limiting behavior is well known in the form:

$$\lim_{x \to 0} (1 + ax)^{\frac{1}{x}} = e^a.$$

Since matrices $\boldsymbol{I}$ and $\boldsymbol{Q}$ commute we can obtain the matrix version of the same expression as follows:

$$\lim_{\delta \to 0} (\boldsymbol{I} + \boldsymbol{Q}\delta)^{\frac{t}{\delta}} = \lim_{k \to \infty} \mathbf{\Pi}(t/k)^k = \lim_{k \to \infty} \left(\boldsymbol{I} + \frac{\boldsymbol{Q}t}{k}\right)^k =$$

$$\lim_{k \to \infty} \sum_{j=0}^{k} \binom{k}{j} \left(\frac{\boldsymbol{Q}t}{k}\right)^j = \lim_{k \to \infty} \sum_{j=0}^{k} \frac{(\boldsymbol{Q}t)^j}{j!} \frac{k!}{k^j\,(k-j)!} = \sum_{j=0}^{\infty} \frac{(\boldsymbol{Q}t)^j}{j!} = e^{\boldsymbol{Q}t}.$$

□

COROLLARY 7.5.– *Given a scaled DPH distribution of size $n$ with representation $(\boldsymbol{\alpha}, \boldsymbol{I} + \boldsymbol{A}\delta)$ and scale factor $\delta$, the limiting behavior as $\delta \to 0$ is the PH($n$) distribution with representation $(\boldsymbol{\alpha}, \boldsymbol{A})$.*

### 7.5.2. The minimum squared coefficient of variation of scaled discrete phase type distributions

Given a scaled DPH distribution with scale factor $\delta$, its mean will be denoted by $m_{\text{sc}}$ while the mean of its unscaled version will be denoted by $m_{\text{u}}$. Accordingly, $m_{\text{sc}} = \delta m_u$. Theorem 6.5 restates theorem 6.4 for scaled DPH distributions.

A simple consequence of theorem 6.5 is the following.

COROLLARY 7.6.– *As $\delta \to 0$, $\frac{m_{\text{sc}}}{\delta}$ tends to infinity and the minimal SCV in* [7.25] *tends to $1/n$.*

Corollary 7.6 shows that the minimal SCV of the DPH class converges to the minimal SCV of the PH class of the same size as $\delta \to 0$. The following corollary presents a stronger convergence result for the case of approximating distributions with low SCV. In the corollary and its proof, the term *best SCV fitting* PH approximation will be used, which refers to the PH approximation that exhibits the same mean and provides the closest approximation for the SCV.

COROLLARY 7.7.– *For distributions whose SCV is lower than the lowest SCV attainable by PH distributions, the best SCV fitting scaled DPH approximation converges, in distribution, to the* best SCV fitting *PH approximation of the same distribution as $\delta$ tends to* 0.

*Proof.* Both the PH and the DPH classes have limits in approximating distributions with low SCV. The best SCV fitting approximation of a distribution with SCV less than these limits is the Erlang($n$) distribution in both the continuous and the discrete case (assuming that $m_{\text{sc}} > n\delta$).

The representation $(\boldsymbol{\alpha}, \boldsymbol{A})$ of the Erlang($n$) distribution with mean $m_{\text{sc}}$, and the representation $(\boldsymbol{\beta}, \boldsymbol{B})$ of the discrete Erlang($n$) distribution with mean $m_{\text{sc}}$ and scale factor $\delta$ are:

$$\boldsymbol{\alpha} = \begin{pmatrix}1 & 0 & \dots & 0\end{pmatrix}, \; \boldsymbol{A} = \begin{pmatrix} -\frac{n}{m_{\text{sc}}} & \frac{n}{m_{\text{sc}}} & 0 & \dots & 0 \\ 0 & -\frac{n}{m_{\text{sc}}} & \frac{n}{m_{\text{sc}}} & \dots & \\ & & \ddots & & \\ 0 & & & \dots & -\frac{n}{m_{\text{sc}}} \end{pmatrix},$$

$$\boldsymbol{\beta} = \begin{pmatrix}1 & 0 & \dots & 0\end{pmatrix}, \; \boldsymbol{B} = \begin{pmatrix} 1-\frac{n\delta}{m_{\text{sc}}} & \frac{n\delta}{m_{\text{sc}}} & 0 & \dots & 0 \\ 0 & 1-\frac{n\delta}{m_{\text{sc}}} & \frac{n\delta}{m_{\text{sc}}} & \dots & \\ & & \ddots & & \\ 0 & & & \dots & 1-\frac{n\delta}{m_{\text{sc}}} \end{pmatrix}.$$

Note that $\boldsymbol{B} = \boldsymbol{I} - \boldsymbol{A}\delta$ and corollary 7.7 follows from corollary 7.5. □

Corollary 7.7 is based on the structural properties of Erlang distributions and its discrete counterpart the negative binomial distribution. When the best fitting scaled DPH and PH distributions are negative binomial and Erlang distributed the convergence at $\delta \to 0$ is provided by corollary 7.7, but this convergence property is not known in general. Instead, in the next section, we provide an extensive experimental study on the relation of the approximating scaled DPH and PH distributions as a function of the scale factor $\delta$.

### 7.5.3. *The optimal scale factor in discrete phase type fitting*

The scale factor, $\delta$, provides a new degree of freedom in fitting, and, furthermore, since the limit of a DPH distribution for $\delta \to 0$ is a PH distribution, the optimization of the scale factor in a fitting problem provides a quantitative way to decide whether a continuous or a discrete approximation performs better in the given problem. Hence, assuming $\delta$ as a decision variable, we can consider the PH and the DPH class as a unique model set in which the choice among DPH or PH classes is given by the optimal value of $\delta$.

Assume $X$ is a continuous random variable whose CDF, $k$-th moment and SCV are $F_X(x)$, $\mathbb{E}(X^k)$ and $SCV_X$, respectively.

To approximate $X$ by a PH distribution, a fitting procedure and a distance measure between $X$ and the approximating PH distributions need to be defined. Then, the fitting algorithm provides the PH distribution that minimizes the chosen distance measure. The minimization of the measure is performed by numerical optimization.

In order to compare the "goodness" of the approximation reached by PH and DPH distributions, in a unified framework, we need to choose a distance measure that is meaningful and applicable both in the continuous as well as in the discrete setting. Here, we use the $D_{\text{scdf}}$ distance measure given in [7.3].

#### 7.5.3.1. *Fitting distributions with low squared coefficient of variation*

The following considerations provide practical upper and lower bounds to guide the choice of a suitable scale factor $\delta$, and are mainly based on the dependence of the minimal coefficient of variation of a scaled DPH distribution on the size $n$ and on the mean $\mathbb{E}(X)$.

Since we only consider DPH distributions with no mass at zero, the mean of any unscaled DPH distribution is greater than 1. This means that $\delta$ should be less than $\mathbb{E}(X)$. However, a more convenient upper bound that exploits the flexibility associated with $n$ phases is given by:

$$\delta \leq \frac{\mathbb{E}(X)}{n-1}. \qquad [7.25]$$

If the SCV of the distribution to be approximated is less than $1/n$, $\delta$ should satisfy the following relation (see theorem 6.1):

$$\delta > \left(\frac{1}{n} - SCV_X\right)\mathbb{E}(X). \qquad [7.26]$$

Let $X$ be an L3 distributed random variable for which $\mathbb{E}(X) = 1$ and $SCV_X = 0.0408$. Table 7.3 reports the lower and upper bounds of $\delta$, with $n = 2, 4, 8, 12$, computed from [7.26] and [7.25].

The CDF and PDF of the approximating PH and DPH distributions of size $n = 10$, with different scale factors $\delta$, are presented in Figure 7.13. When considering the approximate DPH distribution, the $f(x)$ values are calculated at the discrete points $(\delta, 2\delta, 3\delta, \ldots, k\delta, \ldots)$ to which the following mass is assigned:

$$f(k\delta) = 1/\delta(F(k\delta) - F((k-1)\delta)). \qquad [7.27]$$

For the ease of visual interpretation, the points are connected with a line.

When $\delta$ is less than its lower bound defined in [7.26], the required SCV cannot be attained; when $\delta$ becomes too large, the wide separation of the discrete steps increases the approximation error; when $\delta$ is in the proper range (e.g. $n = 10$; $\delta = 0.06$), a reasonably good fit is achieved. This example also suggests that an optimal value of $\delta$ exists that minimizes the chosen distance measure between the L3 distribution and the fitting PH(10) and DPH(10) distributions.

In order to find the optimal $\delta$, Figure 7.14 plots the distance measure $D_{\text{scdf}}$, defined in [7.3], as a function of $\delta$ for various values of the size $n$. A minimum value of $D_{\text{scdf}}$ is attained in the range where the parameters fit the bounds of Table 7.3. As $\delta$ increases, the advantage of having more phases disappears, in accordance with theorem 6.5. The circles in the left part of this figure (as well as in all subsequent figures) indicate the corresponding distance measure $D_{\text{scdf}}$ obtained from PH fitting. As a natural consequence of the observations in Figure 7.13, Figure 7.14 (and the subsequent figures as well) indicates that the distance measure obtained from DPH fitting converges to the distance measure obtained by the PH approximation as $\delta$ tends to 0.

In Figure 7.14, the lowest applied scale factor is 0.01. Lower values of the scale factor render the fitting procedure numerically unstable because the diagonal elements of the transition matrix are close to 1.

| $n$ | Lower bound of $\delta$ equation [7.26] | Upper bound of $\delta$ equation [7.25] |
|---|---|---|
| 4 | 0.2092 | 0.333 |
| 8 | 0.0842 | 0.1428 |
| 12 | 0.0425 | 0.0909 |
| 16 | 0.0217 | 0.0666 |

**Table 7.3.** *Upper and lower bound of $\delta$ for fitting distribution L3*

### 7.5.3.2. *Fitting distributions with high squared coefficient of variation*

We have seen in previous sections that it is beneficial to approximate distributions with a low coefficient of variation by DPH distributions. In this section, we investigate the optimal value of $\delta$ when fitting distributions with a high coefficient of variation.

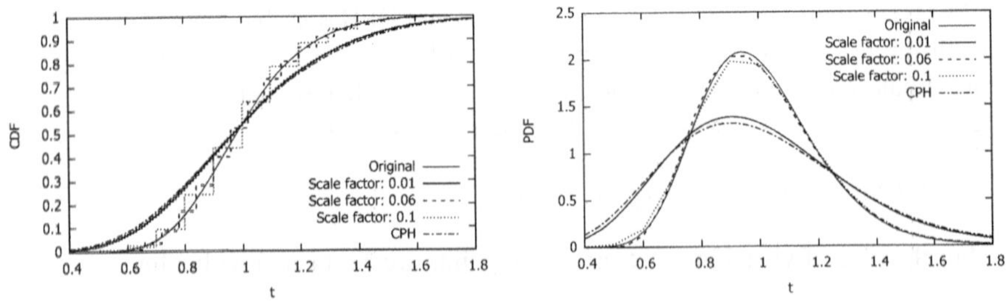


**Figure 7.13.** *Fitting the L3 distribution with PH(10) and DPH(10) distributions*

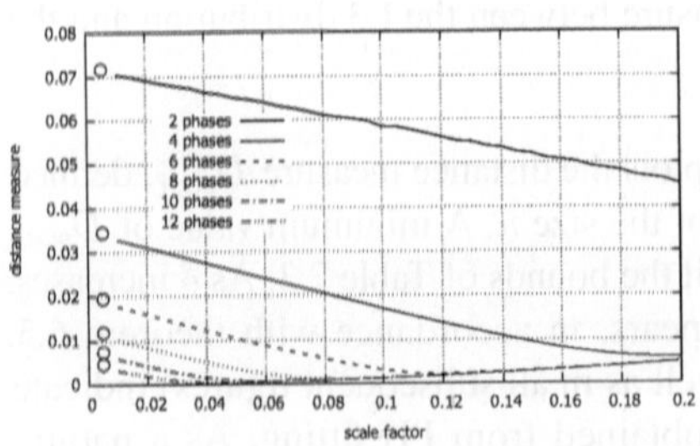


**Figure 7.14.** *$D_{\text{scdf}}$ as the function of $\delta$ for L3 with low SCV*

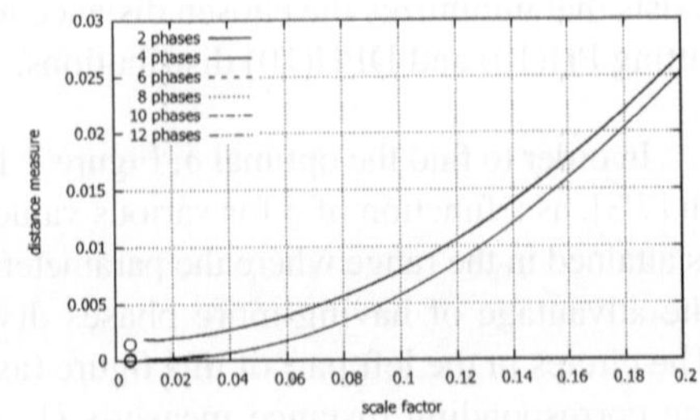


**Figure 7.15.** *$D_{\text{scdf}}$ as the function of $\delta$ for L1 with high SCV*

Let $X$ be an L1 distributed random variable. Then $\mathbb{E}(X) = 1$ and $SCV_X = 24.534$. Figure 7.15 shows the $D_{\text{scdf}}$ measure of the goodness of fit as a function of $\delta$ for various orders $n$ (cases when the number of phases is greater than 2 result in almost the same goodness of fit). The distance measure $D_{\text{scdf}}$ decreases as $\delta \to 0$ indicating that the optimal fitting is achieved by applying PH distributions.

This example suggests that for distributions with $SCV_X > 1/n$ the best fit is obtained by a PH distribution. However, this conclusion might not be true for distributions with finite support; this case is explored in the next section.

#### 7.5.3.3. *Fitting distributions with finite support*

Two features must be considered, namely the SCV and the maximum value of the finite support. It should be stressed that the $D_{\text{scdf}}$ distance measure is not completely appropriate in the case of finite support, since it does not force the approximating PH distribution to have its mass confined on the finite support.

Let $X$ be U2 distributed random variable, then $X \in \{1, 2\}$, $\mathbb{E}(X) = 1.5$ and $SCV_X = 0.0370$. Figure 7.16 shows the $D_{\text{scdf}}$ measure as a function of $\delta$ for various sizes $n$. The figure indicates an optimal $\delta$ value for each size in the plotted range of $\delta$ (except for size 2). This means that the PH approximation is not the best in this case because $SCV_X < 1/n$ for $n < 27$.

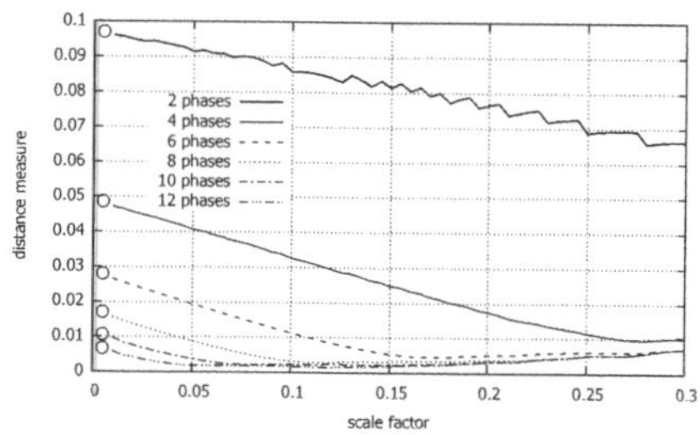


**Figure 7.16.** *$D_{\text{scdf}}$ as the function of $\delta$ for U2*

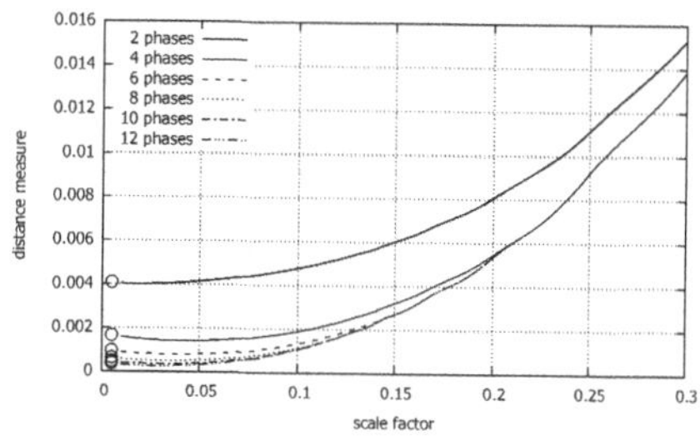


**Figure 7.17.** *$D_{\text{scdf}}$ as the function of $\delta$ for U1*

As a second example, let $X$ be a U1 distributed random variable, then $X \in \{0, 1\}$, $\mathbb{E}(X) = 0.5$ and $SCV_X = 0.333$. Figure 7.17 shows the $D_{\text{scdf}}$ measure as a function of $\delta$ for various sizes $n$. Since in this example $SCV_X > 1/n$ for $n \geq 3$, a PH distribution can attain the coefficient of variation. Nevertheless, the optimal $\delta$ in Figure 7.17, which minimizes the distance measure $D_{\text{scdf}}$ for PH($n$) with size $n > 2$, ranges between $\delta = 0.02$ and $\delta = 0.05$, thus leading to the conclusion that a DPH provides a better fit. This example shows that the coefficient of variation is not the only factor which influences the optimal value of $\delta$. The shape of the distribution plays an important role as well. Our experiments show that a discontinuity in the PDF (or in the CDF) is hard to approximate with a PH distribution, hence in the majority of these cases a DPH distribution provides a better approximation.

Figure 7.18 shows the CDF and the PDF of the U1 distribution, compared with the best fit PH approximations of size $n = 10$, and various scale factors $\delta$. In the case of a DPH approximation, the $f(x)$ values are calculated according to [7.27]. With respect to the $D_{\text{scdf}}$ distance measure, the best approximation is obtained for $\delta = 0.03$, which corresponds to a DPH distribution with infinite support. When $\delta = 0.1$, the approximate distribution has a finite support. Hence, the value $\delta = 0.1$ (for $n = 10$) provides a DPH able to represent the property that the random variable is less than 1.

#### 7.5.3.4. *Approximating exponential distributions*

When it comes to discrete approximation of continuous models, exponential durations are often approximated by a discrete PH distribution with a single phase (see, for example, [HOR 00a]). This approximation, however, is precise only if the scale factor is small enough. Figure 7.19 shows the goodness of fitting as a function of the scale factor for the exponential distribution with rate 1. The figure suggests that it is not worth using more than two phases to approximate exponential distributions.

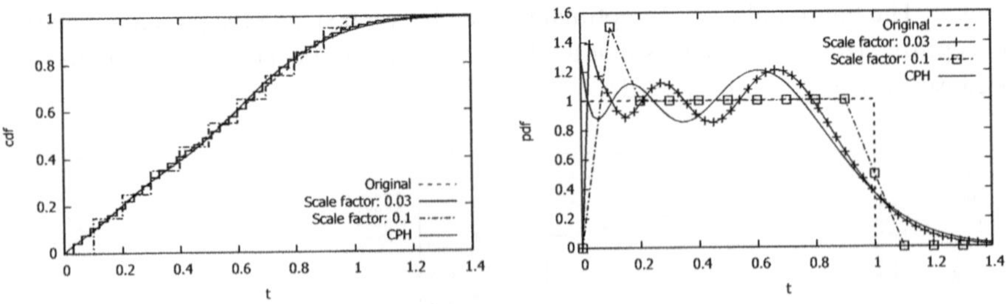


**Figure 7.18.** *Fitting the uniform $(0, 1)$ distribution with PH$(10)$ and DPH$(10)$ distributions*

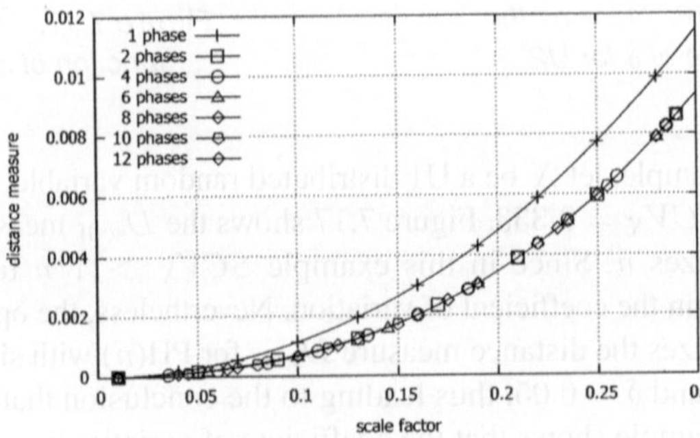


**Figure 7.19.** *$D_{\mathrm{scdf}}$ as the function of $\delta$ for the exponential distribution with rate 1*

### 7.5.4. *Approximating non-Markovian models*

Section 7.5.3 has explored the problem of how to find the best fit among either a DPH or a PH distribution by tuning the scale factor $\delta$. When dealing with a stochastic model of a system that incorporates non-exponential distributions, a well-known solution technique consists of a Markovian approximation of the underlying non-Markovian process by substituting the non-exponential distributions with a best fit PH distribution, and then expanding the state space. A natural question in this case is how to decide between discrete (using DPH) and/or continuous (using PH) approximation in order to minimize the error in the performance measures of interest.

One way to handle this problem is finding the best PH fit for any single distribution and to plug them in the model. In this section, we only consider the case when the PH distributions are either all discrete with the same scale factor, or they are all continuous.

In order to quantitatively evaluate the influence of the scale factor on some performance measures defined at the system level, we consider the example of a preemptive M/G/1/2/2 queue with two classes of jobs. We have chosen this example because accurate analytical solutions are available both in transient condition and in steady state using the methods presented in [GER 00]. The general distribution $G$ is taken from the set of distributions $\{$L1, L3, U1, U2$\}$.

Jobs arrive at the queue with rate $\lambda = 0.5$ in all classes. The service time of a higher priority job is exponentially distributed with parameter $\mu = 1$. The service time distribution of the lower priority job is either L1, L3, U1 or U2. Arrival of a higher priority job preempts the service of the lower priority job being serviced. The policy associated with the preemption of the lower priority job is preemptive repeat different, that is, after the departure of the higher priority customer the service of the low priority customer starts from the beginning with a new service time sample.

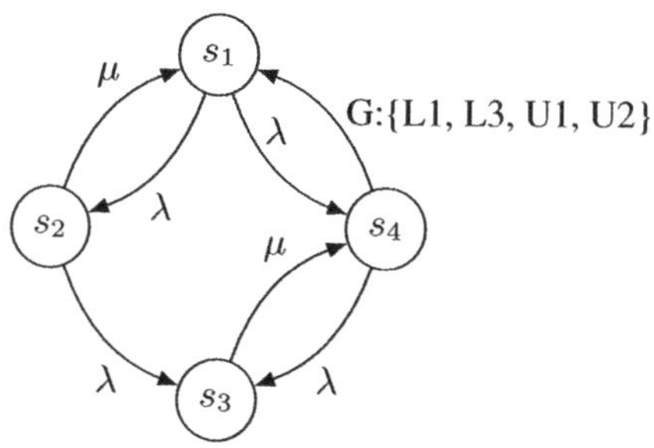


**Figure 7.20.** *The state transitions of the M/G/1/2/2 queue*

The system has four states (Figure 7.20): in state $s_1$, the server is empty; in state $s_2$, a higher priority customer is being serviced with no lower priority customer in the system; in state $s_3$, a higher priority customer is being serviced with a lower priority customer waiting; in state $s_4$, a lower priority job is being serviced (in this case there cannot be a higher priority job in the system).

#### 7.5.4.1. *Steady-state behavior*

Let $p_i$ $(i = 1, \ldots, 4)$ denote the steady-state probability of the M/G/1/2/2 queue obtained from an exact analytical solution.

In order to evaluate the correctness of the PH approximation, we solve the model by substituting the original general distribution (either L1, L3, U1 or U2) and the exponentially distributed inter-arrival and service times with approximating DPH or PH distributions. The fitting was performed applying the $D_{\text{scdf}}$ distance measure. Let $\hat{p}_i$ $(i = 1, \ldots, 4)$ denote the steady-state probability of the resulting M/PH/1/2/2 queue with the PH approximation.

The overall approximation error is measured in terms of the difference between the exact steady-state probabilities $p_i$ and the approximate steady-state probabilities $\hat{p}_i$. Two error measures are defined:

$$\varepsilon_{\text{SUM}} = \sum_i \frac{|p_i - \hat{p}_i|}{p_i} \quad \text{and} \quad \varepsilon_{\text{MAX}} = \max_i \frac{|p_i - \hat{p}_i|}{p_i}.$$

All experiments were performed applying both one- and two-phase approximation for the exponential durations. The evaluated numerical values for $\varepsilon_{\text{SUM}}$ and $\varepsilon_{\text{MAX}}$ are reported in Figures 7.21 and 7.22 and 7.23 and 7.24 for the distribution L3. Since the behavior of $\varepsilon_{\text{MAX}}$ is very similar to the behavior of $\varepsilon_{\text{SUM}}$ in all the cases, for the other cases we report $\varepsilon_{\text{SUM}}$ only (L1: Figures 7.25–7.26, U1: Figures 7.27–7.28, U2: Figures 7.29–7.30).

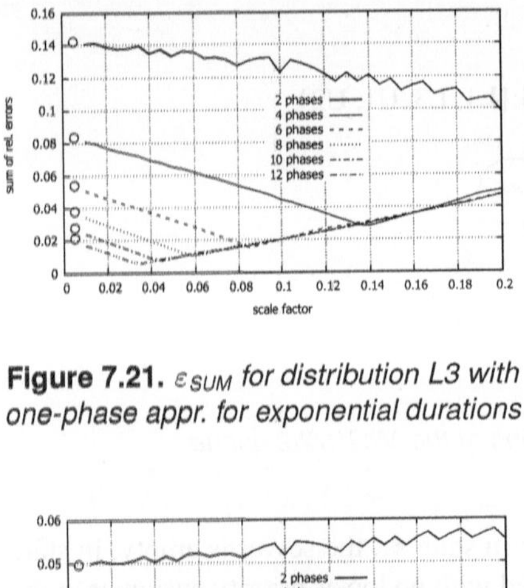


**Figure 7.21.** *$\varepsilon_{SUM}$ for distribution L3 with one-phase appr. for exponential durations*

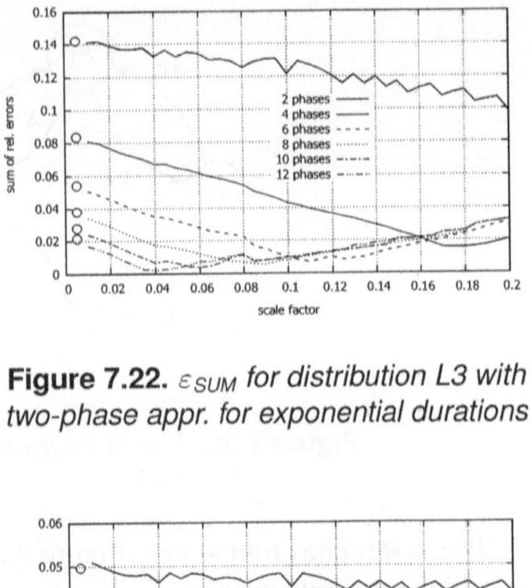


**Figure 7.22.** *$\varepsilon_{SUM}$ for distribution L3 with two-phase appr. for exponential durations*

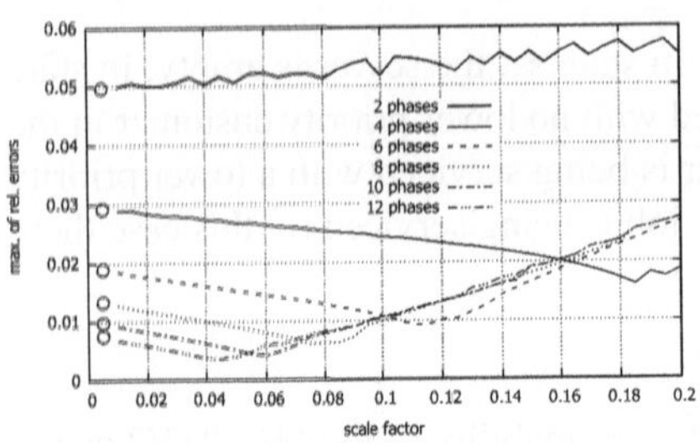


**Figure 7.23.** *$\varepsilon_{MAX}$ for distribution L3 with one-phase appr. for exponential durations*

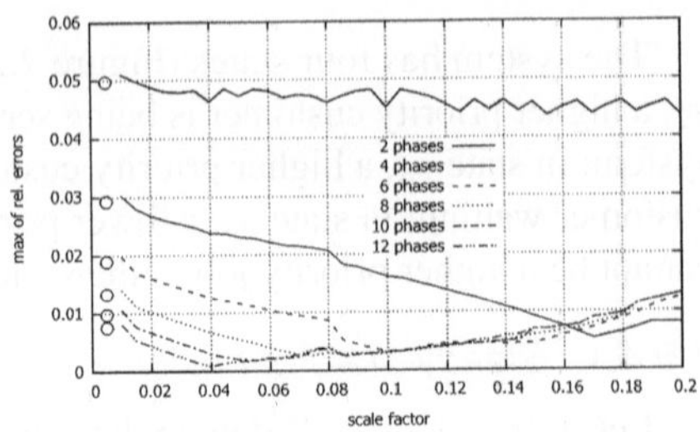


**Figure 7.24.** *$\varepsilon_{MAX}$ for distribution L3 with two-phase appr. for exponential durations*

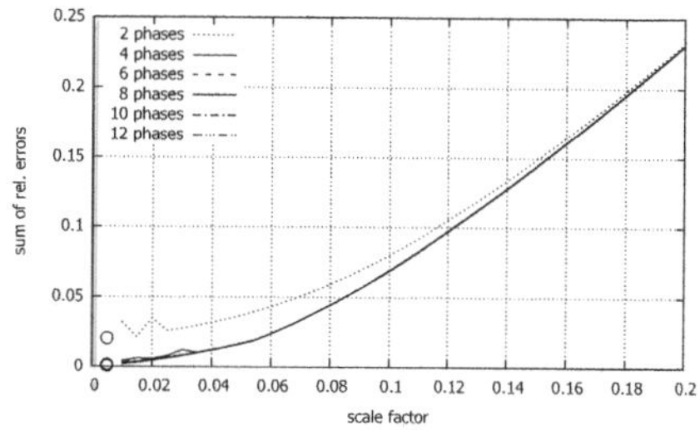


**Figure 7.25.** *$\varepsilon_{SUM}$ for distribution L1 with one-phase appr. for exponential durations*

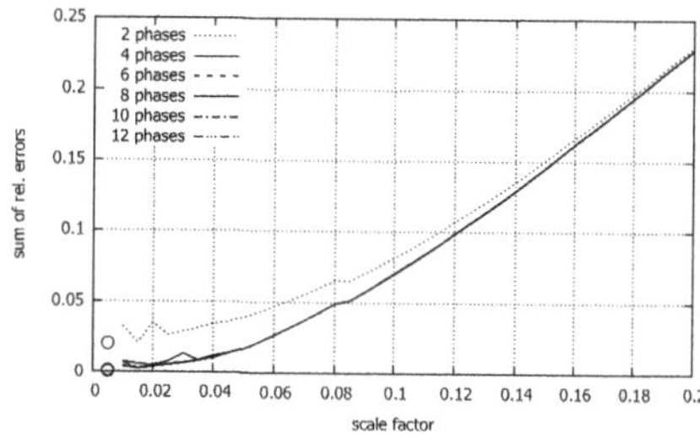


**Figure 7.26.** *$\varepsilon_{SUM}$ for distribution L1 with two-phase appr. for exponential durations*

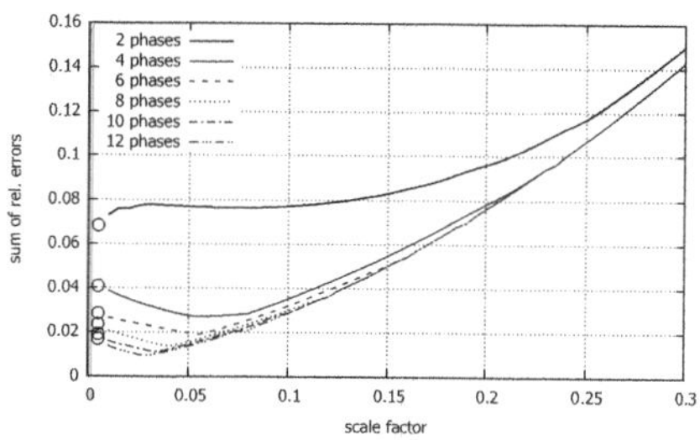


**Figure 7.27.** *$\varepsilon_{SUM}$ for distribution U1 with one-phase appr. for exponential durations*

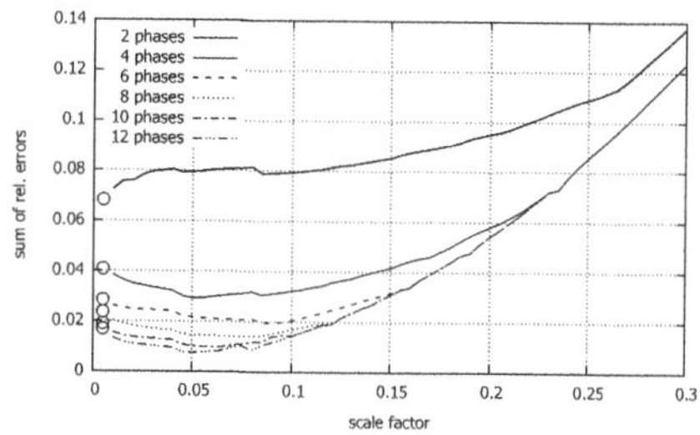


**Figure 7.28.** *$\varepsilon_{SUM}$ for distribution U1 with two-phase appr. for exponential durations*

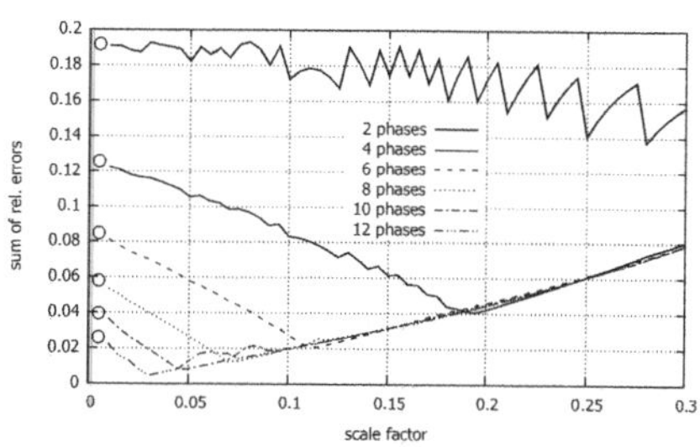


**Figure 7.29.** *$\varepsilon_{SUM}$ for distribution U2 with one-phase appr. for exponential durations*

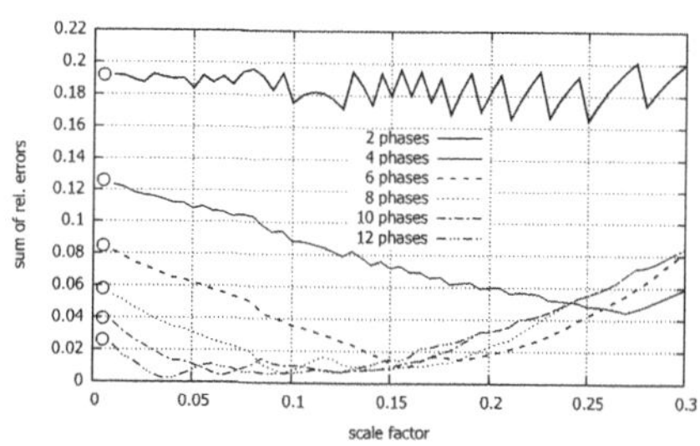


**Figure 7.30.** *$\varepsilon_{SUM}$ for distribution U2 with two-phase appr. for exponential durations*

Figures 7.31 and 7.32 report two cases when neither the service time distribution of the high-priority nor the low-priority jobs are exponentially distributed. In the first case, both the service distributions are L3, while in the second case, higher (lower) priority costumers are served according to distribution L3 (L1).

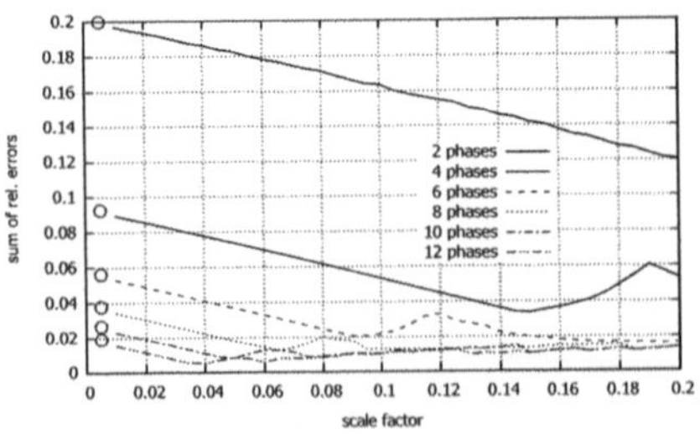


**Figure 7.31.** $\varepsilon_{SUM}$ *for service times L3-L3 with one-phase appr. for exponential durations*

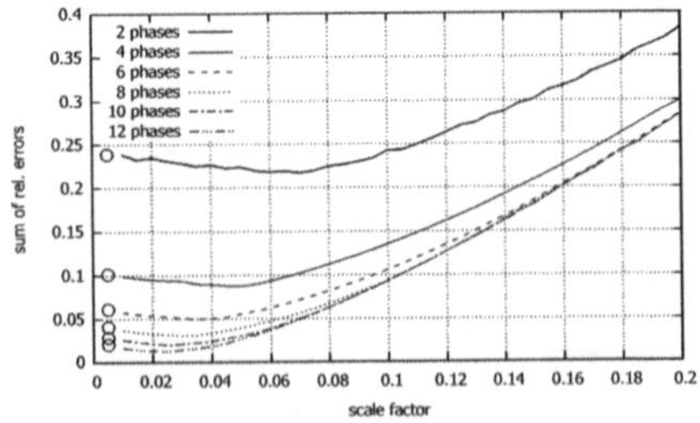


**Figure 7.32.** $\varepsilon_{SUM}$ *for service times L1-L3 with one-phase appr. for exponential durations*

Based on the above experiments, it is rather hard to draw general guidelines for the choice of the scale factor. Some observations, however, can be made:

– If all the optimal scale factors are 0, the approximation error in the evaluation of the performance indices of the global model can be minimized by resorting to continuous PH type approximation. See, for example, the case when the only non-exponential distribution in the model is L1 (Figure 7.25).

– When some of the individual optimal scale factors are 0, while others are not (this is the case, for example, when there are exponential durations and durations with low coefficient of variation in the same model, Figure 7.21), there is an optimal scale factor value (or a range) that gives optimal results for global model. In this case, decreasing the scale factor worsens the precision of the approximation because distributions with low coefficient of variation are approximated worse. On the other hand, increasing the scale factor worsen the approximation because the approximation of the exponential durations is getting less accurate (see Figure 7.21).

– In the cases when one of the individual optimal scale factors is far away from 0, there is a wide range of scale factor values for which the discrete approximation yields better results than the continuous approximation. See, for example, Figure 7.29 where in the case of 10 phases, the discrete approximation gives better results in the range $[0, 0.2]$.

– Applying two-phase discrete PH approximations for the exponential durations slows down the rate at which the approximation error grows with increasing scale factor. See, for example, Figure 7.21 compared to Figure 7.22.

#### 7.5.4.2. *Transient behavior*

In order to investigate the approximation error in the transient behavior, we have considered distribution U2 for the service time of low priority jobs and computed the transient probability of state $s_1$ with two different initial conditions. Figure 7.33

depicts the transient probability of state $s_1$ when the process is started in state $s_1$ with probability 1. Figure 7.34 depicts the transient probability of the state $s_1$ when the service of a lower priority job starts at time 0 (the process starts in $s_4$ with probability 1). All approximations are with DPH distributions of size $n = 10$. Only the DPH approximations are depicted because the PH approximation is very similar to the DPH one with scale factor $\delta = 0.03$. Visual inspection suggests that in the first case (Figure 7.33), the scale factor $\delta = 0.03$, which was optimal for fitting U2 distribution, provides the most accurate results for the transient analysis as well. Instead, in the second case, the approximation with a scale factor $\delta = 0.2$ captures the sharp change in the transient probability better. Moreover, this value of $\delta$ is the only one among the values reported in the figure that results in 0 probability for time points smaller than 1. In other words, the second example depicts the advantage given by DPH distributions to model durations with finite support. This example also suggests that DPH approximation can be of importance when preserving property that the transient probability is zero in an interval of time is crucial (like in modeling time-critical systems) and, hence, DPH approximation can be seen as a bridge between the world of stochastic modeling and the world of functional analysis and model checking [BOB 01].

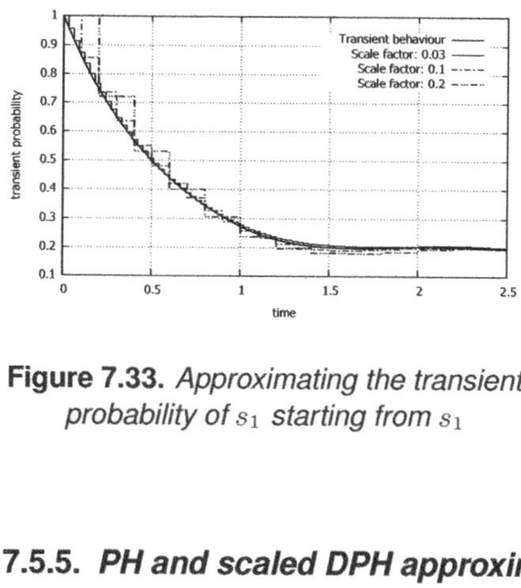


**Figure 7.33.** *Approximating the transient probability of $s_1$ starting from $s_1$*

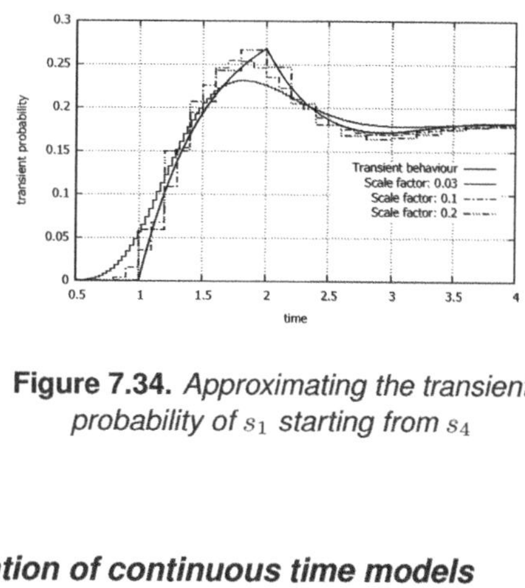


**Figure 7.34.** *Approximating the transient probability of $s_1$ starting from $s_4$*

### 7.5.5. *PH and scaled DPH approximation of continuous time models*

The main message of section 7.5 is that, with the notion of the scale factor $\delta$, we obtain a unified view of DPH and PH distributions. The optimal value $\delta_{opt}$ determines the best distribution in a fitting experiment. When $\delta_{opt} = 0$, the best choice is a PH distribution, while when $\delta_{opt} > 0$, the best choice is a DPH distribution. It has also been shown that the transition from DPH class to PH class is continuous with respect to several properties, like the distance (denoted by $D_{\text{scdf}}$ in 7.3) between the original and the approximate distributions. There are limit theorems for special cases; however, extensive numerical experiments show that the limiting behavior is far more general than the special cases addressed in the theorems.

The numerical examples have also shown that for very small values of $\delta$, the diagonal elements of the transition probability matrix are very close to 1, rendering the DPH fitting procedure numerically unstable.

Advantages and disadvantages of applying DPH distributions with respect to using PH distributions can be summarized as follows.

#### 7.5.5.1. *Advantages of using DPH distributions*

An obvious advantage of the application of DPH distributions is that one can have a closer approximation of distributions with low coefficient of variation. Another important property of the DPH class is that it can capture distributions with finite support and deterministic values. This property allows to capture the periodic behavior of a complex stochastic model, while any PH based approximation of the same model tends to a steady state.

Numerical experiments have also shown that DPH can better approximate distributions with some abrupt or sharp changes in the CDF or in the PDF.

#### 7.5.5.2. *Disadvantages of using DPH distributions*

When applying DPH approximation, coincident events have to be handled, and their consideration may burden the complexity of the analysis significantly. For PH approximation, it is not an issue.

# 8

# Simulation

PH distributions can also be used in simulation studies, where they allow the introduction of realistic event time distributions obtained from measurements. For this application, the efficiency of generating PH-distributed random numbers plays a crucial role. In this chapter, we investigate the efficiency of generating random numbers from continuous PH distributions.

The classic method to generate random samples from a given CDF $F_X(x)$ is the *inverse-function* method. First a uniform random number $U$ is obtained over the interval $(0, 1)$ (this functionality exists in most computational environments). Then in the second step, $X = F_X^{-1}(U)$ provides the random sample whose CDF is $F_X(x)$, that is, $\Pr(X < x) = F_X(x)$.

The inverse-function method can be applied easily to the exponential distribution, where $F_X(x) = 1 - e^{-\lambda x}$. In this case, $X = F_X^{-1}(U)$ simplifies to $X = -\log(1 - U)/\lambda$, which is both simple to implement and fast to evaluate.

However, for general PH distributions, the cumulative distribution function $F_X(x) = 1 - \alpha e^{\mathbf{A}x}\mathbb{1}$ cannot be inverted in an explicit way, thus the inverse-function method is computationally expensive in general.

The next sections introduces several methods to generate PH distributed random samples. Some of them are straightforward, while others apply some non-trivial steps in order to reduce the computational effort.

## 8.1. Generating random samples based on the probabilistic interpretation

Instead of following the inverse-function method and inverting the cumulative distribution function numerically, it is possible to generate PH distributed random samples based on the stochastic interpretation of the underlying Markov chain.

A common feature of all procedures discussed in the rest of the section is that in order to obtain a random sample from a PH distribution, a given number of uniform random variables is necessary and the logarithm function needs to be called a given number of times. These two operations are the computational bottlenecks of all subsequent procedures. Hence, from now on, the efficiency of the procedures will be quantified by

– #uni, the number of uniform random variables required;

– #log, the number of (scalar) logarithms that need to be computed.

### 8.1.1. *Sub-classes of continuous phase type distributions*

Among the sub-classes of PH distributions discussed in section 5.5, there are some sub-classes which are particularly efficient to draw random samples from.

#### 8.1.1.1. *Exponential distribution*

As mentioned above, exponentially distributed samples with parameter $\lambda$ can be obtained as:

$$\text{Exp}(\lambda) \doteq -\frac{1}{\lambda}\log(1-U) \doteq -\frac{1}{\lambda}\log(U), \qquad [8.1]$$

where $U$ is uniformly distributed random variable in $(0, 1)$, and $\doteq$ denotes equality in distribution and we utilized that $U \doteq 1 - U$. With the above introduced metrics, this leads to $\#\text{uni} = 1$ and $\#\text{log} = 1$.

#### 8.1.1.2. *Erlang distribution*

Recognizing that Erlang$(n, \lambda)$ distributions are just the sums of $n$ exponential random variables, one could just repeat [8.1] $n$ times and add up the results, thus $\text{Erl}(n, \lambda) \doteq \sum_{i=1}^{n} - \log(U_i)/\lambda$. This solution needs $\#\text{uni} = n$ uniform random numbers and $\#\text{log} = n$ calls of the logarithm function. The latter, however, can be reduced to 1 by applying the identities of the logarithm, yielding:

$$\text{Erl}(n, \lambda) \doteq -\frac{1}{\lambda}\log\left(\prod_{i=1}^{n} U_i\right). \qquad [8.2]$$

Note that the equality of the rates were essential to apply this trick, thus hypo-exponential distributions with possibly different rates cannot be treated similarly.

#### 8.1.1.3. *Hyper-exponential distribution*

Generating hyper-exponentially distributed samples is simple as well. The first step selects the branch according to initial probability vector $\boldsymbol{\alpha}$, and the corresponding exponential random variable is generated in the second step. Hence, two uniform random numbers and a single logarithm function are necessary, that is, $\#\text{uni} = 2$ and $\#\text{log} = 1$, independent of the number of states.

#### 8.1.1.4. *Hyper-Erlang distribution*

By combining the previous two solutions, one can generate hyper-Erlang distributed random samples efficiently as well, by selecting the branch according to $\boldsymbol{\alpha}$ and generating an Erlang distributed random variable according to [8.2]. For the efficiency metrics, we get $\#\text{log} = 1$ and $\#\text{uni} = 1 + \boldsymbol{\alpha}\boldsymbol{r}^T$ on average, where vector $\boldsymbol{r} = [r_1, \ldots, r_N]$ contains the orders of the contributing Erlang distributions.

### 8.1.2. *The play algorithm*

In the case of more general PH distributions (that do not fall into the classes discussed in section 8.1.1), the most natural way to generate a random number based on the $(\boldsymbol{\alpha}, \boldsymbol{A})$ representation, is to play the evolution of the CTMC until absorption. By play, we mean to simulate the state transitions of the CTMC according to the following basic steps:

– sample the sojourn time in the current state;

– sample the next state visited by the Markov chain.

These steps are repeated until the Markov chain reaches the absorbing state. If the current state is $i$, the sojourn time is an exponentially distributed random variable with parameter $-A_{ii}$, denoted as $\text{Exp}(-A_{ii})$. The probability that the next state is $i$, where $i \neq j$, is $-A_{ij}/A_{ii}$, and the probability of absorption is $1 - \sum_{j \neq i} -A_{ij}/A_{ii}$. Algorithm 8.1 provides the pseudo code of the Play method.

**Algorithm 8.1** Play the Markov chain until absorption

**procedure** PLAY($\boldsymbol{\alpha}, \boldsymbol{A}$)
    clock $\leftarrow 0$
    $i \leftarrow \boldsymbol{\alpha}$-distributed discrete sample for the initial state
    **while** $i \neq$ absorbing state **do**
        $i \leftarrow$ draw a sample for the next state
        clock $\leftarrow$ clock + $\text{Exp}(-A_{ii})$
    **end while**
**return** clock
**end procedure**

If the PH distribution takes $k$ steps till absorption, the efficiency metrics of the Play method are $\#\text{uni} = k+1$ and $\#\text{log} = k$. Instead of repeatedly sampling exponentially distributed samples with the same parameter, Neuts and Pagano [NEU 81b] proposed an elegant solution to reduce #log, called CountPlay. In this procedure, an auxiliary vector, called count, is introduced to store how many times the states of the underlying CTMC have been visited before absorption. The total time to absorption is then obtained as the sum of the corresponding Erlang distributed variables (see algorithm 8.2), because if state $i$ was visited count[$i$] times before absorption, then the time spent in state $i$ is Erlang(count[$i$],$-A_{ii}$) distributed.

**Algorithm 8.2** Count the visits to the states of the Markov chain

**procedure** COUNTPLAY($\boldsymbol{\alpha}, \boldsymbol{A}$)
    count[$i$]$\leftarrow$ 0, $\forall i = 1,\dots,n$
    $i \leftarrow \boldsymbol{\alpha}$-distributed discrete sample for the initial state
    **while** $i \neq$ absorbing state **do**
        count[$i$]$\leftarrow$ count[$i$] + 1
        $i \leftarrow$ draw a sample for the next state
    **end while**
**return** $\sum_{i=1}^{n}$ Erl(count[$i$],$-A_{ii}$)
**end procedure**

In the worst case, when the PH distribution visits all states before the absorption, CountPlay gives $\#\text{log} = n$, which can be significantly less than the number of state transitions before absorption.

We point out that Play and CountPlay are suited to general PH distributions of an arbitrary form, where each phase may have several successor phases. With bi-diagonal APH representations, such as the CF1 form, we can make use of the structural restriction: for each phase, there is exactly one successor phase. Consequently, there is no need to randomly choose the next state. Algorithm 8.3 applies this observation to simplify the Play method.

**Algorithm 8.3** Play the CF1 structure of APH distribution

**procedure** APHPLAY($\boldsymbol{\alpha}, \boldsymbol{A}_m athrmCF1$)
    clock $\leftarrow$ 0
    $i \leftarrow \boldsymbol{\alpha}$-distributed discrete sample for the initial state
    **while** $i \neq$ absorbing state **do**
        clock $\leftarrow$ clock + Exp($-A_{ii}$)
        $i \leftarrow i+1$
    **end while**
**return** clock
**end procedure**

The structural properties of the monocyclic representation (definition 5.16) can be exploited as well. Monocyclic representations consist of FEBs characterized by three parameters: the size $n_i$, the rate $\nu_i$ and the feedback probability $z_i$. The key observation is that the number of times the feedback occurs is geometrically distributed, and given by $\lfloor \log(U)/\log(z_i) \rfloor$, where $U$ is uniformly distributed in $(0, 1)$. If the number of feedbacks is $c$, the time to leave the FEB is Erlang$(c \cdot n_i, \nu_i)$ distributed, which can be efficiently sampled according to [8.2].

In monocyclic representations, however, any states can have positive initial probabilities, even those in the middle of FEBs, which has to be taken into account when generating random numbers. Algorithm 8.4 implements this feature in a computationally efficient way. The efficiency of algorithm 8.4 is due the fact that each $\lambda_i$ intensity is handled in a single step (step 6) according to [8.2], because the number of times the process traverses a phase with intensity $\lambda_i$, $cn_i + \ell$, is computed in advance.

**Algorithm 8.4** Play the monocyclic structure of PH distribution

1: **procedure** MONOCYCLICPLAY($\boldsymbol{\alpha}$, $\boldsymbol{A}_mathrmmonocyclic$)<br>
2: clock $\leftarrow$ 0<br>
3: $i, \ell \leftarrow \boldsymbol{\alpha}$-distributed discrete sample for the initial FEB and the initial state therein<br>
4: **while** $i \neq$ absorbing state **do**<br>
5: $c \leftarrow \lfloor \log(U)/\log(z_i) \rfloor$<br>
6: clock $\leftarrow$ clock + Erl$(cn_i + \ell, \nu_i)$<br>
7: $i \leftarrow i + 1$<br>
8: $\ell \leftarrow n_i$<br>
9: **end while**<br>
10: **return** clock<br>
11: **end procedure**

Some numerical examples are provided in section 8.5.2 comparing the variants of the Play method discussed above.

## 8.2. Representation transformation to improve the Play method

As we saw in section 8.1, generating Erlang, hyper-exponential and PH distributions with many transitions in the generator matrix have different costs. The cost depends on the representation of the distribution, thus for simulation purposes it makes sense to transform the PH distributions to representations that are optimal for random sample generation.

### 8.2.1. *Formulating the optimization problem*

The average cost associated with drawing a random variable from a given PH representation depends on the average number of states traversed up to absorption, denoted by $\overline{k}$. For example, using the Play method the average cost of obtaining a random sample is $\#\text{uni} = \overline{k} + 1, \#\text{log} = \overline{k}$, while in the case of CountPlay it is $\#\text{uni} = n, \#\text{log} = \overline{k}$. In both cases, the average costs depends on $\overline{k}$: the lower $\overline{k}$, the faster the random number generation. Hence, it makes sense to have a closer look at this quantity.

#### 8.2.1.1. *Hyper-exponential and hyper-Erlang distributions*

For hyper-exponential distributions, the average number of state transitions is $\overline{k} = 1$ and for hyper-Erlang distributions, it is $\overline{k} = \boldsymbol{\alpha}\boldsymbol{r}^T$.

#### 8.2.1.2. *APH distribution*

For APH($n$) representations, we have $\overline{k} = \boldsymbol{\alpha}\boldsymbol{\nu}^T$, where $\boldsymbol{\nu} = (n \; n-1 \; \ldots \; 1)$ (thus $\overline{k} = \sum_{i=1}^{n} \alpha_i(n - i + 1)$). That is, the element $i$ of $\boldsymbol{\nu}$ indicates how many phases will be visited before reaching the absorbing state starting from phase $i$.

#### 8.2.1.3. *Monocyclic PH distribution*

Deriving $\overline{k}$ for monocyclic representations is a bit more involved. Given that the representation consists of $m$ FEBs, let us introduce row vector $\boldsymbol{\omega}$ of size $m$, whose $i$th element is the probability of starting from FEB $i$ (e.g. $\omega_1 = \sum_{j=1}^{n_1} \alpha_j$). Furthermore, let $\boldsymbol{\varphi}$ be a row vector of size $m$, whose $i$th element is:

$$\varphi_i = \frac{z_i n_i}{1 - z_i} \sum_{j=i+1}^{m} \frac{n_j}{1 - z_j}$$

(the mean number of steps spent in the $i$th FEB from the first feedback, that is, excluding the steps from the initial state to the feedback state in the first path through the FEB). Let row vector $\boldsymbol{\psi}$ of size $n$ be such that its $i$th element describes how many phases are needed to reach the next FEB (or the absorbing state in the case of the last FEB). For example, if $n_1 = 3$, then $\psi_1 = 3$, $\psi_2 = 2$, $\psi_3 = 1$, $\psi_4 = n_2$, and so on.

Using these notations, the mean number of steps till absorption is $\overline{k} = \boldsymbol{\omega}\boldsymbol{\varphi}^T + \boldsymbol{\alpha}\boldsymbol{\psi}^T$, where $\boldsymbol{\alpha}\boldsymbol{\psi}^T$ contains the number of steps if there is no feedback (i.e. if $z_i = 0$, for $i = 1, \ldots, m$) and $\boldsymbol{\omega}\boldsymbol{\varphi}^T$ contains the additional number of steps due to the loops in the FEB.

#### 8.2.1.4. *General PH distribution*

For a general PH representation, the number of steps until absorption is DPH distributed with representation $(\boldsymbol{\alpha}, \text{diag}\langle 1/A_{ii}\rangle \boldsymbol{A} + \boldsymbol{I})$ and its mean is:

$$\overline{k} = \boldsymbol{\alpha}(\text{diag}\langle 1/A_{ii}\rangle \boldsymbol{A})^{-1}\mathbb{1}. \qquad [8.3]$$

We are interested in the representation with the lowest possible $\overline{k}$. The transformation matrix $\boldsymbol{W}$, leading to the optimal representation, is the solution of the nonlinear optimization problem:

$$\boldsymbol{W} = \arg\min_{\boldsymbol{W}} \gamma(\text{diag}\,\langle 1/G_{ii}\rangle\, \boldsymbol{G})^{-1}\mathbb{1}, \qquad [8.4]$$

subject to

$$\gamma = \boldsymbol{\alpha W}, \boldsymbol{AW} = \boldsymbol{WG}, \boldsymbol{W}\mathbb{1} = \mathbb{1}, \gamma \geq 0, G_{ij} \geq 0, \ \text{for } i \neq j. \qquad [8.5]$$

Hence, we are looking for a PH representation similar to the original one that reaches the absorbing state in as few steps as possible. This nonlinear problem is, however, difficult to solve.

### 8.2.2. *The solution for acyclic phase type distributions*

In the following, we tackle this optimization problem for APH distributions where $(\boldsymbol{\alpha}, \boldsymbol{A})$ is in CF1 form. We choose the CF1 form as our starting point because bi-diagonal forms allow efficient random-variable generation, using the procedure SimplePlay, and because acyclic phase types are commonly given in CF1 form (e.g. by phase-type fitting or moment matching tools). Furthermore, the results in [CUM 82, O'CI 91b] ensure that the results for CF1 are applicable to the whole APH class.

In order to solve the optimization problem, we try to find a bi-diagonal representation $(\boldsymbol{\alpha}^*, \boldsymbol{A}^*)$ for which the average number of traversed states:

$$\overline{k} = \boldsymbol{\alpha}^* \boldsymbol{\nu} = \sum_{i=1}^{n} \alpha_i^*(n - i + 1) \qquad [8.6]$$

is minimal, where $(\boldsymbol{\alpha}^*, \boldsymbol{A}^*)$ denotes the optimal representation.

From the RHS of [8.6], it is obvious that in order to reduce $\overline{k}$, probability mass must be shifted to the higher indices of the initial probability vector $\boldsymbol{\alpha}$. Formally, any probability vector $\boldsymbol{\alpha}'$ which improves $\overline{k}$ must be stochastically larger than $\boldsymbol{\alpha}$.

DEFINITION 8.1.– *The* stochastic ordering *[SZE 95] on the set of stochastic vectors of size n is defined as follows:*

$$\boldsymbol{\alpha} \leq_{st} \boldsymbol{\alpha}' \Leftrightarrow 1 - \sum_{i=1}^{k} \alpha_i \leq 1 - \sum_{i=1}^{k} \alpha_i', \quad \textit{for } k = 1, \ldots, n.$$

At the same time, $(\boldsymbol{\alpha}^*, \boldsymbol{A}^*)$ must represent the same distribution as $(\boldsymbol{\alpha}, \boldsymbol{A})$, and hence the eigenvalues of $\boldsymbol{A}$ and $\boldsymbol{A}^*$ must be same. In the bi-diagonal form, the eigenvalues are the entries of the diagonals. Changing the order of the diagonal elements does not change the eigenvalues, hence a representation where $\boldsymbol{A}^*$ is obtained by re-ordering the diagonal elements is guaranteed to have the same eigenvalues as $(\alpha, \boldsymbol{A})$. Consequently, we consider shifting probability mass to the right by modifying the order of the rates along the matrix diagonal using the following operator. While dealing with re-ordering the diagonal elements in CF1 forms, where the associated bi-diagonal transition matrices are uniquely defined by the order of the eigenvalues, we occasionally represent the transition matrix $\boldsymbol{A}$ by the ordered vector of eigenvalues $\boldsymbol{\Lambda}$. For example:

$$\boldsymbol{\Lambda} = \begin{pmatrix}\lambda_1 & \lambda_2 & \lambda_3 & \lambda_4\end{pmatrix} \text{ denotes } \boldsymbol{A} = \begin{pmatrix} -\lambda_1 & \lambda_1 & & \\ & -\lambda_2 & \lambda_2 & \\ & & -\lambda_3 & \lambda_3 \\ & & & -\lambda_4 \end{pmatrix}, \text{ and}$$

$$\boldsymbol{\Lambda}' = \begin{pmatrix}\lambda_2 & \lambda_4 & \lambda_3 & \lambda_1\end{pmatrix} \text{ denotes } \boldsymbol{A}' = \begin{pmatrix} -\lambda_2 & \lambda_2 & & \\ & -\lambda_4 & \lambda_4 & \\ & & -\lambda_3 & \lambda_3 \\ & & & -\lambda_1 \end{pmatrix}. \qquad [8.7]$$

DEFINITION 8.2.– *The* Swap$(\alpha, \boldsymbol{A}, i)$ operator *exchanges the ith rate with the* $(i+1)$*th rate* $(1 \leq i \leq n-1)$ *in the diagonal of a bi-diagonal representation by swapping the ith and* $(i+1)$*th entry in the vector* $\boldsymbol{\Lambda}$*. The associated similarity transformation matrix* $\boldsymbol{W}$ *has the form:*

$$\boldsymbol{W} = \begin{pmatrix} \ddots & 0 & 0 & 0 & 0 \\ 0 & 1 & 0 & 0 & 0 \\ 0 & w_{i+1,i} & w_{i+1,i+1} & 0 & 0 \\ 0 & 0 & 0 & 1 & 0 \\ 0 & 0 & 0 & 0 & \ddots \end{pmatrix},$$

*where*

$$w_{i+1,i} = \frac{\lambda_i - \lambda_{i+1}}{\lambda_i}, \quad \textit{and} \quad w_{i+1,i+1} = \frac{\lambda_{i+1}}{\lambda_i} \quad \textit{for} \quad 1 \leq i \leq n-1.$$

Where appropriate, we will use the Swap$(\alpha, \boldsymbol{\Lambda}, i)$ notation.

COMMENT 8.3.– *Swap*$(\alpha, \boldsymbol{\Lambda}, i)$ *only involves the ith and* $(i+1)$*th entries of* $\alpha$ *and* $\boldsymbol{\Lambda}$*, which allows for the analytically tractable expressions used in the following. Operators that produce more complex permutations in a single step do not have this property. Furthermore, repeated application of the* Swap *operator is sufficient to generate all permutations of a list [JOH 63].*

Let $(\boldsymbol{\alpha}', \boldsymbol{A}')$ denote the result of applying Swap$(\boldsymbol{\alpha}, \boldsymbol{A}, i)$ on $(\boldsymbol{\alpha}, \boldsymbol{A})$. Since $(\boldsymbol{\alpha}', \boldsymbol{A}')$ is derived by applying a similarity transformation to $(\boldsymbol{\alpha}, \boldsymbol{A})$, both tuples represent the same distribution. Recall from definition 2.10 that:

$$\boldsymbol{\alpha}' = \boldsymbol{\alpha}\boldsymbol{W}, \boldsymbol{A}' = \boldsymbol{W}^{-1}\boldsymbol{A}\boldsymbol{W}.$$

The result of the Swap operation has the following properties:

$$\alpha'_j = \alpha_j, \quad \text{for } \forall j \notin \{i, i+1\}, \qquad [8.8]$$

$$\alpha'_i = \alpha_i + \alpha_{i+1}\frac{\lambda_i - \lambda_{i+1}}{\lambda_i} = \alpha_i + \alpha_{i+1}\left(1 - \frac{\lambda_{i+1}}{\lambda_i}\right), \qquad [8.9]$$

$$\alpha'_{i+1} = \alpha_{i+1}\frac{\lambda_{i+1}}{\lambda_i}, \qquad [8.10]$$

$$\overline{k}(\boldsymbol{\alpha}', \boldsymbol{A}') = \overline{k}(\boldsymbol{\alpha}, \boldsymbol{A}) + \alpha_{i+1}\left(1 - \frac{\lambda_{i+1}}{\lambda_i}\right). \qquad [8.11]$$

Equations [8.8]–[8.11] are valid for any Markovian ($\alpha_i \geq 0$) or non-Markovian ($\alpha_i \in \mathbb{R}$) bi-diagonal representation. Starting from a proper CF1 representation of an APH distribution, the generator remains Markovian after any number of Swap operations, only the initial vector can become negative due to the subtraction in [8.9].

COROLLARY 8.4.– *For Markovian bi-diagonal representations (which can be obtained from CF1 form, by a sequence of* Swap *operations), the* Swap$(\boldsymbol{\alpha}, \boldsymbol{A}, i)$ *operation with adjacent rates* $\lambda_i < \lambda_{i+1}$*, results in* $\overline{k}(\boldsymbol{\alpha}', A') \leq \overline{k}(\boldsymbol{\alpha}, A)$.

*Proof.* For a Markovian bi-diagonal representation $\alpha_{i+1} \geq 0$, using this and $\lambda_i < \lambda_{i+1}$ the statement is a direct consequence of [8.11]. □

Consequently, any Swap$(\boldsymbol{\alpha}, \boldsymbol{A}, i)$ operation where $\lambda_i < \lambda_{i+1}$ and the obtained $\boldsymbol{\alpha}'$ vector is Markovian reduces $\overline{k}$. If all $\lambda_i$ rates can be swapped until $\lambda_i \geq \lambda_{i+1}$ for all $1 \leq i \leq n-1$ and the initial vector remains Markovian, then the obtained representation has minimal $\overline{k}$ cost:

LEMMA 8.5.– *Given a Markovian representation* $(\boldsymbol{\alpha}, \boldsymbol{A})$ *in CF1 form, the representation* $(\boldsymbol{\alpha}^*, \boldsymbol{A}^*)$ *that lists the rates in decreasing order is optimal with respect to* $\overline{k}$ *if* $\boldsymbol{\alpha}^*$ *is a stochastic vector.*

*Proof.* The proof has two steps. First, we show that all bi-diagonal representations are Markovian if $(\boldsymbol{\alpha}^*, \boldsymbol{A}^*)$ is Markovian, and in the second step we show by contradiction that $(\boldsymbol{\alpha}^*, \boldsymbol{A}^*)$ is optimal with respect to $\overline{k}$.

According to property [8.9], a Swap operation applied to a Markovian representation can result in a non-Markovian representation only if a larger rate $\lambda_{i+1}$ is exchanged for a smaller rate $\lambda_i$. Starting from $(\boldsymbol{\alpha}^*, \boldsymbol{A}^*)$, all representations can be obtained by a sequence of Swap operations in which a smaller rate $\lambda_{i+1}$ is exchanged for a larger rate $\lambda_i$. If $(\boldsymbol{\alpha}^*, \boldsymbol{A}^*)$ is Markovian, none of these Swap operations can result in a non-Markovian representation.

Now assume the representation $(\boldsymbol{\alpha}', \boldsymbol{A}')$ with rates:

$$\lambda'_1, \ldots, \lambda'_i, \lambda'_{i+1}, \ldots, \lambda'_n$$

ordered such that $\lambda'_i < \lambda'_{i+1}$ is optimal. Then from [8.11] it follows that by exchanging $\lambda'_i, \lambda'_{i+1}$ using the Swap operator we can obtain a representation $(\boldsymbol{\alpha}'', \boldsymbol{A}'') = \mathsf{Swap}(\boldsymbol{\alpha}', \boldsymbol{A}', i)$ for which $\overline{k}(\boldsymbol{\alpha}'', \boldsymbol{A}'') < \overline{k}(\boldsymbol{\alpha}', \boldsymbol{A}')$. □

The Swap operation may result in a non-stochastic vector $\boldsymbol{\alpha}'$ if $\alpha'_i = \alpha_i + \alpha_{i+1}(1 - \frac{\lambda_{i+1}}{\lambda_i}) < 0$. In this case, $(\boldsymbol{\alpha}', \boldsymbol{A}')$ is a non-Markovian representation of the same APH distribution. This representation is not suitable as input for the Play procedure. Furthermore, both the stochastic ordering and $\overline{k}$ are only defined for stochastic $\boldsymbol{\alpha}$. We must therefore avoid Swap operations that will result in non-stochastic $\boldsymbol{\alpha}'$.

Based on these observations, we propose algorithms for finding the optimal Markovian APH representation when the reversed CF1 form (in which the rates are in decreasing order) is non-Markovian. The algorithms are best discussed as operating on the graph of all permutations of the rate vector $\boldsymbol{\Lambda}$. From each permutation, exactly $n-1$ other permutations can be reached by applying a Swap operation. If the reversed CF1 form is non-Markovian, then some of the permutations have non-Markovian initial vectors.

The most obvious approach proceeds by exploring the complete graph by generating all permutations of $\boldsymbol{\Lambda}$ and minimizing $\overline{k}$ over the subset of permutations for which the associated initial vector is stochastic. This approach is easily implemented, for example, as a modification to the Steinhaus–Johnson–Trotter algorithm for enumerating permutations [JOH 63] and is guaranteed to find the optimum. Unfortunately, since this method explores all $n!$ permutations for an APH($n$) distributions, it is infeasible for larger APH distributions.

Lemma 8.5 provides the basis for the two computationally less expensive algorithms presented in this section. The underlying intuition is as follows: the optimal representation with respect to $\overline{k}$ is somewhere on the Markovian side of the boundary between the Markovian and the non-Markovian representations. Lemma 8.5 implies that the Markovian optimum is along one of the paths from the CF1 to the reversed CF1 form. We thus need to find the (Markovian) point where a path between CF1 and reversed CF1 form crosses the boundary between the Markovian and non-Markovian representations.

Starting with the CF1 form any Swap operation of rate $\lambda_i$ with rate $\lambda_{i+1}$ where $\lambda_i < \lambda_{i+1}$ decreases $\overline{k}$ if the initial vector after the Swap operation is Markovian. Properties [8.10] and [8.11] thus define the order of the Swap operations to search for the optimal Markovian representation without enumerating all permutations and without explicitly computing $\overline{k}$ for the new representation. Algorithm 8.5 follows from this intuition. It is a modified version of the Bubblesort algorithm [KNU 97] that attempts to re-order the rates into the reversed CF1 form.

**Algorithm 8.5** Efficiently re-order the rates into the reversed CF1 form

**procedure** BUBBLESORTOPTIMISE($\boldsymbol{\alpha}, \boldsymbol{\Lambda}_{\text{increasing}}$)
    **for** $i = 1, \ldots, n-1$ **do**
        **for** $j = 1, \ldots, n-1$ **do**
            **if** $\boldsymbol{\Lambda}[j] < \boldsymbol{\Lambda}[j+1] \ \wedge \ (\boldsymbol{\alpha}', \boldsymbol{\Lambda}') := \mathsf{Swap}(\boldsymbol{\alpha}, \boldsymbol{\Lambda}, i)$ is Markovian **then**
                $(\boldsymbol{\alpha}, \boldsymbol{\Lambda}) := (\boldsymbol{\alpha}', \boldsymbol{\Lambda}')$
            **else**
                **break**
            **end if**
        **end for**
    **end for**
**return** $(\boldsymbol{\alpha}, \boldsymbol{\Lambda})$
**end procedure**

Note that the algorithm does not perform Swap operations if the result would be non-Markovian, that is, it does not cross the boundary between Markovian and non-Markovian representations.

Algorithm 8.6 starts from the reversed CF1 form and searches for the point where the path toward the CF1 first crosses the border to the Markovian representations. The path is constructed by swapping pairs of rates such that in the result, the higher rate is to the right (which means that the result is closer to the CF1). The algorithm stops when it encounters a Markovian representation. Otherwise, the procedure terminates when the rates are in increasing order, which is the CF1 form.

### 8.2.3. *A heuristic representation optimization for general continuous phase type distributions*

The CF1 form of APH distributions restricts the complexity of the representation optimization problem defined in [8.4]. In the case of general PH distributions, a similar structural restriction and heuristic optimization can be applied starting from the monocyclic representation of the PH distribution, which is always available according to theorem 5.17.

**Algorithm 8.6** Find the first serial Markovian APH representation

**procedure** FINDMARKOVIANORDER($\boldsymbol{\alpha}, \boldsymbol{\Lambda}_{\text{increasing}}$)
  $(\boldsymbol{\alpha}', \boldsymbol{A}') \leftarrow$ the reversed CF1 form of $(\boldsymbol{\alpha}, \boldsymbol{A})$
  **while** $\exists i \in \{1, \ldots, n-1\} : \alpha'_i < 0$ **do**
    $i \leftarrow \arg\min_i \{\alpha'_i < 0\}$
    $i \leftarrow \max(2, i)$
    **while** $\boldsymbol{\alpha}'$ is not Markovian $\wedge\ \exists k : \boldsymbol{\Lambda}[k] \geq \boldsymbol{\Lambda}[k+1]$ **do**
      $k \leftarrow \arg\min_j \{i - 1 \leq j \leq n - 1 \ : \ \boldsymbol{\Lambda}[j] \geq \boldsymbol{\Lambda}[j+1]\}$
      $(\boldsymbol{\alpha}', \boldsymbol{\Lambda}') \leftarrow \mathsf{Swap}(\boldsymbol{\alpha}', \boldsymbol{\Lambda}', k)$
    **end while**
  **end while**
**return** $(\boldsymbol{\alpha}', \boldsymbol{\Lambda}')$
**end procedure**

The monocyclic representation is a block bi-diagonal representation composed of FEBs, where block level optimization can be achieved in a similar way as the phase level optimization of the CF1 form in the previous section [HOR 16b]; the FEBs can be reordered such that the complexity of random number sampling is minimal.

Unfortunately, due to the complex structure of FEBs, the effect of interchanging the order of two consecutive FEBs is not as straightforward as the case of the CF1 form of acyclic PH distributions. Instead, introducing the Swap operation for the $i$th and $(i+1)$th FEBs of a monocyclic representation and computing the efficiency of the obtained representations by $\overline{k}$ (like in line 4 of algorithm 8.5 for APH) can be applied for finding an optimal order of the FEBs in monocyclic representation.

## 8.3. An acceptance-rejection algorithm

ME distributions do not have a straightforward stochastic interpretation, consequently the methods introduced in the previous sections for generating random variables of Markovian PH representations cannot be used for simulation. To overcome this difficulty, we propose a version of the acceptance-rejection method.

The acceptance-rejection method is a widely used method in simulation [ROB 04]. It consists of two main steps: (1) drawing random samples from an easy-to-compute distribution, and (2) accept the sample with a sample-dependent probability such that the overall probability density of the accepted samples is identical with the required density. The computational complexity of this method depends on the sample efficiency, which is the ratio of the number of accepted and the number of generated samples.

Similar to the case of Markovian models, the representation of ME distributions affects the sample efficiency and the computational complexity. We evaluate the behavior of two particular representations with nice structural properties.

### 8.3.1. *Generating random variables from matrix exponential distributions which have a Markovian generator*

In this section, we consider the case when the representation of the ME distribution has a Markovian generator and a general initial vector that might contain negative elements.

Let $(\boldsymbol{\alpha}, \boldsymbol{A})$ of size $n$ be the representation of the ME distribution such that $\boldsymbol{A}$ is a Markovian generator matrix (off-diagonal elements are non-negative, and the row sums are non-positive). The probability density function can be expressed as a non-convex combination of PH distributions:

$$f(x) = \boldsymbol{\alpha} e^{\boldsymbol{A}x}(-\boldsymbol{A})\mathbb{1} = \sum_{i=1}^{n} \alpha_i \cdot \underbrace{\mathbf{e}_i\, e^{\boldsymbol{A}x}(-\boldsymbol{A})\mathbb{1}}_{g_i(x)}, \qquad [8.12]$$

with $\mathbf{e}_i$ denoting the $i$th unit row vector. Observe that $(\mathbf{e}_i, \boldsymbol{A})$ is a Markovian representation of a PH distribution with pdf $g_i(x)$; consequently, $\int_0^\infty g_i(x)dx = 1$.

To cope with the negative coefficients, we apply an acceptance–rejection method to generate a random variable as follows. The set of coefficients of the density function is divided into $\mathcal{A}_+$ and $\mathcal{A}_-$, such that $i \in \mathcal{A}_+$ if $\alpha_i \geq 0$ and $i \in \mathcal{A}_-$ otherwise. This way, $f(x)$ is separated into a positive part $(f_+(x))$ and a negative part $(f_-(x))$:

$$f(x) = \underbrace{\sum_{i\in\mathcal{A}_+} \alpha_i \cdot g_i(x)}_{f_+(x)} + \underbrace{\sum_{i\in\mathcal{A}_-} \alpha_i \cdot g_i(x)}_{f_-(x)}, \qquad [8.13]$$

where $f_+(x) \geq 0,\ \forall x \geq 0$ and $f_-(x) \leq 0,\ \forall x \geq 0$.

Multiplying with the inverse of $q^* = \sum_{j\in\mathcal{A}_+} \alpha_j$, where $q^* \geq 1$, the positive part gets normalized and we get:

$$\hat{f}_+(x) = \frac{1}{q^*} \sum_{i\in\mathcal{A}_+} \alpha_i \cdot g_i(x), \qquad [8.14]$$

which is a valid PH distribution with Markovian representation $\left(\sum_{i\in\mathcal{A}_+} \alpha_i/q^*\, \mathbf{e}_i, \boldsymbol{A}\right)$, where the initial vector is non-negative and normalized. With these notations and definitions, the acceptance–rejection based method to generate random numbers from $(\boldsymbol{\alpha}, \boldsymbol{A})$ is formalized by algorithm 8.7.

**Algorithm 8.7** Generating ME distributed samples which have a Markovian generator

**procedure** ACCEPTREJECTSAMPLING($\boldsymbol{\alpha}, \boldsymbol{A}$)
  **while** true **do**
    $I \leftarrow$ discrete random sample with distribution $\sum_{i \in \mathcal{A}_+} \alpha_i / q^* \, \mathbf{e}_i$
    $R \leftarrow$ random sample with PDF $g_I(x)$ by any PH sampling method
      // $R$ is $\hat{f}_+(x)$ distributed random sample
    $p_{\text{accept}}(R) \leftarrow \frac{f_+(R)+f_-(R)}{f_+(R)}$
    $U \leftarrow$ uniform$(0, 1)$ sample
    **if** $U < p_{\text{accept}}(R)$ **then**
      **return** $R$
    **end if**
  **end while**
**end procedure**

THEOREM 8.6.– *Algorithm 8.7 provides a random number with PDF $f(x)$, and the number of required samples is geometrically distributed with PMF $(1-q^*)^{n-1}q^*$.*

*Proof.* Let $f^*(x)$ be the probability density of the sample generated by algorithm 8.7. In accordance with the proof of the acceptance–rejection method, we are going to show that $f^*(x) = f(x)$. The probability density that the first step of the algorithm results in sample $R$ is $\hat{f}_+(R)$. The probability density that sample $R$ is the accepted can be computed as:

$$f^*(R) = \frac{\hat{f}_+(R) p_{\text{accept}}(R)}{\displaystyle\int_x \hat{f}_+(x) p_{\text{accept}}(x) dx}$$

$$= \frac{\dfrac{f_+(R)}{q^*} \dfrac{f_+(R)+f_-(R)}{f_+(R)}}{\displaystyle\int_x \dfrac{f_+(x)}{q^*} \dfrac{f_+(x)+f_-(x)}{f_+(x)} dx} = \frac{\frac{1}{q^*} f(R)}{\frac{1}{q^*} \int_x f(x) dx} = f(R).$$

The steps of the iterative procedure are independent. The probability of accepting a sample is $\int_x \hat{f}_+(x) p_{\text{accept}}(x) dx = 1/q^*$. □

### 8.3.2. *Generating matrix exponential distributed random variables using feedback Erlang blocks*

As shown in section 8.3.1, there are several representations from which it is efficient to draw random numbers. In this section, we present two general representations with special structures composed of FEBs.

#### 8.3.2.1. *Hyper-feedback-Erlang representation*

DEFINITION 8.7.– A hyper-feedback-Erlang (Hyper-FE) distribution *is defined by an initial probability vector* $\boldsymbol{\alpha}$ *and a transient generator* $\boldsymbol{A}$ *which has the following special structure (see Figure 8.1):*

$$\boldsymbol{A} = \begin{pmatrix} \boldsymbol{M_1} & & & \\ & \boldsymbol{M_2} & & \\ & & \ddots & \\ & & & \boldsymbol{M_J} \end{pmatrix}, \qquad [8.15]$$

*where matrices* $\boldsymbol{M_j}$ *of size* $n_j m_j \times n_j m_j$ *are the sub-generators of* $m_j$ *identical concatenated FEBs with parameters* $(\lambda_j, n_j, z_j)$*:*

$$\boldsymbol{M_j}(\lambda_j, n_j, z_j, m_j) = \left(\begin{array}{cccc|c|cccc} -\lambda_j & \lambda_j & & & & & & & \\ & \ddots & \ddots & & & & & & \\ & & -\lambda_j & \lambda_j & & & & & \\ z_j\lambda_j & & & -\lambda_j & (1-z_j)\lambda_j & & & & \\ \hline & & & & \ddots & \ddots & & & \\ \hline & & & & & -\lambda_j & \lambda_j & & \\ & & & & & & \ddots & \ddots & \\ & & & & & & & -\lambda_j & \lambda_j \\ & & & & & z_j\lambda_j & & & -\lambda_j \end{array}\right). \qquad [8.16]$$

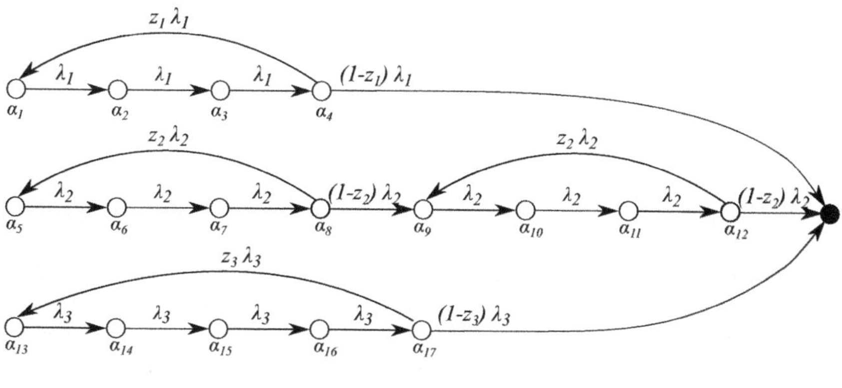


**Figure 8.1.** *Structure of the Hyper-FE distribution with FEB size* $n_1 = 4, n_2 = 4, n_3 = 5$ *and eigenvalue multiplicity* $m_1 = 1, m_2 = 2, m_3 = 1$

The structure of matrix $\boldsymbol{M}_j(\lambda_j, n_j, z_j, m_j)$ in [8.16] and Figure 8.1 (where the multiplicity of the second eigenvalue is two) emphasizes the fact that to maintain the multiplicity of the eigenvalues which are larger than one, the Hyper-FE structure should contain as many concatenated FEBs as the multiplicity of the related eigenvalue.

To transform a general non-Markovian representation $(\boldsymbol{\tau}, \boldsymbol{T})$ into a similar representation with the structure of matrix $\boldsymbol{A}$ in [8.15], we can apply a slightly modified version of algorithm 5.3. First, the FEBs are generated from the eigenvalues in the same way as in Step 1 of algorithm 5.3, than $\boldsymbol{A}$ is generated from these FEBs according to [8.15] (which is the only difference compared to algorithm 5.3). The initial vector associated with matrix $\boldsymbol{A}$ is computed with the help of the transformation matrix $\boldsymbol{W}$ in the same way as in Step 2 of algorithm 5.3. That is, $\boldsymbol{W}$ is the solution of $\boldsymbol{T}\boldsymbol{W} = \boldsymbol{W}\boldsymbol{A}, \boldsymbol{W}\mathbb{1} = \mathbb{1}$ and $\boldsymbol{\alpha} = \boldsymbol{\tau}\boldsymbol{W}$.

Vector $\boldsymbol{\alpha}$ is composed of sub-vectors according to the underlying $\boldsymbol{M}_j$ block structure of $\boldsymbol{A}$ and the vector element associated with state $i$ of block $j$ is denoted by $\alpha_{j,i}$. Similar to [8.12], the probability density function can then be expressed as:

$$f(x) = \boldsymbol{\alpha} e^{\boldsymbol{A}x}(-\boldsymbol{A})\mathbb{1} = \sum_{j=1}^{J} \sum_{i=1}^{n_j m_j} \alpha_{j,i} \cdot \underbrace{\mathbf{e}_i \, e^{\boldsymbol{M}_j x}(-\boldsymbol{M}_j)\mathbb{1}}_{g_{j,i}(x)}, \qquad [8.17]$$

where $J$ denotes the total number of distinct real eigenvalues and complex conjugate eigenvalue pairs. Note that $(\mathbf{e}_i, \boldsymbol{M}_j)$ is a Markovian representation for $g_{k,i}(x)$, from which it is efficient to draw random numbers since it is composed by FEBs.

We obtain a random variable with density $g_{k,i}(x)$ denoted by $R_{g_{k,i}}$ using:

$$L_{j,i} = n_j \, m_j - i + 1 + \sum_{\ell=\lceil i/n_j \rceil}^{m_j} n_j \cdot \left\lfloor \frac{\log U_\ell}{\log z_j} \right\rfloor, \qquad [8.18]$$

$$R_{g_{j,i}} = -\frac{1}{\lambda_j} \log \prod_{\ell=1}^{L_{j,i}} U_\ell.$$

In this expression, $L_{j,i}$ corresponds to the number of steps (exponential distributions) taken before absorption. $n_j \, m_j - i + 1$ is the number of steps that is taken without feedback, while the sum represents the number of steps due to feedback: $\lfloor \log U_\ell / \log z_j \rfloor$ is the geometrically distributed random variable for the number of feedback loops and $n_j$ is the number of steps for a feedback loop. The efficiency of computing $R_{g_{k,i}}$ according to [8.18] is due to the separation of the two steps, sampling the number of phases with rate $\lambda_j$ and sampling an Erlang($L_{j,i}, \lambda_j$) distribution.

In the case when $\alpha_{j,i} \geq 0, \forall i, j$, generating a random variable from $f(x)$ is simple: draw a discrete random sample from distribution $\boldsymbol{\alpha}$ for the starting point of the Hyper-FE structure, and draw a $g_{j,i}(x)$ distributed random number according to [8.18].

However, in case the initial vector has negative elements, we apply the acceptance–rejection method to generate a random variable as described in section 8.3.1. Utilizing the Hyper-FE structure of $\boldsymbol{A}$ allows for efficient generation of the random variable in the third line of algorithm 8.7.

In each iteration of the algorithm, before accepting a sample there is exactly one logarithm function computed to obtain a sample from an Erlang($L_{j,i}$) distribution and $(m_j - \lceil i/n_j \rceil + 1)$ logarithm functions are computed to draw the number of times a feedback loop is traversed in the FEBs. Note that it is not necessary to evaluate $\log z_j$ every time, since it can be pre-calculated before starting the algorithm. The total number of logarithms evaluated is:

$$\#\text{ilog} = \sum_{j=1}^{J} \sum_{i=1}^{n_j m_j} \alpha_{j,i} \cdot (2 + m_j - \lceil i/n_j \rceil). \qquad [8.19]$$

For the average number of uniformly distributed random samples required in one iteration before accepting the sample, we get:

$$\#\text{iuni} = \sum_{j=1}^{J} \sum_{i=1}^{n_j m_j} \alpha_{j,i}$$

$$\cdot \left( \underbrace{\left(1 + m_j - \lceil i/n_j \rceil\right)}_{\text{to evaluate } L_{j,i}} + \underbrace{\left(n_j m_j - i + 1 + (1 + m_j - \lceil i/n_j \rceil) n_j/(1 - z_j)\right)}_{E(L_{j,i}) \text{ uniforms required by } R_{g_{j,i}}} \right).$$

Taking into consideration that the mean number of rejections until a sample is accepted is $1/q^*$, we have basic operations:

$$\#\text{log} = \frac{\#\text{ilog}}{q^*}, \quad \#\text{uni} = \frac{\#\text{iuni}}{q^*}. \qquad [8.20]$$

### 8.3.2.2. *Hypo-feedback-Erlang representation*

DEFINITION 8.8.– *A hypo-feedback-Erlang (Hypo-FE) distribution is defined by an initial probability vector* $\boldsymbol{\alpha}$ *and a transient generator* $\boldsymbol{A}$ *which has the following special structure (see Figure 8.2):*

$$\boldsymbol{A} = \begin{pmatrix} \boldsymbol{M}_1 & \boldsymbol{M}_1' & & \\ & \boldsymbol{M}_2 & \boldsymbol{M}_2' & \\ & & \ddots & \\ & & & \boldsymbol{M}_J \end{pmatrix}, \qquad [8.21]$$

*where matrices $\boldsymbol{M_j}$ are defined in* [8.16] *and:*

$$\boldsymbol{M}'_j = (-\boldsymbol{M}_j)\mathbb{1} \cdot \mathbf{e}_1 \,. \tag{8.22}$$

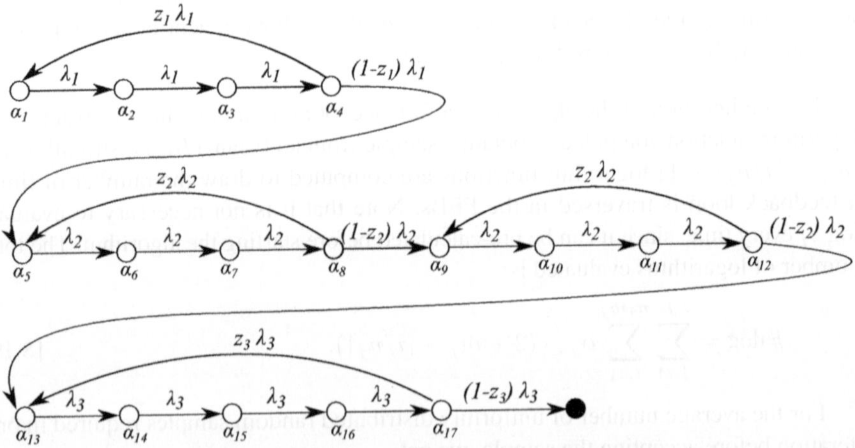


**Figure 8.2.** *Structure of the Hypo-FE distribution*

Apart from the ordering of the FEBs, definition 8.8 is equivalent with definition 5.16. The main difference is that definition 5.16 does not contain restrictions on the order of the FEBs, while definition 8.8 requires that the FEBs associated with the same eigenvalue (representing a higher multiplicity of the eigenvalue) are directly connected.

Algorithm 5.3 can transform any general non-Markovian representation $(\boldsymbol{\tau}, \boldsymbol{T})$ into a similar representation with the structure of matrix $\boldsymbol{A}$ according to [8.21].

The main benefit of using Hypo-FE distribution is that the Erlang distribution sampling associated with the FEBs of a monocyclic representation can be unified for the FEBs associated with the same eigenvalue:

$$L_{j,i} = n_j m_j - i + 1 + \sum_{\ell=\lceil i/n_j \rceil}^{m_j} n_j \cdot \left\lfloor \frac{\log U_\ell}{\log z_j} \right\rfloor , \tag{8.23}$$

$$R_{g_{j,i}} = -\frac{1}{\lambda_j} \log \prod_{\ell=1}^{L_{j,i}} U_\ell + \sum_{r=j+1}^{J} (-1) \frac{1}{\lambda_r} \log \prod_{\ell=1}^{L_{r,1}} U_\ell . \tag{8.24}$$

The only difference compared to Hyper-FE structure is that after traversing the initially selected block ($j$), all consecutive $\boldsymbol{M}_{j+1}, \ldots, \boldsymbol{M}_J$ blocks are traversed until absorption.

The cost of generating a random sample from the Hypo-FE structure is calculated similar to the Hyper-FE case. The final expressions including the cost of sample rejections are:

$$\#\log = \frac{1}{q^*}\sum_{j=1}^{J}\sum_{i=1}^{n_j m_j} \alpha_{j,i}\cdot\left((2+m_j-\lceil i/n_j\rceil)+\sum_{r=j+1}^{J}(1+m_r)\right), \qquad [8.25]$$

$$\begin{aligned}\#\text{uni} =& \frac{1}{q^*}\sum_{j=1}^{J}\sum_{i=1}^{n_j m_j}\alpha_{j,i}\cdot\Bigg[\underbrace{(1+m_j-\lceil i/n_j\rceil)}_{\text{to evaluate } L_{j,i}} \\ &+\underbrace{\left(n_j m_j - i + 1 + (1+m_j-\lceil i/n_j\rceil)n_j/(1-z_j)\right)}_{E(L_{j,i})\text{ uniforms required by first term of } R_{g_{j,i}}} \\ &+\underbrace{\sum_{r=j+1}^{J}\left(m_r + n_r m_r + n_r m_r/(1-z_r)\right)\Bigg]}_{\text{uniforms required by the sum in } R_{g_{j,i}}}.\end{aligned} \qquad [8.26]$$

It might appear that generating Hypo-FE distributed sample is more expensive compared to Hyper-FE distributed one due to the additional sum appearing in $R_{g_{j,i}}$ in [8.23]. However, the initial vectors ($\alpha$) of the two representations are different; consequently, the mean number of rejections $1/q^*$ are different as well.

Numerical experiments indicate that either Hyper-FE or Hypo-FE can perform better in specific examples.

## 8.4. Overview of simulation methods

Many of the subclasses of PH distributions can be efficiently sampled, as detailed in section 8.1.1. The canonical forms of APH distributions are commonly not optimal for random sampling, because a reordering of the transition rates in the CF1 form might result in Markovian representations with lower sampling cost, as detailed in section 8.2.2. The effect of interchanging a larger transition rate with a lower one is determined by [8.11].

PH distributions have infinitely many Markovian and non-Markovian representations, which allow different numerical procedures to generate PH distributed random samples. The structure-based methods for PH distributions without specific structure, for example, the Play and CountPlay methods, are commonly less efficient than transforming the representation into a special structure and generate a sample based on that representation [HOR 16b]. Some of the potential structures used are based on FEBs.

Unfortunately, the sample generation efficiency of FEB-based representations does not exhibit simple relations such as [8.11] for APH distributions. Different serial and parallel compositions of FEBs can provide efficient representations, but the most efficient representation can be found only by exhaustive search.

In the case when the FEB-based representation of a PH distribution is such that the associated initial vector contains negative elements, the acceptance–rejection method of section 8.3 can be applied to generate random samples.

## 8.5. Numerical examples

### 8.5.1. *Optimizing acyclic phase type representations*

In this section, we demonstrate the effect of reordering the phases in CF1 representation of acyclic PH distributions. The cost of the representation, $\overline{k}$, is computed according to [8.6]. To simplify the notation, instead of the generator matrices in CF1 form, we present the vector $\mathbf{\Lambda}$ whose elements are the ordered list of eigenvalues as defined in [8.7].

– Generalized Erlang distribution:

If the initial vector of a CF1 APH distribution is $e_1$, that is, the Markov chain starts from phase 1 and visits all states before absorption, the reordering of the phases has no effect on the initial vector and consequently no effect on the cost of sample generation. For example, if $\mathbf{\Lambda} = \begin{pmatrix}1 & 2 & 3 & 4\end{pmatrix}$ and $\boldsymbol{\alpha} = \begin{pmatrix}1 & 0 & 0 & 0\end{pmatrix}$, then for any reordering of the eigenvalues, $\boldsymbol{\alpha}' = \begin{pmatrix}1 & 0 & 0 & 0\end{pmatrix}$ and $\overline{k} = 4$.

– APH with Markovian initial vector for eigenvalues with decreasing order:

In the CF1 form, the eigenvalues are increasingly ordered. If the order of the eigenvalues can be reversed such that the associated initial vector remains Markovian, then the obtained representation is optimal according to corollary 8.4. For $\mathbf{\Lambda} = \begin{pmatrix}1 & 2 & 3 & 4\end{pmatrix}$ and $\boldsymbol{\alpha} = \begin{pmatrix}0.7 & 0.15 & 0.09 & 0.06\end{pmatrix}$, the average sampling cost is $\overline{k} = 3.49$. The representation of this APH distribution with decreasing order of eigenvalues is $\mathbf{\Lambda}' = \begin{pmatrix}4 & 3 & 2 & 1\end{pmatrix}, \boldsymbol{\alpha}' = \begin{pmatrix}0.46 & 0.12 & 0.18 & 0.24\end{pmatrix}$, whose average cost is $\overline{k}' = 2.8$. Since $\boldsymbol{\alpha}'$ is Markovian, this representation provides the optimal average cost.

– APH with non-Markovian initial vector for eigenvalues with decreasing order:

Unfortunately, the representation with decreasing eigenvalues might be non-Markovian. For example, for $\mathbf{\Lambda} = \begin{pmatrix}1 & 2 & 3 & 4\end{pmatrix}, \boldsymbol{\alpha} = \begin{pmatrix}0.5 & 0.4 & 0.05 & 0.05\end{pmatrix}$, the representation with decreasing rates is $\mathbf{\Lambda}' = \begin{pmatrix}4 & 3 & 2 & 1\end{pmatrix}, \boldsymbol{\alpha}' = \begin{pmatrix}-0.6 & 1.4 & 0 & 0.2\end{pmatrix}$, which is non-Markovian. In this case, algorithm 8.6 can be used to find a Markovian representation with minimal cost, which is $\mathbf{\Lambda}'' = \begin{pmatrix} 2 & 4 & 3 & 1 \end{pmatrix}, \boldsymbol{\alpha}'' = \begin{pmatrix} 0.1 & 0.7 & 0 & 0.2 \end{pmatrix}$, with $\overline{k}(\boldsymbol{\alpha}'', \mathbf{\Lambda}'') = 2.7$.

### 8.5.2. *Optimizing general continuous phase type representations*

In this section, we demonstrate the effect of using optimized representations instead of using Play and CountPlay methods with general PH representation. The following methods are considered in the comparison:

– the Play method (without representation optimization);

– the CountPlay method (without representation optimization);

– the *monocyclic representation* defined in definition 5.16, assuming increasing transition rate of the FEBs;

– *monocyclic representation* with optimized ordering of the FEBs.

The execution time of generating a single random variable using any of the above methods is dominated by the elementary operations of drawing a uniformly distributed pseudo-random number and of computing a logarithm. The exact cost of these operations depends on hardware and software specifics. In the numerical examples, the presented methods will be evaluated using a combined cost function that depends on the ratio $L$ of the costs of these operations:

$$n_{\text{Total}} = n_{\text{Uni}} + L \cdot n_{\text{Log}}, \qquad [8.27]$$

where the computational complexity of generating a uniformly distributed pseudo-random number is $L$ times larger than the computational complexity of computing the logarithm of a number.

In our investigation, we use the PH distribution with representation:

$$\boldsymbol{\alpha} = \begin{pmatrix} 0.40535 & 0.0914074 & 0.403507 & 0.099735 \end{pmatrix},$$

$$\mathbf{A} = \begin{pmatrix} -8.69773 & 1.1017 & 2.227 & 4.6353 \\ 2.10619 & -10.1553 & 4.61562 & 1.50111 \\ 1.46425 & 4.261 & -12.0265 & 4.40511 \\ 3.85212 & 5.72213 & 0.704262 & -14.1558 \end{pmatrix}.$$

In this PH distribution, the average number of state transitions up to absorption is:

$$n^* = \boldsymbol{\alpha}(\text{diag}\langle 1/a_{ii}\rangle \mathbf{A})^{-1}\mathbb{1} = 5.54692.$$

According to the Play method, at each transition an exponentially distributed sojourn time spent in the current phase is generated, followed by drawing the next phase to visit. Generating exponentially distributed random variables involves generating a uniform sample and a logarithm operation. To obtain the next phase (according to the appropriate discrete distribution), a uniform sample is needed. Furthermore, another uniform sample is required to obtain the initial phase of the distribution. Thus, assuming $L = 1$, we have:

$$n_{\text{Uni}} = 1 + 2n^* = 12.0938,$$

$$n_{\text{Log}} = n^* = 5.54692,$$

$$n_{\text{Total}} = n_{\text{Uni}} + L \cdot n_{\text{Log}} = 17.6408.$$

According to the CountPlay method, at each transition only the next phase is drawn and the number of visits to the current phase is incremented by one. Thus, assuming $L = 1$, we have:

$$n_{\text{Uni}} = 1 + 2n^* = 12.0938,$$

$$n_{\text{Log}} = \tilde{n} = 2.85703,$$

$$n_{\text{Total}} = n_{\text{Uni}} + L \cdot n_{\text{Log}} = 14.9508,$$

where $\tilde{n}$ is the average number of visited phases out of the all phases $n$.

Next, the monocyclic representation of PH($\alpha$, $A$) is constructed (see Figure 8.3). The number of logarithms required depends on how many states are traversed till the absorption. Note that drawing Erlang distributed random variables needs only a single logarithm operation, and the geometrically distributed variable representing the number of times the feedback is taken needs an additional logarithm as well. The number of logarithms required is calculated as:

$$n_{\text{Log}} = 0.527017 \cdot 3 + 0.13063 \cdot 2$$
$$+ (0.102815 + 0.129809 + 0.109729) \cdot 1 + 1 = 3.18466.$$

Now we calculate the number of uniform samples required. One uniform is required for exponential distributions and $n$ uniforms are needed for Erlang($n$) distributions. The number of times the feedback is taken is geometrically distributed with parameter $z = 0.115691/14.9325 = 0.00774763$, thus the mean number of times the feedback is taken is $z/(z-1) = 0.007808$. An additional uniform sample is needed to determine the initial phase of the PH distribution as well. This yields:

$$n_{\text{Uni}} = 0.527017 \cdot 5 + 0.13063 \cdot 4 + 0.102815 \cdot 3$$
$$+ 0.129809 \cdot 2 + 0.109729 \cdot 1 + 0.007808 \cdot 3 + 1 = 5.85882.$$

Thus, the total cost of the (ordered) monocyclic representation is $n_{\text{Total}} = 9.04349$, which is only about half the cost of the Play method.

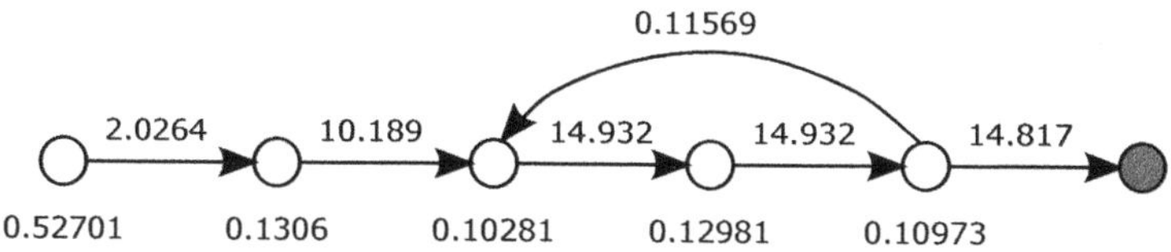


**Figure 8.3.** *Monocyclic representation of PH($\alpha$, $A$)*

In the hope of reducing the cost even further, we can try to change the order of FEBs. As there are three FEBs in this example, there are six different orderings to check. Figure 8.4 depicts the ordering that has been found to be optimal. Note that an Erlang(3) distribution has been appended to maintain the Markovian property of the initial distribution according to Step (3) of the monocyclic transformation procedure algorithm 5.3. The cost of this structure is $n_{\text{Total}} = 6.06503$. At first sight, it seems to be contradictory that the size of the PH is increased to reduce the cost of generating random variables from it, but Figure 8.4 shows that the initial probabilities closer to the absorbing state are increased, reducing the average cost of the representation.

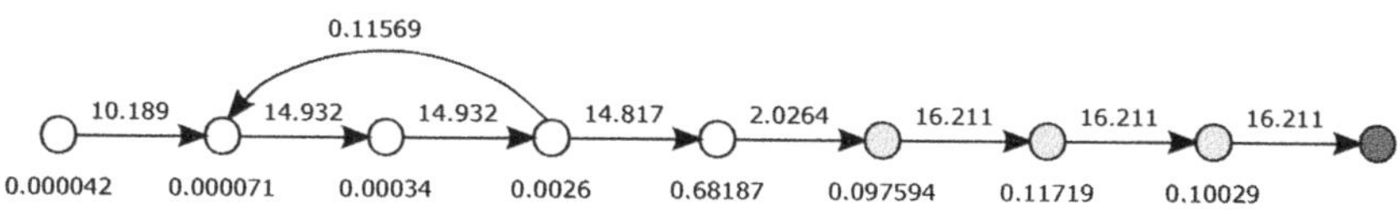


**Figure 8.4.** *Representation of PH($\alpha$, $A$) with optimal FEBs ordering*

Our findings on the execution speeds of the optimization methods are as follows:

– Drawing PH variables using the Play and the CountPlay methods does not require preliminary representation transformation.

– The transformation to the monocyclic representation is fast. It terminates promptly with PHs having less than 100 phases, and takes a few seconds with 100–1000 phases.

– Finding the optimal order of FEBs in terms of the cost function requires an exhaustive search, which is not a scalable approach due to the combinatorial explosion (i.e. for $m$ FE blocks, $m!$ orderings must be considered).

– The monocyclic representation and its optimized version allow more efficient generation of PH distributed samples than general vector–matrix representation. This property is verified for a large number of randomly generated PH distributions in [HOR 16b].

9

# Continuous Phase Type Distribution Based Random Variables

PH distributions are extensively studied and many of their beneficial properties are discussed in the previous chapters. There are application fields where the statistical properties of the quantities of interest differ from the ones of PH distributions (e.g. quantities of interest have finite support). The first part of this section focuses on the case where PH distributions are used to generate distributions with specific statistical properties and makes use of the beneficial properties of PH distribution in their analysis. The second part of this section is devoted to the use of PH distributions in multivariate statistical modeling.

## 9.1. Functions of continuous phase type distributions

There are many papers dealing with random variables generated from PH distributions. For example, Ramaswami and Viswanath [RAM 14] introduce distributions with finite support, Ahn et al. [AHN 12] address distributions with heavy tail, and Shanthikumar [SHA 85] and Ahn and Ramaswami [AHN 05] deal with distributions with support on the entire real axis. A common feature of these distributions is that they are generated from PH distributed random variables and an appropriate function, which transforms the PH distribution to a distribution with the desired statistical properties.

Before the individual discussion of the different cases, we follow a general approach and discuss functions of PH distributions as proposed in [PAV 22].

Let $X$ be a PH($\alpha, \boldsymbol{A}$) distributed random variable and $g(\cdot) : \mathbb{R}^+ \to \mathbb{S} \subseteq \mathbb{R}$ a function with finite derivatives on $\mathbb{R}^+$. First, we discuss the case when $g(\cdot)$ is an invertible bijective function on $\mathbb{R}^+$ and later we generalize to the case when $g(\cdot)$ is non-invertible.

### 9.1.1. *Bijective functions of PH distributions*

If $g'(x) > 0$ at $x$, then for an infinitesimally small $\Delta$, $g(x+\Delta) = g(x) + \Delta g'(x) + \sigma(\Delta) > g(x)$ and the density of $Y = g(X)$ at $y = g(x)$ (i.e. $x = g^{-1}(y)$) can be computed as:

$$\begin{aligned} f_Y(y) &= \lim_{\Delta\to 0} \frac{1}{\Delta g'(g^{-1}(y)) + \sigma(\Delta)} \mathrm{P}(y \le Y < y + \Delta g'(g^{-1}(y)) + \sigma(\Delta)) \\ &= \lim_{\Delta\to 0} \frac{1}{\Delta g'(x) + \sigma(\Delta)} \mathrm{P}(g(x) \le Y < g(x+\Delta)) \\ &= \lim_{\Delta\to 0} \frac{1}{\Delta g'(x) + \sigma(\Delta)} \mathrm{P}(x \le X < x+\Delta) \\ &= \frac{1}{g'(x)} f_X(x) = \frac{1}{g'(g^{-1}(y))} f_X(g^{-1}(y)) \,. \end{aligned} \qquad [9.1]$$

If $g'(x) < 0$, then the same steps give:

$$\begin{aligned} f_Y(y) &= \lim_{\Delta\to 0} \frac{1}{\Delta |g'(g^{-1}(y))| + \sigma(\Delta)} \mathrm{P}(y + \Delta g'(g^{-1}(y)) + \sigma(\Delta) \le Y < y) \\ &= \frac{1}{|g'(x)|} f_X(x) = \frac{1}{|g'(g^{-1}(y))|} f_X(g^{-1}(y)), \end{aligned}$$

from which we conclude that in general:

$$f_Y(y) = \frac{1}{|g'(g^{-1}(y))|} f_X(g^{-1}(y)) \,. \qquad [9.2]$$

### 9.1.2. *General functions of PH distributions*

If $g(\cdot)$ is not invertible, but there is a countable $\mathbb{O}_1, \mathbb{O}_2, \ldots, \mathbb{O}_N$ disjoint partition (finite if $N < \infty$ or infinite if $N = \infty$) of $\mathbb{R}^+$ such that $\mathbb{O}_i \cap \mathbb{O}_j = \emptyset$ for $i \neq j$ and $\bigcup_{i=1}^N \mathbb{O}_i = \mathbb{R}^+$ and $g(\cdot)$ is invertible in $\mathbb{O}_i$ for $\forall i \le N$, then we can generalize the result of the invertible case as follows.

Let $I_i(y) = \mathcal{I}_{\{\exists x \in \mathbb{O}_i : g(x) = y\}}$ be the indicator that $y$ is an image from $\mathbb{O}_i$, that is, there exist $x \in \mathbb{O}_i$ such that $g(x) = y$ and let $g_i^{-1}(y)$ be the inverse of $y$ in $\mathbb{O}_i$.

Since the density of $Y$ at $y$ can have contributions from all $\mathbb{O}_i$ intervals, we have:

$$f_Y(y) = \sum_{i=1}^{N} \frac{I_i(y)}{|g_i'(g_i^{-1}(y))|} f_X(g_i^{-1}(y)). \qquad [9.3]$$

### 9.1.3. *The considered transformation functions*

Different $g(\cdot)$ transformation functions applied for the PH distributed random variable $X$ result in different $f_Y(y)$ densities for $Y = g(X)$.

#### 9.1.3.1. *The* mod *function*

The cases when:

$$Y_1 = a + X \mod (b-a) \text{ and } Y_2 = b - X \mod (b-a) \qquad [9.4]$$

are considered in [RAM 14] and are referred to as *PH distribution with finite support.* In these cases, $\mathbb{S} = (a,b)$, $N = \infty$, $\mathbb{O}_i = [(i-1)(b-a), i(b-a))$ and $I_i(y) = 1$ for $\forall y \in \mathbb{S}$ and $i \geq 0$.

When $X$ is PH($\boldsymbol{\alpha}, \boldsymbol{A}$) distributed, the density function of $Y_1$ and $Y_2$ is:

$$f_{Y_1}(y) = \sum_{i=0}^{\infty} \boldsymbol{\alpha} e^{\boldsymbol{A}(b-a)(i-1)} e^{\boldsymbol{A}(y-a)} \boldsymbol{a} = \boldsymbol{\alpha} \left(\boldsymbol{I} - e^{\boldsymbol{A}(b-a)}\right)^{-1} e^{\boldsymbol{A}(y-a)} \boldsymbol{a}, \qquad [9.5]$$

$$f_{Y_2}(y) = \sum_{i=0}^{\infty} \boldsymbol{\alpha} e^{\boldsymbol{A}(b-a)(i-1)} e^{\boldsymbol{A}(b-y)} \boldsymbol{a} = \boldsymbol{\alpha} \left(\boldsymbol{I} - e^{\boldsymbol{A}(b-a)}\right)^{-1} e^{\boldsymbol{A}(b-y)} \boldsymbol{a}, \qquad [9.6]$$

for $a < y < b$, and otherwise they are zero.

To investigate the properties of the obtained density functions, we first assume that $\boldsymbol{A}$ is diagonalizable and $\boldsymbol{A} = \boldsymbol{B}^{-1} \boldsymbol{\Lambda} \boldsymbol{B}$ is the Jordan decomposition of $\boldsymbol{A}$. In this case, according to [2.11], the density of $X$ is $\boldsymbol{\alpha} e^{\boldsymbol{A}t} \boldsymbol{a} = -\sum_{i=1}^{n} c_i \lambda_i e^{\lambda_i t}$, where $n$ is the size of $\boldsymbol{A}$ and $\lambda_i$ is its $i$th eigenvalue, from which:

$$f_{Y_1}(y) = \boldsymbol{\alpha} \left(\boldsymbol{I} - e^{\boldsymbol{A}(b-a)}\right)^{-1} e^{\boldsymbol{A}(y-a)} \boldsymbol{a} = -\sum_{i=1}^{n} \frac{c_i \lambda_i e^{\lambda_i (y-a)}}{1 - e^{\lambda_i (b-a)}}, \qquad [9.7]$$

$$f_{Y_2}(y) = \boldsymbol{\alpha} \left(\boldsymbol{I} - e^{\boldsymbol{A}(b-a)}\right)^{-1} e^{\boldsymbol{A}(b-y)} \boldsymbol{a} = -\sum_{i=1}^{n} \frac{c_i \lambda_i e^{\lambda_i (b-y)}}{1 - e^{\lambda_i (b-a)}}. \qquad [9.8]$$

Since $\boldsymbol{A}$ is the generator of a PH distribution, its eigenvalues have negative real parts. When $\lambda_i$ and $c_i$ are real and $c_i > 0$ for $\forall i$, then $f_{Y_1}(y)$ is the sum of exponentially decreasing functions in $(a,b)$ and $f_{Y_2}(y)$ is the sum of exponentially increasing functions. Consequently, $f_{Y_1}(y)$ is monotone decreasing and $f_{Y_2}(y)$ is monotone increasing. If $\boldsymbol{A}$ has some complex conjugate pairs of eigenvalues, then the exponential decay is altered by a trigonometric function as shown in Figure 2.2.

For the interval $(a,b)$, $Y_1$ and $Y_2$ can have polynomial PDF as well, as demonstrated below.

According to theorem 2.17, the dominant eigenvalue of $\boldsymbol{A}$, denoted by $\hat{\lambda}$, is real and negative. If $\hat{\lambda}$ tends to 0, then the related term of the density function of $Y_1$ is:

$$\lim_{\hat{\lambda}\to 0} -\frac{\hat{c}\hat{\lambda}e^{\hat{\lambda}(y-a)}}{1-e^{\hat{\lambda}(b-a)}} = \frac{\hat{c}}{b-a}, \qquad [9.9]$$

which is independent of $y$. That is, when the dominant eigenvalue of $\boldsymbol{A}$ tends to 0 and $|\hat{c}| \gg |c_i|$ for $\forall i$ apart from the dominant eigenvalue, then both $f_{Y_1}(y)$ and $f_{Y_2}(y)$ tend to a constant function in $(a,b)$, and consequently, both $Y_1$ and $Y_2$ tend to be uniformly distributed.

When $\boldsymbol{A}$ is non-diagonalizable, according to [2.12], the density of $X$ is $\boldsymbol{\alpha} e^{\boldsymbol{A}t}\boldsymbol{a} = \sum_{i=1}^{\#\lambda}\sum_{j=1}^{\#\lambda_i} c_{ij}(-\lambda_i)^j t^{j-1} e^{\lambda_i t}$, from which:

$$f_{Y_1}(y) = \boldsymbol{\alpha}\left(\boldsymbol{I} - e^{\boldsymbol{A}(b-a)}\right)^{-1} e^{\boldsymbol{A}(y-a)}\boldsymbol{a} = \sum_{i=1}^{\#\lambda}\sum_{j=1}^{\#\lambda_i} d_{ij}(-\lambda_i)^j (y-a)^{j-1} e^{\lambda_i(y-a)}, \qquad [9.10]$$

$$f_{Y_2}(y) = \boldsymbol{\alpha}\left(\boldsymbol{I} - e^{\boldsymbol{A}(b-a)}\right)^{-1} e^{\boldsymbol{A}(b-y)}\boldsymbol{a} = \sum_{i=1}^{\#\lambda}\sum_{j=1}^{\#\lambda_i} d_{ij}(-\lambda_i)^j (b-y)^{j-1} e^{\lambda_i(b-y)} \qquad [9.11]$$

where the $d_{ij}$ coefficients depend on $a, b, \lambda_i, c_{i1}, \ldots, c_{i\#\lambda_i}$, but are independent of $y$.

Assuming that the multiplicity of the dominant eigenvalue is $\#\hat{\lambda} > 1$ and $\hat{\lambda}$ tends to 0,

$$\lim_{\hat{\lambda}\to 0} f_{Y_1}(y) = \frac{\hat{c}(y-a)^{\#\hat{\lambda}-1}}{(\#\hat{\lambda}-1)!(b-a)} + f_{\text{rem}}(y) \qquad [9.12]$$

where $f_{\text{rem}}(y)$ contains lower order polynomials of $(y-a)$ and additional terms associated with the non-dominant eigenvalues of $\boldsymbol{A}$. That is, when the dominant eigenvalue of $\boldsymbol{A}$ tends to 0 and $f_{\text{rem}}(y)$ is negligible, then the dominant term of $f_{Y_1}(y)$ is $(y-a)^{\#\hat{\lambda}-1}$. Similarly, when the multiplicity of the dominant eigenvalue is $\#\hat{\lambda} > 1$ and the dominant eigenvalue tends to 0:

$$\lim_{\hat{\lambda}\to 0} f_{Y_2}(y) = \frac{\hat{c}(b-y)^{\#\hat{\lambda}-1}}{(\#\hat{\lambda}-1)!(b-a)} + f_{\text{rem}}(y), \qquad [9.13]$$

where $f_{\text{rem}}(y)$ contains lower order polynomials of $(b - y)$ and additional terms associated with the non-dominant eigenvalues of $\boldsymbol{A}$. When $f_{\text{rem}}(y)$ is negligible, then the dominant term of $f_{Y_2}(y)$ is $(b - y)^{\#\hat{\lambda}-1}$.

To conclude this part, we note that PH distributions with finite support generated by the mod function can exhibit density functions with exponential polynomial and purely polynomial terms.

#### 9.1.3.2. *The* exp *function*

The case when $Y = e^X$ is considered in [AHN 12]. In this case, the $g(x) = e^x$ transform function is invertible and the density function of $Y = e^X$ is:

$$f_Y(y) = \frac{1}{y}\boldsymbol{\alpha} e^{\boldsymbol{A}\log(y)}\boldsymbol{a} \qquad [9.14]$$

for $y > 1$.

Similar to the previous point, we first assume that $\boldsymbol{A}$ is diagonalizable, the Jordan decomposition of $\boldsymbol{A}$ is $\boldsymbol{A} = \boldsymbol{B}^{-1}\boldsymbol{\Lambda}\boldsymbol{B}$ and investigate the properties of $f_Y(y)$ using [2.11]:

$$f_Y(y) = \frac{1}{y}\boldsymbol{\alpha} e^{\boldsymbol{A}\log(y)}\boldsymbol{a} = -\sum_{i=1}^{n} c_i\lambda_i y^{\lambda_i - 1}, \qquad [9.15]$$

which is a mixture of Pareto density functions, when the eigenvalues of $\boldsymbol{A}$ are real. If $\boldsymbol{A}$ has complex conjugate eigenvalue pairs the polynomial decay of the Pareto density is altered by a trigonometric function similar to Figure 2.2.

The *shape* parameters of the Pareto density functions are characterized by the eigenvalues of $\boldsymbol{A}$ and the smallest shape parameter is associated with the dominant eigenvalue.

When $\boldsymbol{A}$ is non-diagonalizable, the density of $X$ has the form of [2.12] and the density of $Y$ is:

$$f_Y(y) = \frac{1}{y}\boldsymbol{\alpha} e^{\boldsymbol{A}\log(y)}\boldsymbol{a} \qquad [9.16]$$

$$= -\sum_{i=1}^{\#\lambda} d_{i1}\lambda_i y^{\lambda_i - 1} - \sum_{i=1}^{\#\lambda}\sum_{j=2}^{\#\lambda_i} d_{ij}\frac{\log(y)^{j-2}}{(j-2)!}(j - 1 + \lambda_i\log(y))y^{\lambda_i - 1}$$

where the $d_{ij}$ coefficients depend on $c_{i1}, \ldots, c_{i\#\lambda_i}$, but are independent of $y$. The terms with $j = \#\lambda_i$ result in the slowest decay term associated with eigenvalue $\lambda_i$ but, due to the slow increase of $\log(y)$, the difference between the decay of the $j = 1$ and the $j = \#\lambda_i$ terms is negligible.

#### 9.1.3.3. *The* log *function*

The transformation function $g(x) = \log(x)$ results in a distribution with support on the entire real axis. In this case,

$$f_Y(y) = e^y \boldsymbol{\alpha} e^{\boldsymbol{A} e^y} \boldsymbol{a}. \tag{9.17}$$

When $\boldsymbol{A}$ is diagonalizable with Jordan decomposition $\boldsymbol{A} = \boldsymbol{B}^{-1} \boldsymbol{\Lambda} \boldsymbol{B}$, the PDF can be written as:

$$f_Y(y) = e^y \boldsymbol{\alpha} e^{\boldsymbol{A} e^y} \boldsymbol{a} = -e^y \sum_{i=1}^{n} c_i \lambda_i e^{\lambda_i e^y}. \tag{9.18}$$

The obtained distribution is light tailed to the right and heavy tailed to the left as demonstrated in Figure 9.1.

There are many other transformation functions that map $\mathbb{R}^+$ to $\mathbb{R}$ but the most common way to generate distribution on $\mathbb{R}$ from PH distributed random variables is based on a different method, which is summarized in the next section.

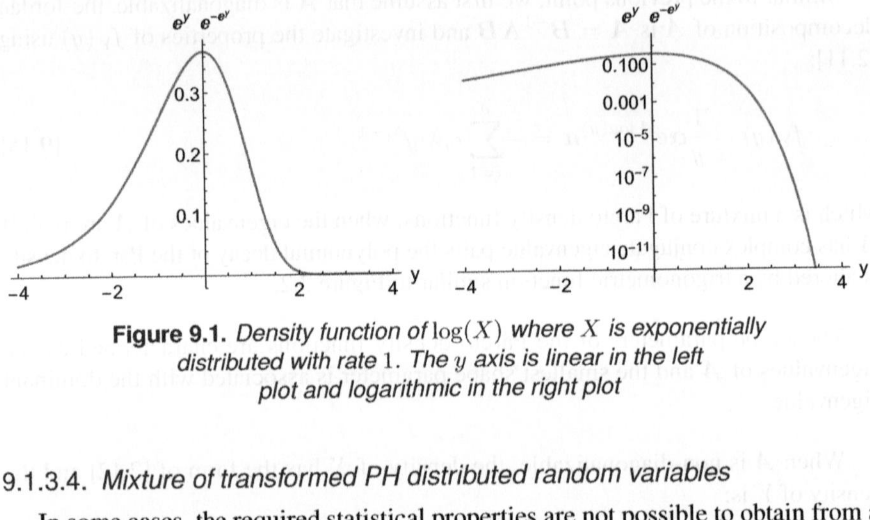


**Figure 9.1.** *Density function of* $\log(X)$ *where* $X$ *is exponentially distributed with rate* 1. *The* $y$ *axis is linear in the left plot and logarithmic in the right plot*

#### 9.1.3.4. *Mixture of transformed PH distributed random variables*

In some cases, the required statistical properties are not possible to obtain from a single transformed PH distributed random variable: instead, a mixture of such variables are applied.

Let $X_k$ be a PH($\boldsymbol{\alpha}_k, \boldsymbol{A}_k$) distributed random variable, $g^{(k)}(\cdot)$ be an $\mathbb{R}^+ \to \mathbb{S}_k \subseteq \mathbb{R}$ function with finite derivatives on $\mathbb{R}^+$ for $k \in \{1, \ldots, K\}$ and $\Theta$ a discrete random variable on $\{1, \ldots, K\}$ with distribution $P(\Theta = k) = p_k$ for $k \in \{1, \ldots, K\}$. The mixture of the PH distributed random variables is obtained as:

$$Y = \sum_{k=1}^{K} \mathcal{I}_{\{\Theta = k\}} g^{(k)}(X_k). \tag{9.19}$$

Assuming general $g^{(k)}(\cdot)$ transform functions as in [9.3], the density function of $Y$ is:

$$f_Y(y) = \sum_{k=1}^{K} p_k \sum_{i=1}^{N} \frac{I_i^{(k)}(y)}{\left|g_i^{(k)\prime}\left(g_i^{(k)^{-1}}(y)\right)\right|} f_{X_k}\left(g_i^{(k)^{-1}}(y)\right). \quad [9.20]$$

The properties of this distribution can be obtained from a similar analysis of the $g^{(k)}(\cdot)$ transform functions as in the previous section.

In the literature, these mixtures are most commonly applied with $K = 2$. For example, the *bilateral PH distribution*, which was introduced in [SHA 85] and its basic properties are obtained in [AHN 05], uses $K = 2$, $g^{(1)}(x) = x$ and $g^{(2)}(x) = -x$, from which its density function is:

$$f_Y(y) = \begin{cases} p_1 \boldsymbol{\alpha}_1 e^{\boldsymbol{A}_1 y} \boldsymbol{a}_1 & \text{if } y \geq 0, \\ p_2 \boldsymbol{\alpha}_2 e^{-\boldsymbol{A}_2 y} \boldsymbol{a}_2 & \text{if } y < 0. \end{cases} \quad [9.21]$$

Further properties of this distribution are studied in [HOR 07]. Another example for such mixtures is the mixture of $Y_1$ and $Y_2$ (defined in [9.4]), which is investigated in [RAM 14].

## 9.2. Multivariate phase type distributions

### 9.2.1. *Subset-based multivariate phase type distributions*

The concept of generating a random variable based on a background Markov chain, $X(t)$, with $n$ transient and one absorbing state can be extended to multiple random variables. The first proposal in this direction is in [ASS 84], where subsets of the state space of the background Markov chain denoted by $\Gamma_j$, are introduced and random variables $T_k$ for $j = 1, \ldots, J$ are defined as:

$$T_j = \min\{t : X(t) \in \Gamma_j, t \geq 0\},$$

where $\bigcap_{j=1}^{J} \Gamma_j$ contains the absorbing state and the initial state of the Markov chain is in $\bigcap_{j=1}^{J} \Gamma_j^c$ (which should be non-empty).

In the bivariate case, when $J = 2$, the joint distribution of $T_1$ and $T_2$ can be obtained from the following division of the state space:

|  | $\Gamma_1^c$ | $\Gamma_1$ |
|---|---|---|
| $\Gamma_2^c$ | $S_1$ | $S_2$ |
| $\Gamma_2$ | $S_3$ | $S_4$ |

,

that is,

$$\mathcal{S}_1 = \Gamma_1^c \cap \Gamma_2^c, \mathcal{S}_2 = \Gamma_1 \cap \Gamma_2^c, \mathcal{S}_3 = \Gamma_1^c \cap \Gamma_2, \mathcal{S}_4 = \Gamma_1 \cap \Gamma_2.$$

The associated block structure of the infinitesimal generator matrix is:

$$\boldsymbol{Q} = \begin{pmatrix} \boldsymbol{Q}_{11} & \boldsymbol{Q}_{12} & \boldsymbol{Q}_{13} & \boldsymbol{Q}_{14} \\ \boldsymbol{Q}_{21} & \boldsymbol{Q}_{22} & \boldsymbol{Q}_{23} & \boldsymbol{Q}_{24} \\ \boldsymbol{Q}_{31} & \boldsymbol{Q}_{32} & \boldsymbol{Q}_{33} & \boldsymbol{Q}_{34} \\ \boldsymbol{Q}_{41} & \boldsymbol{Q}_{42} & \boldsymbol{Q}_{43} & \boldsymbol{Q}_{44} \end{pmatrix}.$$

To obtain positive $T_1$, $T_2$ variables, $X(0) \in \mathcal{S}_1$ has to hold, which is assumed also in [ASS 84].

Let $\boldsymbol{\alpha}_1$ denote the initial distribution in $\mathcal{S}_1$, where $\boldsymbol{\alpha}_1 \mathbb{1} = 1$.

We denote the time of the first departure from $\mathcal{S}_1$ by $\gamma_1$. $\gamma_1$ is PH($\boldsymbol{\alpha}_1, \boldsymbol{Q}_{11}$) distributed. At the first departure from $\mathcal{S}_1$, the background Markov chain can move:

– to $\mathcal{S}_2$, which means that $T_1 < T_2$;

– or to $\mathcal{S}_3$, which means that $T_2 < T_1$;

– or to $\mathcal{S}_4$, which means that $T_1 = T_2$.

In the first case $T_1 = \gamma_1$ and the remaining time of $T_2$ is PH distributed with representation:

$$\boldsymbol{\alpha}_{\mathcal{S}_1 \cup \mathcal{S}_2} = \left(\boldsymbol{0} \ -\boldsymbol{\alpha}_1 \boldsymbol{Q}_{11}^{-1} \boldsymbol{Q}_{12}\right), \text{ and } \boldsymbol{Q}_{\mathcal{S}_1 \cup \mathcal{S}_2} = \begin{pmatrix} \boldsymbol{Q}_{11} & \boldsymbol{Q}_{12} \\ \boldsymbol{Q}_{21} & \boldsymbol{Q}_{22} \end{pmatrix}.$$

That is, the joint distribution of $T_1$ and $T_2$ satisfies:

$$\frac{d}{dt_1}\frac{d}{dt_2} \mathrm{P}(T_1 < T_2, T_1 < t_1, T_2 < t_2) = \left(\boldsymbol{0} \ \boldsymbol{\alpha}_1 e^{\boldsymbol{Q}_{11} t_1} \boldsymbol{Q}_{12}\right) e^{\boldsymbol{Q}_{\mathcal{S}_1 \cup \mathcal{S}_2}(t_2 - t_1)} \boldsymbol{q}_{\mathcal{S}_1 \cup \mathcal{S}_2},$$

where $\boldsymbol{q}_{\mathcal{S}_1 \cup \mathcal{S}_2} = -\boldsymbol{Q}_{\mathcal{S}_1 \cup \mathcal{S}_2} \mathbb{1}$. The case $T_2 < T_1$ can be computed similarly.

In the last case, when $T_1 = T_2 = \gamma_1$, we have:

$$\frac{d}{dt} \mathrm{P}(T_1 = T_2 < t) = \boldsymbol{\alpha}_1 e^{\boldsymbol{Q}_{11} t} \boldsymbol{Q}_{14} \mathbb{1}.$$

### 9.2.2. *Reward-based multivariate phase type distributions*

Reward-based multivariate phase type (MVPH) distributions with severe structural restrictions were considered in [ASS 84]. Here, we follow the more flexible MVPH distribution model proposed in [KUL 89] where a reward matrix $\boldsymbol{R}$ is assigned to a transient Markov chain with initial distribution $\boldsymbol{\alpha}$ and transient generator $\boldsymbol{A}$. As long as the background Markov chain stays in state $i$, reward of type $j$ is accumulated at rate $r_{ij} \geq 0$ where $r_{ij}$ is element $i, j$ of matrix $\boldsymbol{R}$. That is,

$$\frac{d}{dt} Y_j(t) = r_{ij} \quad \text{if } X(t) = i \in \mathcal{S}$$

where $X(t)$ is the state of the background Markov chain, $Y_j(t)$ is the amount of type $j$ reward at time $t$ and $\mathcal{S}$ is the set of transient states.

In this MVPH model, $T_j$ is the amount of type $j$ reward accumulated until absorption, that is, during a PH($\boldsymbol{\alpha}, \boldsymbol{A}$) distributed time. This class of MVPH distributions is commonly denoted as MPH*($\boldsymbol{\alpha}, \boldsymbol{A}, \boldsymbol{R}$).

Unfortunately, in general, the distribution of MPH*($\boldsymbol{\alpha}, \boldsymbol{A}, \boldsymbol{R}$) cannot be expressed as nicely as the subset-based MVPH distributions. It still can be described however in the multiple Laplace transform domain. Let the initial state dependent $J$-fold Laplace transform of $T_1, \ldots, T_J$ be:

$$V_i^*(v_1, \ldots, v_J) = \mathbb{E}\Big(e^{-\sum_{j=1}^{J} v_j T_j} | X(0) = i\Big). \qquad [9.22]$$

We compute $V_i^*(v_1, \ldots, v_J)$ utilizing that the sojourn time in state $i$ is exponentially distributed with parameter $\vartheta_i = -a_{ii}$, and the next state visited after state $i$ is state $k$ with probability $a_{ik}/\vartheta_i$, where $a_{ik}$ is element $i, k$ of matrix $\boldsymbol{A}$. Additionally, the probability that after visiting state $i$, the background Markov chain moves to the absorbing state is $a_i/\vartheta_i$, where $a_i = -\sum_{k \in \mathcal{S}} a_{ik}$.

$$\begin{aligned} &V_i^*(v_1, \ldots, v_J) \\ &= \int_{x=0}^{\infty} \vartheta_i e^{-\vartheta_i x} e^{-\sum_{j=1}^{J} v_j r_{ij} x} dx \cdot \left( \sum_{k \in \mathcal{S}, k \neq i} \frac{a_{ik}}{\vartheta_i} \mathbb{E}\Big(e^{-\sum_{j=1}^{J} v_j T_j} | X(0) = k\Big) + \frac{a_i}{\vartheta_i} \right) \\ &= \frac{\vartheta_i}{\vartheta_i + \sum_{j=1}^{J} v_j r_{ij}} \cdot \left( \sum_{k \in \mathcal{S}, k \neq i} \frac{a_{ik}}{\vartheta_i} V_k^*(v_1, \ldots, v_J) + \frac{a_i}{\vartheta_i} \right) \end{aligned}$$

where the first term contains the reward accumulated during the first visit to state $i$ and the second term contains the reward accumulated after the first transition of the background Markov chain. Multiplying with the denominator of the first term, we obtain:

$$\sum_{j=1}^{J} v_j r_{ij} V_i^*(v_1, \ldots, v_J) = \sum_{k \in \mathcal{S}} a_{ik} V_k^*(v_1, \ldots, v_J) + a_i. \qquad [9.23]$$

Introducing column vector $\boldsymbol{V}^*(v_1, \ldots, v_J)$ and diagonal matrix $\boldsymbol{D}$:

$$\boldsymbol{V}^*(\underline{v}) = \begin{pmatrix} V_1^*(v_1, \ldots, v_J) \\ \vdots \\ V_J^*(v_1, \ldots, v_J) \end{pmatrix}, \boldsymbol{D}(\underline{v}) = \begin{pmatrix} \sum_{j=1}^J v_j r_{1j} & & \\ & \ddots & \\ & & \sum_{j=1}^J v_j r_{1N} \end{pmatrix},$$

and using matrix $\boldsymbol{A}$ and column vector $\boldsymbol{a} = -\boldsymbol{A}\mathbb{1}$, [9.23] can be written in matrix form as:

$$\boldsymbol{D}(\underline{v})\boldsymbol{V}^*(\underline{v}) = \boldsymbol{A}\boldsymbol{V}^*(\underline{v}) + \boldsymbol{a}, \qquad [9.24]$$

from which

$$\boldsymbol{V}^*(\underline{v}) = (\boldsymbol{D}(\underline{v}) - \boldsymbol{A})^{-1}\boldsymbol{a} \qquad [9.25]$$

is the $J$-fold Laplace transform description of the joint distribution of $T_1, \ldots, T_J$.

The subset-based MVPH distribution with initial probability vector $\boldsymbol{\alpha}$, generator matrix $\boldsymbol{A}$ and subsets $\Gamma_1, \ldots, \Gamma_J$ can be equivalently described by MPH*$(\boldsymbol{\alpha}, \boldsymbol{A}, \boldsymbol{R})$ simply with $r_{ij} = \mathcal{I}_{\{i \in \Gamma_j^c\}}$.

### 9.2.3. *Linear projection based multivariate multivariate phase type distributions*

Following the characterization of multivariate Gaussian distributions, Nielsen [NIE 22] proposed another definition for MVPH distributions.

DEFINITION 9.1.– *[NIE 22, Definition 2] $T_1, \ldots, T_J$ has a MVPH distribution if $\sum_{j=1}^J w_j T_j$ has a univariate PH distribution for all $w_j \geq 0$.*

The characterization of MVPH and MPH* distributions is still an open research problem. It is known that any MPH*$(\boldsymbol{\alpha}, \boldsymbol{A}, \boldsymbol{R})$ distribution is also an MVPH distribution since $\sum_{j=1}^J w_j T_j$ is an MPH*$(\boldsymbol{\alpha}, \boldsymbol{A}, \boldsymbol{R}\boldsymbol{w})$ distribution where $\boldsymbol{w}$ is a column vector composed from the $w_j$ parameters. That is, the obtained model is a MPH* with a single reward, which is the weighted sum of the original reward rates of MPH*$(\boldsymbol{\alpha}, \boldsymbol{A}, \boldsymbol{R})$ with weights $w_j$.

When the background Markov chain spends an exponentially distributed time in state $i$ with parameter $\vartheta_i$ and accumulates reward at rate $\rho_i$, the reward accumulated in state $i$ is exponentially distributed with parameter $\vartheta_i/\rho_i$. The same amount of reward is accumulated in state $i$ if the background Markov chain spends an exponentially distributed time in state $i$ with parameter $\vartheta_i/\rho_i$ and accumulates reward at rate 1.

Utilizing this identity,

$$\mathrm{MPH}^*(\boldsymbol{\alpha}, \boldsymbol{A}, \boldsymbol{R}\boldsymbol{w}) \equiv \mathrm{PH}(\boldsymbol{\alpha}, \boldsymbol{G}\boldsymbol{A}), \tag{9.26}$$

where $\boldsymbol{G} = \mathrm{diag}\left\langle \frac{1}{\sum_{j=1}^{J} r_{ij} w_j} \right\rangle$.

Whether MPH* is a true subset of MVPH is an open reaseach problem [NIE 22].

# APPENDICES

# Appendix 1

## Description of Related Software Tools

Here, we provide a list of related tools with a brief description.

**BuTools** [HOR 17, BUT 15] provides a collection of computational methods that are useful for Markovian and non-Markovian matrix analytic performance analysis. It provides packages to obtain, analyze, transform and minimize discrete and continuous PH distributions and Markovian arrival processes (MAPs); to fit empirical measurement data and to evaluate the result; to obtain many performance measures of various Markovian queuing systems; and to solve block-structured Markov chains.

**PhFit** [HOR 02] implements PH fitting procedures that allow for approximating distributions or sets of samples not only by continuous but by discrete PH distributions as well. The implemented algorithms separate the fitting of the body and the tail part of the distribution that results in satisfactory fitting also for heavy-tailed distributions. Moreover, PhFit allows the user to choose the distance measure according to which the fitting is performed.

**KPC-Toolbox** [CAS 08] is a library of MATLAB functions for fitting an empirical dataset into a Markov model such as a PH distribution or a MAP using exact and approximate moment matching techniques. As special cases, it includes hyper-exponential, Erlang, hypo-exponential, generalized Erlang, and acyclic PH distributions.

**SMCSolver** [BIN 06] is a package for solving structured Markov chains that can arise, for example, from applying PH distributions in queues. It contains the most advanced algorithms for solving QBD, M/G/1 and G/M/1 problems. It is provided in two versions: a Matlab toolbox and a Fortran 95 version with a user-friendly graphical interface.

**HyperStar** [REI 13] is an extensible tool with a graphical user-interface that enables simple and efficient fitting of mixtures of special cases of PH distributions (e.g. Erlang distributions) to data sets.

**EMPHT** [LAU 19] is an efficient implementation of the EM algorithm (as described in [ASM 96]) for fitting PH distributions to data.

# Appendix 2

## Acronyms

**APH** acyclic phase type

**BE** Bernstein expolynomial

**BP** Bernstein polynomial

**CCDF** complementary cumulative density function

**CDF** cumulative distribution function

**CTMC** continuous time Markov chain

**DPH** discrete phase type

**DTMC** discrete time Markov chain

**EM** expectation maximization

**FEB** feedback Erlang block

**GHErD** generalized hyper-Erlang distribution

**Hyper-FE** hyper-feedback-Erlang

**Hypo-FE** hypo-feedback-Erlang

**LHS** left-hand side

**LT** Laplace transform

**MD** moment distance

**ME** matrix exponential

**MECO** mixture of Erlang distributions of common order

**MG** matrix geometric

**MVPH** multivariate phase type

**PDF** probability density function

**PH** continuous phase type

**PMF** probability mass function

**RE** relative entropy

**RHS** right-hand side

**RV** random variable

**SCV** squared coefficient of variation

# References


[AHN 05] AHN S., RAMASWAMI V., "Bilateral phase type distributions", *Stochastic Models*, vol. 21, nos. 2–3, pp. 239–259, 2005.

[AHN 12] AHN S., KIM J.H., RAMASWAMI V., "A new class of models for heavy tailed distributions in finance and insurance risk", *Insurance: Mathematics and Economics*, vol. 51, no. 1, pp. 43–52, 2012.

[ALD 87] ALDOUS D., SHEPP L., "The least variable phase type distribution is Erlang", *Stochastic Models*, vol. 3, pp. 467–473, 1987.

[ALM 21] ALMOUSA S.A.-D., TELEK M., "Enhanced optimization of high order concentrated matrix-exponential distributions", *Annales Mathematicae et Informaticae*, vol. 53, pp. 5–19, 2021.

[ASM 96] ASMUSSEN S., NERMAN O., OLSSON M., "Fitting phase-type distributions via the EM algorithm", *Scandinavian Journal of Statistics*, vol. 23, no. 4, pp. 419–441, 1996.

[ASS 84] ASSAF D., LANGBERG N.A., SAVITS T.H. et al., "Multivariate phase-type distributions", *Operations Research*, vol. 32, no. 3, pp. 688–702, 1984.

[BAR 65] BARLOW R.E., PROSCHAN F., *Mathematical Theory of Reliability*, Siam, Philadelphia, 1965.

[BEN 04] BENVENUTI L., FARINA L., "Eigenvalue regions for positive systems", *Systems & Control Letters*, vol. 51, no. 3, pp. 325–330, 2004.

[BIN 06] BINI D.A., MEINI B., STEFFÉ S. et al., "Structured Markov chains solver: Software tools", *Proceedings from the 2006 Workshop on Tools for Solving Structured Markov Chains, SMCtools '06*, Association for Computing Machinery, New York, 2006.

[BLA 17] BLADT M., NIELSEN B.F., *Matrix-Exponential Distributions in Applied Probability*, Springer, New York, 2017.

[BOB 94] BOBBIO A., TELEK M., "A benchmark for PH estimation algorithms: Results for Acyclic-PH", *Stochastic Models*, vol. 10, no. 3, pp. 661–677, 1994.

[BOB 01] BOBBIO A., HORVÁTH A., "Petri nets with discrete phase type timing: A bridge between stochastic and functional analysis", *Proc. of 2nd Workshop on Models for Time-Critical Systems*, Aalborg, 2001.

[BOB 04] BOBBIO A., HORVÁTH A., TELEK M., "The scale factor: A new degree of freedom in phase type approximation", *Performance Evaluation*, vol. 56, nos. 1–4, pp. 121–144, 2004.

[BOB 05] BOBBIO A., HORVÁTH A., TELEK M., "Matching three moments with minimal acyclic phase type distributions", *Stochastic Models*, vol. 21, pp. 303–326, 2005.

[BOL 06] BOLCH G., GREINER S., DE MEER H. et al., *Queueing Networks and Markov Chains*, John Wiley & Sons, New York, 2006.

[BRO 60a] BROCKMEYER E., "A survey of A. K. Erlang's mathematical works", *Acta Polytech. Scandinav.*, vol. 287, pp. 101–126, 1960.

[BRO 60b] BROCKMEYER E., HALSTRØM H., JENSEN A., *The Life and Works of A.K. Erlang*, Acta Polytechnica Scandinavica/Danish Academy of Technical Sciences, Copenhagen, 1960.

[BUC 10] BUCHHOLZ P., KEMPER P., KRIEGE J., "Multi-class Markovian arrival processes and their parameter fitting", *Performance Evaluation*, vol. 67, no. 11, pp. 1092–1106, 2010.

[BUC 13] BUCHHOLZ P., TELEK M., "On minimal representation of rational arrival processes", *Annals of Operations Research*, vol. 202, pp. 35–58, 2013.

[BUC 14] BUCHHOLZ P., KRIEGE J., FELKO I., *Input Modeling with Phase-Type Distributions and Markov Models: Theory and Applications*, Springer, Cham, 2014.

[BUR 56] BURKE P.J., "The output of a queuing system", *Operations Research*, vol. 4, no. 6, pp. 699–704, 1956.

[BUT 15] BUTOOLS, A collection of useful tools for Markovian performance evaluation, available at: http://webspn.hit.bme.hu/butools.

[CAS 08] CASALE G., ZHANG E.Z., SMIRNI E., "KPC-Toolbox: Simple yet effective trace fitting using Markovian arrival processes", *2008 Fifth International Conference on Quantitative Evaluation of Systems*, IEEE Computer Society in Los Alamitos, California, pp. 83–92, 2008.

[COX 55] COX D., "A use of complex probabilities in the theory of stochastic processes", *Proceedings of the Cambridge Phylosophical Society*, vol. 51, pp. 313–319, 1955.

[CUM 82] CUMANI A., "On the canonical representation of homogeneous Markov processes modeling failure-time distributions", *Microelectronics and Reliability*, vol. 22, pp. 583–602, 1982.

[DUF 07] DUFRESNE D., "Fitting combinations of exponentials to probability distributions", *Applied Stochastic Models in Business and Industry*, vol. 23, no. 1, pp. 23–48, 2007.

[ÉLT 06] ÉLTETŐ T., RÁCZ S., TELEK M., "Minimal coefficient of variation of matrix exponential distributions", *2nd Madrid Conference on Queueing Theory*, Madrid, 2006.

[ÉLT 08] ÉLTETŐ T., VADERNA P., "Finding upper-triangular representations for phase-type distributions with 3 distinct real poles", *Annals of Operations Research*, vol. 160, pp. 139–172, 2008.

[FAC 03] FACKRELL M.W., Characterization of matrix-exponential distributions, PhD Thesis, School of Applied Mathematics, The University of Adelaide, 2003.

[FEL 98] FELDMANN A., WHITT W., "Fitting mixtures of exponentials to long-tail distributions to analyze network performance models", *Performance Evaluation*, vol. 31, no. 3, pp. 245–279, 1998.

[GER 00] GERMAN R., *Performance Analysis of Communication Systems: Modeling with Non-Markovian Stochastic Petri Nets*, John Wiley and Sons, Chichester, 2000.

[GRA 83] GRAGG W.B., LINQUIST A., "On the partial realization problem", *Linear Algebra and Applications*, vol. 50, pp. 277–319, 1983.

[HE 05] HE Q., ZHANG H., "A note on unicyclic representations of phase type distributions", *Stochastic Models*, vol. 21, no. 2–3, pp. 465–483, 2005.

[HE 14] HE Q., *Fundamentals of Matrix-Analytic Methods*, Springer, New York, 2014.

[HE 19] HE Q., HORVÁTH G., HORVÁTH I. et al., "Moment bounds of PH distributions with infinite or finite support based on the steepest increase property", *Advances in Applied Probability (AAP)*, vol. 51, no. 1, pp. 168–183, 2019.

[HOR 00a] HORVÁTH A., PULIAFITO A., SCARPA M. et al., "Analysis and evaluation of non-Markovian stochastic Petri nets", *Proc. of 11th Performance TOOLS*, Springer, Schaumberg, IL, pp. 171–187, 2000.

[HOR 00b] HORVÁTH A., TELEK M., "Approximating heavy tailed behavior with phase type distributions", *Proc. of 3rd International Conference on Matrix-Analytic Methods in Stochastic Models*, Leuven, 2000.

[HOR 02] HORVÁTH A., TELEK M., "PhFit: A general phase-type fitting tool", *Computer Performance Evaluation: Modeling Techniques and Tools*, Springer, Berlin, Heidelberg, pp. 82–91, 2002.

[HOR 07] HORVÁTH A., TELEK M., "On the properties of acyclic bilateral phase type distributions", *International Workshop on Tools for solving Structured Markov Chains, SMCtools*, Nantes, 2007.

[HOR 09] HORVÁTH G., TELEK M., "On the canonical representation of phase type distributions", *Performance Evaluation*, vol. 66, no. 8, pp. 396–409, 2009.

[HOR 14] HORVÁTH I., TELEK M., "A heuristic procedure for compact Markov representation of PH distributions", *Proceedings of the 8th International Conference on Performance Evaluation Methodologies and Tools, VALUETOOLS '14*, pp. 195–202, 2014.

[HOR 15] HORVÁTH I., TELEK M., "A constructive proof of the phase-type characterization theorem", *Stochastic Models*, vol. 31, no. 2, pp. 316–350, 2015.

[HOR 16a] HORVÁTH A., SCARPA M., TELEK M., "Phase type and matrix exponential distributions in stochastic modeling", in FIONDELLA L., PULIAFITO A. (eds). *Principles of Performance and Reliability Modeling and Evaluation: Essays in Honor of Kishor Trivedi on his 70th Birthday*, Springer, Cham, pp. 3–25, 2016.

[HOR 16b] HORVÁTH G., REINECKE P., TELEK M. et al., "Heuristic representation optimization for efficient generation of PH-distributed random variates", *Annals of Operations Research*, vol. 239, no. 2, pp. 643-665, 2016.

[HOR 16c] HORVÁTH I., SÁFÁR O., TELEK M. et al., "Concentrated matrix exponential distributions", *EPEW*, 2016.

[HOR 17] HORVÁTH G., TELEK M., "BuTools 2: A rich toolbox for Markovian performance evaluation", *Proc. of 10th EAI International Conference on Performance Evaluation Methodologies and Tools*, ACM, pp. 137–142, 2017.

[HOR 20] HORVÁTH G., HORVÁTH I., TELEK M., "High order concentrated matrix-exponential distributions", *Stochastic Models*, vol. 36, no. 2, pp. 176–192, 2020.

[ITO 97] ITO H., "A new statement about the theorem determining the region of eigenvalues of stochastic matrices", *Linear Algebra and Its Applications*, vol. 267, pp. 241–246, 1997.

[JOH 63] JOHNSON S.M., "Generation of permutations by adjacent transposition", *Mathematics of Computation*, vol. 17, no. 83, pp. 282–285, 1963.

[JOH 89] JOHNSON M., TAAFFE M., "Matching moments to phase distributions: Mixtures of Erlang distributions of common order", *Stochastic Models*, vol. 5, no. 4, pp. 711–743, 1989.

[KAR 51] KARPELEVICH F.I., "On the characteristic roots of matrices with non-negative elements", *Izvestiya Akad. Nauk SSSR. Ser. Mat.*, vol. 15, pp. 361–383, 1951.

[KEL 78] KELLOGG R., STEPHENS A., "Complex eigenvalues of a non-negative matrix with a specified graph", *Linear Algebra and Its Applications*, vol. 20, no. 2, pp. 179–187, 1978.

[KLE 75] KLEINROCK L., *Queuing Systems, Volume 1: Theory*, Wiley Interscience, New York, 1975.

[KNU 97] KNUTH D.E., *The Art of Computer Programming*, volume 3, Addison-Wesley, Upper Saddle River, NJ, 1997.

[KOL 11] KOLOSSVÁRY I., TELEK M., "Explicit evaluation of ME(3) membership", *8th International Conference on Quantitative Evaluation of SysTems (QEST),* Centre for Telematics and Information Technology (CTIT), University of Twente, Aachen, pp. 11–13, 2011.

[KUL 89] KULKARNI V.G., "A new class of multivariate phase type distributions", *Operations Research*, vol. 37, no. 1, pp. 151–158, 1989.

[KUL 11] KULKARNI V.G., *Modeling and Analysis of Stochastic Systems*, 2nd edition, Chapman & Hall, London, 2011.

[LAK 19] LAKATOS L., SZEIDL L., TELEK M., *Introduction to Queueing Systems with Telecommunication Applications,* 2nd edition, Springer, Cham, 2019.

[LAU 19] LAUB P., "A Julia port of the EMpht.c program", available at: https://github.com/Pat-Laub/EMpht.jl, 2019.

[LIE 90] VAN DE LIEFVOORT A., The moment problem for continuous distributions, Technical Report, University of Missouri, Kansas City, WP-CM-1990-02, 1990.

[LOR 53] LORENTZ G., *Bernstein Polynomials*, University of Toronto Press, Toronto, 1953.

[MÉS 22] MÉSZÁROS A., TELEK M., "Concentrated matrix exponential distributions with real eigenvalues", *Probability in the Engineering and Informational Sciences*, vol. 36, no. 4, pp. 1171–1187, 2022.

[MOC 99] MOCANU S., COMMAULT C., "Sparse representations of phase-type distributions", *Commun. Stat., Stochastic Models*, vol. 15, no. 4, pp. 759–778, 1999.

[NEU 75] NEUTS M.F., "Probability distributions of phase type", *Liber Amicorum Prof. Emeritus H. Florin*, University of Louvain, pp. 173–206, 1975.

[NEU 81a] NEUTS M.F., *Matrix-Geometric Solutions in Stochastic Models: An Algorithmic Approach*, Dover, New York, 1981.

[NEU 81b] NEUTS M.F., PAGANO M.E., "Generating random variates from a distribution of phase type", *WSC '81: Proceedings of the 13th Conference on Winter Simulation*, IEEE Press, Piscataway, NJ, pp. 381–387, 1981.

[NEU 81c] NEUTS M.F., *Matrix-Geometric Soluitons in Stochastic Models: An Algorithmic Approach*, The Johns Hopkins University Press, Baltimore/London, 1981.

[NIE 22] NIELSEN B.F., "Characterisation of multivariate phase type distributions", *Queueing Systems*, vol. 100, pp. 229–231, 2022.

[O'CI 89] O'CINNEIDE C.A., "On non-uniqueness of representations of phase-type distributions", *Communications in Statistics – Stochastic Models*, vol. 5, no. 2, pp. 247–259, 1989.

[O'CI 90] O'CINNEIDE C.A., "Characterization of phase-type distributions", *Stochastic Models*, vol. 6, no. 1, pp. 1–57, 1990.

[O'CI 91a] O'CINNEIDE C.A., "Phase-type distributions and majorization", *Annals of Applied Probability*, vol. 1, no. 2, pp. 219–227, 1991.

[O'CI 91b] O'CINNEIDE C.A., "Phase-type distributions and invariant polytopes", *Advances in Applied Probability*, vol. 23, no. 3, pp. 515–535, 1991.

[O'CI 93] O'CINNEIDE C.A., "Triangular order of triangular phase-type distributions", *Stochastic Models*, vol. 9, pp. 507–529, 1993.

[O'CI 99] O'CINNEIDE C.A., "Phase-type distributions: open problems and a few properties", *Communications in Statistics – Stochastic Models*, vol. 15, no. 4, pp. 731–758, 1999.

[PAP 13] PAPP J., TELEK M., "Canonical representation of discrete phase type distributions of order 2 and 3", *Proc. of UK Performance Evaluation Workshop, UKPEW 2013*, Loughborough, 2013.

[PAV 22] PAVITHRA C.R., DEEPAK T., "Parameter estimation and computation of the Fisher information matrix for functions of phase type random variables", *Computational Statistics & Data Analysis*, vol. 167, p. 107362, 2022.

[PUL 11] PULUNGAN R., "The order of phase-type distributions", *Proceedings of the 6th SEAMS-UGM Conference*, Yogyakarta, pp. 65–78, 2011.

[RAM 14] RAMASWAMI V., VISWANATH N., "Phase type distributions with finite support", *Stochastic Models*, vol. 30, no. 4, pp. 576–597, 2014.

[REI 13] REINECKE P., KRAUSS T., WOLTER K., "Phase-type fitting using HyperStar", in BALSAMO M.S., KNOTTENBELT W.J., MARIN A. (eds), *Computer Performance Engineering*, Springer, Berlin, Heidelberg, pp. 164–175, 2013.

[REI 14] REINECKE P., TELEK M., "Does a given vector-matrix pair correspond to a PH distribution?", *Performance Evaluation*, vol. 80, pp. 40–51, 2014.

[ROB 04] ROBERT C., CASELLA G., *Monte Carlo Statistical Methods*, Springer, New York, 2004.

[ROS 19] ROSS S.M., *Introduction to Probability Models*, 12th edition, Academic Press, London, 2019.

[SHA 85] SHANTHIKUMAR J., "Bilateral phase type distributions", *Naval Res. Log. Quarterly.*, vol. 32, p. 119, 1985.

[STI 94] STIELTJES T., "Recherches sur les fractions continues", *Annales de la Faculté des Sciences de Toulouse*, vol. 2, pp. 1–122, 1894.

[STO 83] STOYAN D., DALEY D., *Comparison Methods for Queues and Other Stochastic Models*, John Wiley & Sons, New York, 1983.

[SZE 95] SZEKLI R., *Stochastic Ordering and Dependence in Applied Probability*, Springer, New York, 1995.

[TEL 00] TELEK M., "The minimal coefficient of variation of discrete phase type distributions", *3rd International Conference on Matrix-Analytic Methods in Stochastic Models*, Leuven, 2000.

[TEL 02] TELEK M., HEINDL A., "Matching moments for acyclic discrete and continuous phase-type distributions of second order", *International Journal of Simulation Systems, Science & Technology*, vol. 3, no. 3–4, pp. 47–57, 2002.

[TEL 07] TELEK M., HORVÁTH G., "A minimal representation of Markov arrival processes and a moments matching method", *Performance Evaluation*, vol. 64, no. 9–12, pp. 1153–1168, 2007.

[THÜ 06] THÜMMLER A., BUCHHOLZ P., TELEK M., "A novel approach for phase-type fitting with the EM algorithm", *IEEE Trans. on Dependable and Secure Computing*, vol. 3, no. 3, pp. 245–258, 2006.

[TRI 01] TRIVEDI K.S., *Probability and Statistics with Reliability, Queuing and Computer Science Applications*, 2nd edition, John Wiley & Sons, Hoboken, NJ, 2001.

[VAN 78] VAN LOAN C., "Computing integrals involving the matrix exponential", *IEEE Transactions on Automatic Control*, vol. 23, no. 3, pp. 395–404, 1978.

[VER 99] VERSCHELDE J., "Algorithm 795: PHCpack: A general-purpose solver for polynomial systems by homotopy continuation", *ACM Transactions on Mathematical Software (TOMS)*, vol. 25, no. 2, pp. 251–276, 1999.

[YAO 02] YAO R., "A proof of the steepest increase conjecture of a phase-type density", *Stochastic Models*, vol. 18, no. 1, pp. 1–6, 2002.

# Index

Other titles from

iSTE

in

Networks and Telecommunications

## 2023

RAVALIMINOARIMALALASON Toky Basilide, RANDIMBINDRAINIBE Falimanana
*Queues Applied to Telecoms: Courses and Exercises*

TONG Van Van, SOUIHI Sami, TRAN Hai-Anh, MELLOUK Abdelhamid
*Troubleshooting for Network Operators: The Road to a New Paradigm with Encrypted Traffic*
*(New Generation Networks Set – Volume 3)*

## 2022

BANNOUR Fetia, SOUIHI Sami, MELLOUK Abdelhamid
*Software-Defined Networking: Extending SDN Control to Large-Scale Networks*
*(New Generation Networks Set – Volume 2)*

BENAROUS Leila, BITAM Salim, MELLOUK Abdelhamid
*Security in Vehicular Networks: Focus on Location and Identity Privacy*
*(New Generation Networks Set – Volume 1)*

## 2021

LAUNAY Frédéric
*NG-RAN and 5G-NR: 5G Radio Access Network and Radio Interference*

## 2020

GONTRAND Christophe
*Digital Communication Techniques*

PUJOLLE Guy
*Software Networks: Virtualization, SDN, 5G and Security (2nd edition revised and updated)*
*(Advanced Network Set – Volume 1)*

## 2019

LAUNEY Frédéric, PEREZ André
*LTE Advanced Pro: Towards the 5G Mobile Network*
*Harmonic Concept and Applications*

TOUNSI Wiem
*Cyber-Vigilance and Digital Trust: Cyber Security in the Era of Cloud Computing and IoT*

## 2018

ANDIA Gianfranco, DURO Yvan, TEDJINI Smail
*Non-linearities in Passive RFID Systems: Third Harmonic Concept and Applications*

BOUILLARD Anne, BOYER Marc, LE CORRONC Euriell
*Deterministic Network Calculus: From Theory to Practical Implementation*

LAUNAY Frédéric, PEREZ André
*LTE Advanced Pro: Towards the 5G Mobile Network*

PEREZ André
*Wi-Fi Integration to the 4G Mobile Network*

## 2017

BENSLAMA Malek, BENSLAMA Achour, ARIS Skander
*Quantum Communications in New Telecommunications Systems*

HILT Benoit, BERBINEAU Marion, VINEL Alexey, PIROVANO Alain
*Networking Simulation for Intelligent Transportation Systems: High Mobile Wireless Nodes*

LESAS Anne-Marie, MIRANDA Serge
*The Art and Science of NFC Programming*
*(Intellectual Technologies Set – Volume 3)*

## 2016

AL AGHA Khaldoun, PUJOLLE Guy, ALI-YAHIYA Tara
*Mobile and Wireless Networks*
*(Advanced Network Set – Volume 2)*

BATTU Daniel
*Communication Networks Economy*

BENSLAMA Malek, BATATIA Hadj, MESSAI Abderraouf
*Transitions from Digital Communications to Quantum Communications: Concepts and Prospects*

CHIASSERINI Carla Fabiana, GRIBAUDO Marco, MANINI Daniele
*Analytical Modeling of Wireless Communication Systems*
*(Stochastic Models in Computer Science and Telecommunication Networks Set – Volume 1)*

EL FALLAH SEGHROUCHNI Amal, ISHIKAWA Fuyuki, HÉRAULT Laurent, TOKUDA Hideyuki
*Enablers for Smart Cities*

PEREZ André
*VoLTE and ViLTE*

## 2015

BENSLAMA Malek, BATATIA Hadj, BOUCENNA Mohamed Lamine
*Ad Hoc Networks Telecommunications and Game Theory*

BENSLAMA Malek, KIAMOUCHE Wassila, BATATIA Hadj
*Connections Management Strategies in Satellite Cellular Networks*

BERTHOU Pascal, BAUDOIN Cédric, GAYRAUD Thierry, GINESTE Matthieu
*Satellite and Terrestrial Hybrid Networks*

CUADRA-SANCHEZ Antonio, ARACIL Javier
*Traffic Anomaly Detection*

LE RUYET Didier, PISCHELLA Mylène
*Digital Communications 1: Source and Channel Coding*

PEREZ André
*LTE and LTE Advanced: 4G Network Radio Interface*

PISCHELLA Mylène, LE RUYET Didier
*Digital Communications 2: Digital Modulations*

## 2014

ANJUM Bushra, PERROS Harry
*Bandwidth Allocation for Video under Quality of Service Constraints*

BATTU Daniel
*New Telecom Networks: Enterprises and Security*

BEN MAHMOUD Mohamed Slim, GUERBER Christophe, LARRIEU Nicolas, PIROVANO Alain, RADZIK José
*Aeronautical Air−Ground Data Link Communications*

BITAM Salim, MELLOUK Abdelhamid
*Bio-inspired Routing Protocols for Vehicular Ad-Hoc Networks*

CAMPISTA Miguel Elias Mitre, RUBINSTEIN Marcelo Gonçalves
*Advanced Routing Protocols for Wireless Networks*

CHETTO Maryline
*Real-time Systems Scheduling 1: Fundamentals*
*Real-time Systems Scheduling 2: Focuses*

EXPOSITO Ernesto, DIOP Codé
*Smart SOA Platforms in Cloud Computing Architectures*

MELLOUK Abdelhamid, CUADRA-SANCHEZ Antonio
*Quality of Experience Engineering for Customer Added Value Services*

OTEAFY Sharief M.A., HASSANEIN Hossam S.
*Dynamic Wireless Sensor Networks*

PEREZ André
*Network Security*

PERRET Etienne
*Radio Frequency Identification and Sensors: From RFID to Chipless RFID*

REMY Jean-Gabriel, LETAMENDIA Charlotte
*LTE Standards*
*LTE Services*

TANWIR Savera, PERROS Harry
*VBR Video Traffic Models*

VAN METER Rodney
*Quantum Networking*

XIONG Kaiqi
*Resource Optimization and Security for Cloud Services*

## 2013

ASSING Dominique, CALÉ Stéphane
*Mobile Access Safety: Beyond BYOD*

BEN MAHMOUD Mohamed Slim, LARRIEU Nicolas, PIROVANO Alain
*Risk Propagation Assessment for Network Security: Application to Airport Communication Network Design*

BERTIN Emmanuel, CRESPI Noël
*Architecture and Governance for Communication Services*

BEYLOT André-Luc, LABIOD Houda
*Vehicular Networks: Models and Algorithms*

BRITO Gabriel M., VELLOSO Pedro Braconnot, MORAES Igor M.
*Information-Centric Networks: A New Paradigm for the Internet*

DEUFF Dominique, COSQUER Mathilde
*User-Centered Agile Method*

DUARTE Otto Carlos, PUJOLLE Guy
*Virtual Networks: Pluralistic Approach for the Next Generation of Internet*

FOWLER Scott A., MELLOUK Abdelhamid, YAMADA Naomi
*LTE-Advanced DRX Mechanism for Power Saving*

JOBERT Sébastien *et al.*
*Synchronous Ethernet and IEEE 1588 in Telecoms: Next Generation Synchronization Networks*

MELLOUK Abdelhamid, HOCEINI Said, TRAN Hai Anh
*Quality-of-Experience for Multimedia: Application to Content Delivery Network Architecture*

NAIT-SIDI-MOH Ahmed, BAKHOUYA Mohamed, GABER Jaafar, WACK Maxime
*Geopositioning and Mobility*

PEREZ André
*Voice over LTE: EPS and IMS Networks*

## 2012

AL AGHA Khaldoun
*Network Coding*

BOUCHET Olivier
*Wireless Optical Communications*

DECREUSEFOND Laurent, MOYAL Pascal
*Stochastic Modeling and Analysis of Telecoms Networks*

DUFOUR Jean-Yves
*Intelligent Video Surveillance Systems*

EXPOSITO Ernesto
*Advanced Transport Protocols: Designing the Next Generation*

JUMIRA Oswald, ZEADALLY Sherali
*Energy Efficiency in Wireless Networks*

KRIEF Francine
*Green Networking*

PEREZ André
*Mobile Networks Architecture*

## 2011

BONALD Thomas, FEUILLET Mathieu
*Network Performance Analysis*

CARBOU Romain, DIAZ Michel, EXPOSITO Ernesto, ROMAN Rodrigo
*Digital Home Networking*

CHABANNE Hervé, URIEN Pascal, SUSINI Jean-Ferdinand
*RFID and the Internet of Things*

GARDUNO David, DIAZ Michel
*Communicating Systems with UML 2: Modeling and Analysis of Network Protocols*

LAHEURTE Jean-Marc
*Compact Antennas for Wireless Communications and Terminals: Theory and Design*

PALICOT Jacques
*Radio Engineering: From Software Radio to Cognitive Radio*

PEREZ André
*IP, Ethernet and MPLS Networks: Resource and Fault Management*

RÉMY Jean-Gabriel, LETAMENDIA Charlotte
*Home Area Networks and IPTV*

TOUTAIN Laurent, MINABURO Ana
*Local Networks and the Internet: From Protocols to Interconnection*

## 2010

CHAOUCHI Hakima
*The Internet of Things*

FRIKHA Mounir
*Ad Hoc Networks: Routing, QoS and Optimization*

KRIEF Francine
*Communicating Embedded Systems / Network Applications*

## 2009

CHAOUCHI Hakima, MAKNAVICIUS Maryline
*Wireless and Mobile Network Security*

VIVIER Emmanuelle
*Radio Resources Management in WiMAX*

## 2008

CHADUC Jean-Marc, POGOREL Gérard
*The Radio Spectrum*

GAÏTI Dominique
*Autonomic Networks*

LABIOD Houda
*Wireless Ad Hoc and Sensor Networks*

LECOY Pierre
*Fiber-optic Communications*

MELLOUK Abdelhamid
*End-to-End Quality of Service Engineering in Next Generation Heterogeneous Networks*

PAGANI Pascal *et al.*
*Ultra-wideband Radio Propagation Channel*

## 2007

BENSLIMANE Abderrahim
*Multimedia Multicast on the Internet*

PUJOLLE Guy
*Management, Control and Evolution of IP Networks*

SANCHEZ Javier, THIOUNE Mamadou
*UMTS*

VIVIER Guillaume
*Reconfigurable Mobile Radio Systems*

Printed and bound by CPI Group (UK) Ltd, Croydon, CR0 4YY
13/09/2026
14961024-0004